INTERNATIONAL UNION OF CRYSTALLOGRAPHY
BOOK SERIES

IUCr Monographs on Crystallography

1 *Accurate molecular structures*
A. Domenicano, I. Hargittai, editors

2 *P. P. Ewald and his dynamical theory of X-ray diffraction*
D. W. J. Cruickshank, H. J. Juretschke, N. Kato, editors

3 *Electron diffraction techniques, Vol. 1*
J. M. Cowley, editor

4 *Electron diffraction techniques, Vol. 2*
J. M. Cowley, editor

5 *The Rietveld method*
R. A. Young, editor

6 *Introduction to crystallographic statistics*
U. Shmueli, G. H. Weiss

7 *Crystallographic instrumentation*
L. A. Aslanov, G. V. Fetisov, G. A. K. Howard

8 *Direct phasing in crystallography*
C. Giacovazzo

9 *The weak hydrogen bond*
G. R. Desiraju, T. Steiner

10 *Defect and microstructure analysis by diffraction*
R. L. Snyder, J. Fiala and H. J. Bunge

11 *Dynamical theory of X-ray diffraction*
A. Authier

12 *The chemical bond in inorganic chemistry*
I. D. Brown

13 *Structure determination from powder diffration data*
W. I. F. David, K. Shankland, L. B. McCusker, Ch. Baerlocher, editors

14 *Polymorphism in molecular crystals*
J. Bernstein
15 *Crystallography of modular materials*
G. Ferraris, E. Makovicky, S. Merlino
16 *Diffuse X-ray scattering and models of disorder*
T. R. Welberry
17 *Crystallogrphy of the polymethylene chain: an inquiry into the structure of waxes*
D. L. Dorset
18 *Crystalline molecular complexes and compounds: structure and principles*
F. H. Herbstein
19 *Molecular aggregation: structure analysis and molecular simulation of crystals and liquids*
A. Gavezzotti
20 *Aperiodic crystals: from modulated phases to quasicrystals*
T. Janssen, G. Chapuis, M. de Boissieu
21 *Incommensurate crystallography*
S. van Smaalen

IUCr Texts on Crystallography

1 *The solid state*
A. Guinier, R. Julien
4 *X-ray charge densities and chemical bonding*
P. Coppens
5 *The basics of crystallography and diffraction, second edition*
C. Hammond
6 *Crystal structure analysis: principles and practice*
W. Clegg, editor
7 *Fundamentals of crystallography, second edition*
C. Giacovazzo, editor
8 *Crystal structure refinement: a crystallographer's guide to SHELXL*
P. Müller, editor
9 *Theory and techniques of crystal structure determination*
U. Shmueli
10 *Advanced structural inorganic chemistry*
Wai-Kee Li, Gong-Du Zhou, Thomas Mak

Incommensurate Crystallography

SANDER VAN SMAALEN
Laboratory of Crystallography, University of Bayreuth

OXFORD
UNIVERSITY PRESS

OXFORD
UNIVERSITY PRESS

Great Clarendon Street, Oxford OX2 6DP

Oxford University Press is a department of the University of Oxford.
It furthers the University's objective of excellence in research, scholarship,
and education by publishing worldwide in
Oxford New York

Auckland Cape Town Dar es Salaam Hong Kong Karachi
Kuala Lumpur Madrid Melbourne Mexico City Nairobi
New Delhi Shanghai Taipei Toronto
With offices in

Argentina Austria Brazil Chile Czech Republic France Greece
Guatemala Hungary Italy Japan Poland Portugal Singapore
South Korea Switzerland Thailand Turkey Ukraine Vietnam

Published in the United States
by Oxford University Press Inc., New York

First published 2007

British Library Cataloguing in Publication Data
Data available

Library of Congress Cataloging in Publication Data
Data available

Printed in Great Britain
on acid-free paper by Biddles Ltd., King's Lynn, Norfolk

ISBN 978–0–19–857082–0 (Hbk)

1 3 5 7 9 10 8 6 4 2

to Jan L. de Boer

PREFACE

Translational symmetry is firmly established as the key property characterizing the crystalline state of matter. This is founded in the successes of translational symmetry in explaining physical properties of crystalline compounds and Bragg reflections in their diffraction patterns, as well as in the many substances that possess periodic atomic structures. An alternative and perhaps better characterization of the crystalline state is by long-range order of the atomic structure. The past fifty years have seen a continuous increase of the awareness that long-range order can be achieved differently than by translational symmetry. Modulated crystals and composite crystals have atomic structures that can be described as variations on periodic structures, while quasicrystals differ from crystals with translational symmetry in a more fundamental way. This book presents the methods of structural analysis of incommensurately modulated crystals and incommensurate composite crystals.

The crystallography of aperiodic crystals employs many concepts that are routinely applied to periodic crystals. The present text has been written under the assumption that the reader is familiar with concepts like space group symmetry, Bragg reflections and vector calculus. This assumption is motivated by the recognition that readers interested in aperiodic crystals will often have a background in the solid state sciences, and by the fact that many books are available that deal with the crystallography of translational symmetric structures at both introductory and advanced levels.

This book aims at providing a complete overview of the different aspects of aperiodic crystallography. Chapter 1 gives an introduction to the special structural features of incommensurate crystals. The atomic structures and diffraction are presented in relation to the structures and diffraction of periodic crystals, while it is shown that space group symmetry is lost. The second part (Chapters 2–5) gives the theoretical background of the superspace theory for the description of the atomic structures and symmetry of aperiodic crystals. The theory is introduced at an elementary level (Chapters 2 and 3), mainly concentrating on the simplest case of an incommensurately modulated crystal with a one-dimensional (1D) modulation wave. Chapter 4 then provides generalizations of the theory, while Chapter 5 develops the relation between incommensurate crystals and superstructures. The third part (Chapters 6–10) presents methods of structural analysis of incommensurate crystals. Chapters 6 and 7 introduce the quantitative description of diffraction (structure factors) and structure refinements. Chapter 8 discusses the electron density as it can be studied by Fourier maps, while the maximum entropy method is presented as a method to determine the precise shapes of modulation functions from electron densities in superspace. Methods of structure solution are introduced in Chapters 9 and 10. Finally, Chapter 11 presents superspace as a tool for crystal chemical considerations.

It is my intention that this book may serve two groups of readers. Scientists who feel a need to understand the special structural features of aperiodic crystals should be able to obtain this knowledge from Chapters 1–6. Crystallographers who want to determine the atomic structures of modulated and composite crystals from diffraction data will find the tools in Chapters 7–11.

Bayreuth: Sander van Smaalen

CONTENTS

ACKNOWLEDGEMENTS

I am greatly indebted to Andreas Schönleber for critical reading of the manuscript. I thank all colleagues who provided me with original electronic files of pictures from their publications, or who have allowed me to use one of their unpublished figures.

Figures and diagrams have been produced with CORALDRAW12 and PHOTO-PAINT12 by Coral corporation, JANA2000 by Petricek *et al.* (2000), MISTEK, and DIAMOND Version 3 by Crystal Impact.

1

STRUCTURE AND DIFFRACTION OF APERIODIC CRYSTALS

1.1 Introduction to aperiodic order

1.1.1 *Lattice periodic structures*

Crystals are usually defined as solid materials in which the atoms are arranged according to a space lattice. The atomic structures of these periodic crystals are completely characterized by three basis vectors of the translational symmetry and the coordinates of the atoms in one unit cell (Fig. 1.1). The position of atom μ with respect to the origin of the unit cell is

$$\mathbf{x}^0(\mu) = x_1^0(\mu)\,\mathbf{a}_1 + x_2^0(\mu)\,\mathbf{a}_2 + x_3^0(\mu)\,\mathbf{a}_3\,, \tag{1.1}$$

where $(x_1^0(\mu), x_2^0(\mu), x_3^0(\mu))$ are relative coordinates with respect to the basis vectors $\{\mathbf{a}_1,\ \mathbf{a}_2,\ \mathbf{a}_3\}$. Consequently, all possible positions within one unit cell are described by coordinates with $0 \leqslant x_i^0(\mu) < 1$ $(i = 1, 2, 3)$. Translational symmetry is characterized by the lattice $\Lambda = \{\mathbf{a}_1, \mathbf{a}_2, \mathbf{a}_3\}$ with lattice vectors

$$\mathbf{L} = l_1\,\mathbf{a}_1 + l_2\,\mathbf{a}_2 + l_3\,\mathbf{a}_3\,, \tag{1.2}$$

where l_i $(i = 1, 2, 3)$ are integers. Translational symmetry implies that equivalent atoms are found at positions that differ from each other by a lattice vector. The positions of these atoms are given by

$$\bar{\mathbf{x}} = \mathbf{L} + \mathbf{x}^0(\mu)\,, \tag{1.3}$$

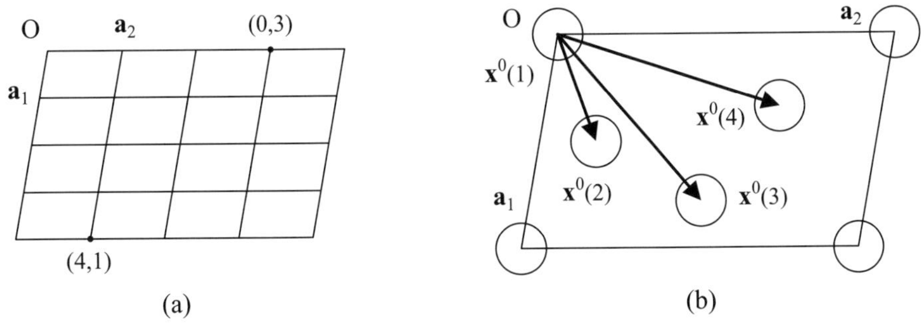

FIG. 1.1. (a) Lattice with basis vectors $\mathbf{a}_1$ and $\mathbf{a}_2$. Lattice vectors $\mathbf{L} = (0, 3)$ and $(4, 1)$ are indicated. (b) One unit cell with atoms at different positions $\mathbf{x}^0(\mu)$ for $\mu = 1, \cdots, 4$.

with relative coordinates $(\bar{x}_1, \bar{x}_2, \bar{x}_3)$. The atomic structure of an infinite crystal is obtained by varying $\mathbf{L}$ over all lattice vectors for each of the atoms $\mu = 1, \cdots, N$ in the unit cell.

For the theory of the structure, symmetry and diffraction of periodic crystals I refer to the many textbooks on this subject, *e.g.* by Giacovazzo *et al.* (2002). In order to facilitate the presentation of the theory of aperiodic crystals, subscripts $i = 1, 2, 3$ are used to indicate the three space directions. Accordingly, $\{\mathbf{a}_1, \mathbf{a}_2, \mathbf{a}_3\}$, (x_1, x_2, x_3) and $(h_1 h_2 h_3)$ are identified with the commonly employed symbols $\{\mathbf{a}, \mathbf{b}, \mathbf{c}\}$, (x, y, z) and $(h\, k\, l)$, respectively.

1.1.2 *Modulated structures*

Before the solid state sciences became dominated by translational symmetry, the crystalline state of matter was defined by long-range order of the atomic structures. Indeed, periodicity implies long-range order, and compounds with a periodic atomic structure are crystalline. However, crystalline order can be achieved in different ways than by translational symmetry, resulting in aperiodic crystals. Presently, three alternatives to translational symmetry are known: incommensurately modulated crystals, incommensurate composite crystals (Section 1.1.3) and quasicrystals (Section 1.1.4).

Modulated structures can be derived from structures with translational symmetry. For example, consider a periodic structure with one atom at the origin of the unit cell [Fig. 1.2(a)]. A twofold superstructure corresponding to a doubled $\mathbf{a}_2$ axis can be formed by displacements of the atoms by equal amounts, alternatingly into the directions $\pm\mathbf{a}_2$ [Fig. 1.2(b)]. Obviously, the new structure has long-range order as well as translational symmetry, as expressed by a unit cell of twice the volume of the original unit cell. Alternatively, a superstructure can be described as a commensurately modulated structure. The displacements of the atoms are obtained as the values of a periodic function (modulation function) that has a periodicity equal to the superlattice period, *i.e.* a period equal to $2\mathbf{a}_2$ in the present example. It is now a small step to consider a modulation wave of which the period does not match any integral number of lattice translations. Such a modulation wave is said to be incommensurate with the periodic basic structure [Fig. 1.2(c)]. Because of the incommensurability, the resulting structure does not have translational symmetry, whatever size of supercell would be proposed. Yet, long-range order is preserved, provided that the shapes and amplitudes of the modulation waves are known. The determination of modulation waves, in addition to the positions $\mathbf{x}^0(\mu)$ of the atoms in the unit cell of the basic structure [eqn (1.1)], is one of the tasks of incommensurate crystallography.

Displacements of atoms from the positions in the basic structure towards the positions in the superstructure can have any direction and magnitude. Therefore three independent modulation functions are required to describe the displacement of one atom, that can be considered as one longitudinal wave [displacement along $\mathbf{a}_2$ in the example—Fig. 1.2(c)] and two independent transversal waves [displacements along $\mathbf{a}_1$ and $\mathbf{a}_3$—Fig. 1.2(d)]. The modulation function

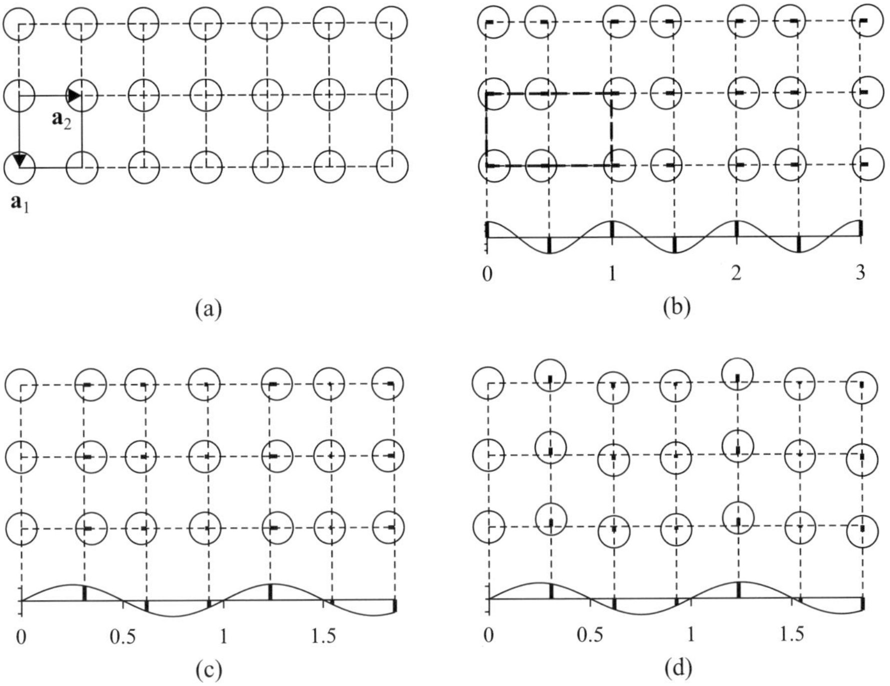

FIG. 1.2. Crystal structures with displacement modulations. (a) Basic structure. (b) Twofold superstructure with the supercell indicated by heavy dashed lines. (c) Incommensurate longitudinal modulation. (d) Incommensurate transversal modulation. Lattices of the periodic basic structures are indicated by dashed grids. Circles denote atoms that are shifted out of lattice periodic positions by varying amounts given by the heavy bars. Numbers count periods of the modulation waves $\mathbf{u}(\bar{x}_4)$ [eqn (1.7)].

$\mathbf{u}^\mu$ of atom μ thus is a vector function with components $(u_1^\mu, u_2^\mu, u_3^\mu)$ along the three basis vectors $\{\mathbf{a}_1, \mathbf{a}_2, \mathbf{a}_3\}$.

Modulation functions are wave functions. They are characterized by a wave vector $\mathbf{q}$ that specifies the direction of the wave and its wavelength. The components of $\mathbf{q}$ are given with respect to the basis vectors of the reciprocal lattice of the basic structure, $\Lambda^* = \{\mathbf{a}_1^*, \mathbf{a}_2^*, \mathbf{a}_3^*\}$, according to

$$\mathbf{q} = \sigma_1\, \mathbf{a}_1^* + \sigma_2\, \mathbf{a}_2^* + \sigma_3\, \mathbf{a}_3^* \,. \tag{1.4}$$

Incommensurability is ensured by one, two or all three components σ_i ($i = 1, 2, 3$) being irrational numbers. The example of Fig. 1.2 has σ_2 irrational, while $\sigma_1 = \sigma_3 = 0$. The modulation is one-dimensional if it is described by a single modulation wave vector, irrespective of the number of non-zero or irrational components of this vector. Wave functions $\mathbf{u}(\bar{x}_4)$ are periodic functions of the argument

$$\bar{x}_4 = t + \mathbf{q}\cdot\bar{\mathbf{x}}\,. \tag{1.5}$$

The argument is denoted by $\bar{x}_4$ in regard to the superspace theory presented in Chapter 2. $\mathbf{q}\cdot\bar{\mathbf{x}}$ is the scalar product of the vectors $\mathbf{q}$ and $\bar{\mathbf{x}}$. With components of $\mathbf{q}$ with respect to the reciprocal lattice, and with components of the vector $\bar{\mathbf{x}}$ with respect to the direct lattice [eqn (1.3)], the scalar product can be evaluated towards

$$\mathbf{q}\cdot\bar{\mathbf{x}} \;=\; \sigma_1\,\bar{x}_1 + \sigma_2\,\bar{x}_2 + \sigma_3\,\bar{x}_3\,. \tag{1.6}$$

The parameter t is a real number describing the initial phase of the wave. It plays a central role in the superspace analysis of aperiodic crystals, but for now it can be set to any fixed value, *e.g.* $t = 0$. The only restrictions on modulation functions are that they are periodic functions,

$$\mathbf{u}^{\mu}(\bar{x}_4 + 1) = \mathbf{u}^{\mu}(\bar{x}_4)\,. \tag{1.7}$$

The displacement of atom μ out of the basic structure position $\bar{\mathbf{x}}$ then is

$$\mathbf{u}^{\mu}(\bar{x}_4) = u_1^{\mu}(\bar{x}_4)\,\mathbf{a}_1 + u_2^{\mu}(\bar{x}_4)\,\mathbf{a}_2 + u_3^{\mu}(\bar{x}_4)\,\mathbf{a}_3\,. \tag{1.8}$$

Examples of functions $u_i^{\mu}(\bar{x}_4)$ are given at the bottom of Figs. 1.2(b),(c),(d). The position $\mathbf{x}$ of atom μ in unit cell $\mathbf{L}$ of the basic structure is the sum of the basic structure position [eqn (1.3)] and the modulation function [eqn (1.8)],

$$\mathbf{x} = \bar{\mathbf{x}} + \mathbf{u}^{\mu}(\bar{x}_4)\,, \tag{1.9}$$

with the argument of the modulation function given by eqn (1.5).

Any periodic function can be written as a Fourier series,

$$u_i^{\mu}(\bar{x}_4) = \sum_{n=1}^{\infty} A_i^n(\mu)\,\sin(2\pi n\bar{x}_4) + B_i^n(\mu)\,\cos(2\pi n\bar{x}_4)\,. \tag{1.10}$$

The Fourier amplitudes $\mathbf{A}^n(\mu) = [A_1^n(\mu),\ A_2^n(\mu),\ A_3^n(\mu)]$ and $\mathbf{B}^n(\mu) = [B_1^n(\mu),\ B_2^n(\mu),\ B_3^n(\mu)]$ define the modulation functions of atom μ. They are the structural parameters of the modulation that need to be determined in the structural analysis of incommensurately modulated crystals. Equation (1.10) shows that arbitrary shapes of modulation functions require an infinite number of parameters, although only a finite number of them can experimentally be determined. This problem is solved by considering the obvious property, that modulation functions must be bounded, *i.e.* modulation functions must have finite values for all values of their arguments, and they should not exceed a value of the order of the interatomic distances. From this property it is easy to derive that the magnitudes of the Fourier amplitudes need to decrease faster than $\frac{1}{n}$ in the limit of large n. So, Fourier amplitudes will become zero for the orders n beyond some maximum order $n_{\max}$, and good approximations to modulation functions can always be obtained by a finite number of Fourier amplitudes. In practice, only the first-order ($n = 1$) or a few low-order Fourier coefficients can be determined from the scattering information.

Modulations are not restricted to displacive modulations, but they may also apply to variations of other structural properties. In compounds with partial occupancies of atomic sites, vacancy ordering may lead to incommensurately modulated structures. An example is pyrrhotite $Fe_{1-x}S$, where a fraction x of the iron sites is not occupied by an atom. These vacancies order according to a modulation wave that describes a variation of occupational probability of the iron sites. The modulation wave vector is $\mathbf{q} = (0, 0, 2x)$, that will be commensurate or incommensurate, depending on the composition x of the compound (Yamamoto and Nakazawa, 1982). For $x = 1/8$ a fourfold superstructure results ($N = 1/(2x) = 4$), with complete vacancy ordering: in every second layer one out of four iron sites is empty (Fig. 1.3). For $x = 1/12$ a sixfold superstructure ($N = 6$) has fully occupied layers alternating with double layers in which one out of four sites has the probability one half that it is a vacancy. For $x = 0.09$ ($N = 5.54$) the modulation is incommensurate, as expressed by fractional probabilities for sites to be vacancies. A true superstructure cannot be found, but for $N = 5.5$ an elevenfold superstructure provides a reasonable approximation to the incommensurate modulation. With the exception of the 4C structure (Fig. 1.3), all compounds have remaining disorder, as expressed by fractional occupation probabilities. This is commonly observed for occupational modulation waves, and it is the fundamental difference to displacement modulations, for which any shape of modulation wave defines a fully ordered structure.

Modulations involving the alternate occupancy of a single crystallographic site with atoms of different chemical elements are defined in a similar way as vacancy ordering. Systematic variations of the occupational probabilities from one to the next unit cell may result in incommensurately modulated structures or superstructures, depending on the period of the modulation wave. An example is the ordering of Cu and Au atoms in the binary alloy AuCu (Yamamoto, 1982*a*). Other modulation types include the incommensurate ordering of the directions of atomic magnetic moments in the rare earth metals (Koehler, 1972). Furthermore atomic displacement parameters (ADPs), as they occur in the Debye–Waller factor, may be modulated. This happens often in conjunction to displacement or occupational modulation waves.

1.1.3 *Incommensurate composite crystals*

Incommensurate composite crystals are based on two, interpenetrating lattice-periodic structures that are mutually incommensurate. Their building principle is easily visualized for layered compounds. Crystal structures of compounds like graphite and NbS_2 can be considered as the stacking of layers, with each layer being a few atoms thick. Within the layers, atoms form a connected net of strong chemical bonds, while neighbouring layers interact through much weaker

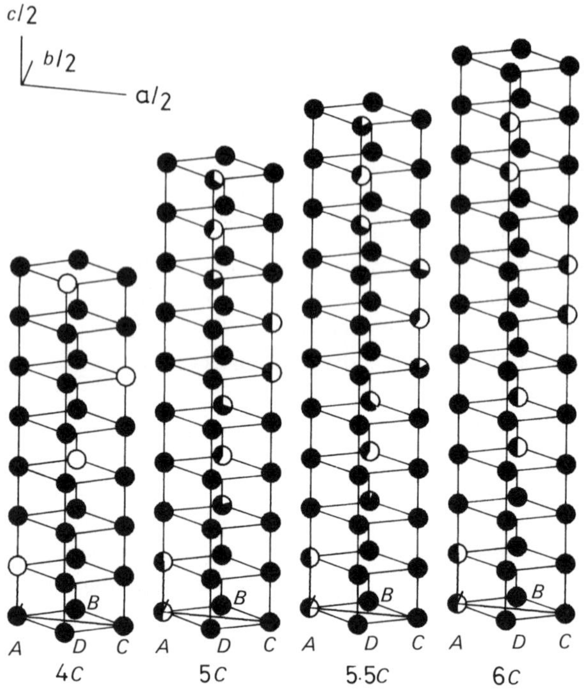

FIG. 1.3. Vacancy ordering in $Fe_{1-x}S$ for four different compositions x. Only iron atoms are shown. The period of the superstructure is indicated by NC with $N = 1/(2x)$ for $x = \frac{1}{8}, \frac{1}{10}$, 0.09 and $\frac{1}{12}$, respectively. The fraction of the area of a circle that is white defines the probability that the corresponding site is a vacancy. Reprinted with permission from Yamamoto and Nakazawa (1982), copyright (1982) IUCr.

Van der Waals bonds. Two independent lattice periodicities parallel to the layers are mainly determined by the strong bonds, while the third lattice periodicity is given by the repeat distance along the stack direction. Accordingly, many layered compounds have a crystal structure with translational symmetry.

A variation on layered compounds is provided by the alternate stacking of two chemically distinct types of layers. As an example consider the inorganic misfit layer compound $[LaS]_{1.14}[NbS_2]$ (Wiegers, 1996). Layers of composition NbS_2 alternate with layers LaS that have a structure equal to a two-atom thick slice of the rock-salt type structure [Fig. 1.4(a),(b)]. The set of layers NbS_2 has approximate three-dimensional translational symmetry, with empty space in their unit cell at positions where the LaS layers are located. The latter form a structure with approximate three-dimensional translational symmetry in a similar way. Along the stack direction the lattices of NbS_2 and LaS share a common period, that encompasses one layer of each type. However, parallel to the lay-

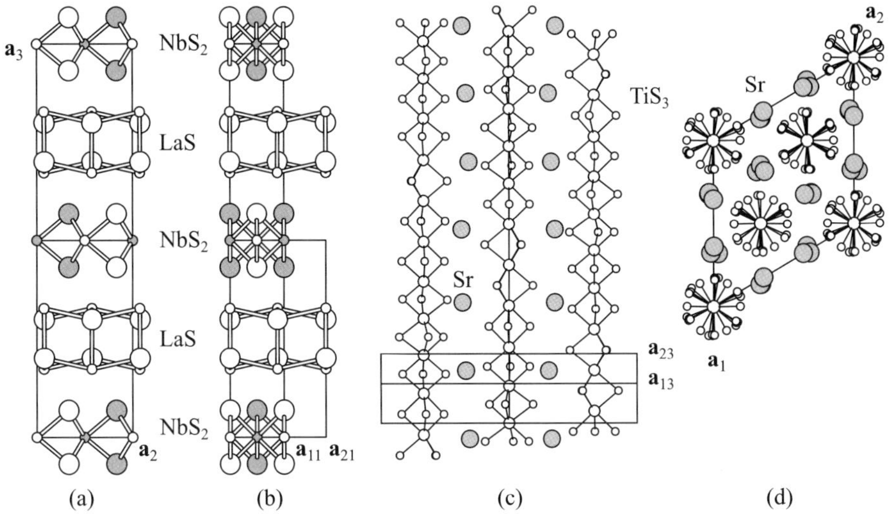

FIG. 1.4. (a) Projection along the mutually incommensurate axes $\mathbf{a}_{\nu 1}$ of the two subsystems $\nu = 1, 2$ of the crystal structure of $[LaS]_{1.14}[NbS_2]$. (b) Projection along $\mathbf{a}_2 = \mathbf{a}_{12} = \mathbf{a}_{22}$, showing the incommensurate character of this compound. Adapted from Jobst and van Smaalen (2002). Large circles denote sulfur atoms and small circles depict metal atoms. White and grey circles denote atoms with different values of the projected coordinate. (c) Two types of chains in $[Sr]_{1+x}[TiS_3]$ $(1 + x = 2a_{13}/a_{23} = 1.132)$. a_{13} is the lattice parameter a_3 of the Sr subsystem and a_{23} is the lattice parameter a_3 of the TiS_3 subsystem. (d) Projection along the chains, showing the two periodicities common to the subsystems: $\mathbf{a}_1 = \mathbf{a}_{11} = \mathbf{a}_{21}$ and $\mathbf{a}_2 = \mathbf{a}_{12} = \mathbf{a}_{22}$. Large circles denote metal atoms, small circles represent sulfur atoms (Palatinus *et al.*, 2005*a*).

ers NbS_2 and LaS have their own lattice constants, as they are determined by the strong chemical bonds within the layers, and the periodicities parallel to the layers are mutually incommensurate. At this point it is noticed that incommensurate composite crystals are not formed by combining arbitrary selected layers. Some relation between the chemically distinct layers must exist. In the present example, one of the lattice parameters of the compound NbS_2 is almost equal to the lattice parameter of rock-salt type LaS. In the composite crystal $[LaS]_{1.14}[NbS_2]$ these two lattice parameters are exactly equal, as is the result from small distortions of the individual layers. Therefore, the two lattices characterizing $[LaS]_{1.14}[NbS_2]$ share two lattice parameters [Fig. 1.4(a)], while along the third direction they are mutually incommensurate [Fig. 1.4(b)].

All atoms that belong to one of the approximate lattices define a subsystem of the composite crystal. They are enclosed in square brackets in the chemical formula. $[LaS]_{1.14}[NbS_2]$ is composed of two subsystems with compositions LaS and NbS_2, respectively. The seeming non-stoichiometry of composite crystals is explained by their incommensurability. The unit cell of subsystem NbS_2 contains an integral number ($Z_1 = 4$) of formula units NbS_2, as well as that the unit cell of subsystem LaS contains an integral number ($Z_2 = 8$) of formula units LaS. Because the volume of the unit cell of LaS is larger than the volume of the unit cell of NbS_2 by an irrational factor, only a fraction of the eight formula units LaS fits into one unit cell of NbS_2. The fraction appears to be 0.57, and for each formula unit NbS_2 this gives $(0.57 \times 8)/4 = 1.14$ formula units LaS, thus providing the composition indicated in the chemical formula.

Other possibilities to build incommensurate composite crystals include arrangements with collinear columns of atoms, each with their own periodicity along the chain direction, as in the example of $[Sr]_{1+x}[TiS_3]$ ($x \approx 0.1$) [Fig. 1.4(c)] (Onoda *et al.*, 1993; Gourdon *et al.*, 2000). Two independent directions perpendicular to the chains represent periodicities common to the subsystems Sr and TiS_3, as is easily viewed in a projection of the structure along the chain direction [Fig. 1.4(d)]. A third building principle is provided by framework compounds, where channels in the framework are filled with columns of atoms, that have a repeat distance different from the periodicity the framework structure. Because framework and filler atoms occupy different parts of space, the periodicities perpendicular to the channel/chain directions are equal to the two subsystems. Examples are the urea/alkane inclusion compounds (Yeo and Harris, 1997) and the inorganic compound $[Cr_7Se_{12}][Eu_3CrSe_3]_{0.4050}[Eu_3Se]_{0.5447}$ with composition $Eu_{1-p}Cr_2Se_{4-p}$ ($p = 0.284$) (Fig. 1.5; Brouwer and Jellinek 1977). The first subsystem with composition Cr_7Se_{12} forms a framework of edge-sharing and face-sharing $CrSe_6$ octahedra. Hexagonal channels are filled with columns of Eu_3CrSe_3, while trigonal channels contain Eu_3Se. The framework, the hexagonal columns and the trigonal columns each have their own periodicity along the channel axis, and $Eu_{1-p}Cr_2Se_{4-p}$ is one of the few known composite crystals with three subsystems.

An incommensurate composite crystal represents a single thermodynamic phase, and interactions between the subsystems are intrinsic to its existence. Because of the incommensurability, each subsystem attains an incommensurately modulated structure, where the other subsystems act as incommensurate external potentials for the subsystem under consideration. Therefore, incommensurate composite crystals can be considered as the intergrowth of two or more incommensurately modulated structures. For example, displacive modulations in $[Sr]_{1+x}[TiS_3]$ have a relatively large amplitudes, resulting in wavy chains of Sr atoms in Fig. 1.4(c), and in multiple positions in the projection of the structure onto the basal plane [Fig. 1.4(d)]. Accordingly, atomic structures of composite crystals can be described by specifying the basic structure coordinates $\mathbf{x}^0(\mu)$ [eqn (1.1)] and modulation functions $\mathbf{u}^\mu(\bar{x}_4)$ or their Fourier coefficients $\mathbf{A}^n(\mu)$ and $\mathbf{B}^n(\mu)$ [eqns (1.8) and (1.10)] for each atom μ. The coordinates then refer to the basis vectors of the basic structure of the subsystem to which atom μ belongs to. In order to distinguish the subsystems an additional parameter has been intro-

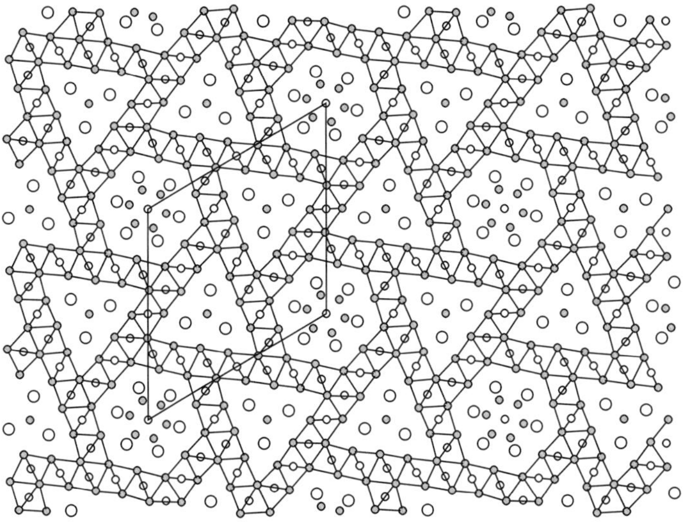

FIG. 1.5. Crystal structure of $[Cr_7Se_{12}][Eu_3CrSe_3]_{0.4050}[Eu_3Se]_{0.5447}$ projected along the channel axis. $a_1 = a_2 = 21.41$, $a_{13} = 3.446$, $a_{23} = 5.672$ and $a_{33} = 4.218$ Å. Grey circles represent Se, large circles are Eu and small circles are Cr. Coordinates from Brouwer and Jellinek (1977).

duced, which has values $\nu = 1, 2, \cdots$ for the first, second and further subsystems, and which will be used as an additional subscript or superscript as appropriate. The approximate translational symmetry of subsystem ν is characterized by the lattice

$$\Lambda_\nu = \{\mathbf{a}_{\nu 1}, \mathbf{a}_{\nu 2}, \mathbf{a}_{\nu 3}\}\,. \tag{1.11}$$

The modulation wave vector of subsystem ν is [eqn (1.4)]

$$\mathbf{q}_\nu = \sigma_1^\nu\,\mathbf{a}_{\nu 1}^* + \sigma_2^\nu\,\mathbf{a}_{\nu 2}^* + \sigma_3^\nu\,\mathbf{a}_{\nu 3}^*\,, \tag{1.12}$$

and the argument of the modulation functions of subsystem ν is given by [eqn (1.5)]

$$\bar{x}_{\nu 4} = t_\nu + \mathbf{q}_\nu \cdot \bar{\mathbf{x}}\,. \tag{1.13}$$

The parameters for different subsystems are not independent, as was already seen in this section for the lattices Λ_ν. The relations between t_1 and t_2 as well as the relations between $\mathbf{q}_1$ and $\mathbf{q}_2$ can be derived from the superspace description, that will be presented in Chapter 4.

1.1.4 *Quasicrystals*

A distinct property of quasicrystals is that their diffraction patterns exhibit fivefold and other non-crystallographic point symmetries. The discovery of quasicrystals questioned one of the fundamental theorems of crystallography, that

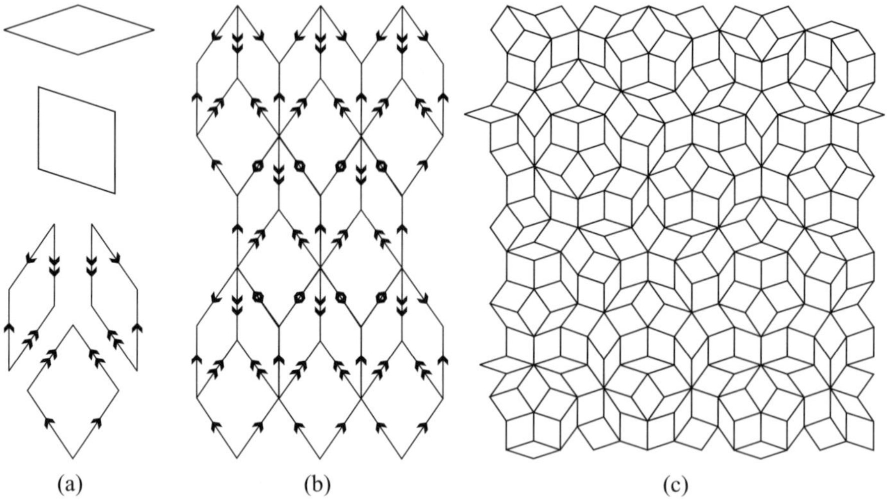

FIG. 1.6. (a) Fat and skinny rhombuses and the formation of a hexagonal unit cell from two skinny and one fat rhombus. (b) Two-dimensional periodic tiling that violates the matching rules. (c) Two-dimensional Penrose tiling that fulfills the matching rules. Penrose tiling drawn with QUASI.C by E. Weeks.

colloquially states that crystals can only have twofold, threefold, fourfold and sixfold rotational symmetries. However, the precise formulation of this theorem is that three-dimensional translational symmetry admits crystallographic point symmetries only. Indeed quasicrystals lack translational symmetry, and rotational symmetries are allowed according to any point group in three-dimensional space (Shechtman *et al.*, 1984).

The building principles of periodic crystals can be understood from the consideration of a three-dimensional lattice [Fig. 1.1(a)]. In the mathematical theory of tilings, space is said to be covered by a periodic tiling based on a single prototile (the unit cell). Important properties of tilings are that different tiles do not overlap and that there are no gaps between the tiles. Based on this definition a plethora of non-periodic tilings can be designed (Grünbaum and Shephard, 1987). However, most of them lack the fundamental properties of crystal structures: (i) long-range order, and (ii) homogeneity. The latter property represents the requirement that the atomic structure of a finite patch of the crystal does not depend on the location where this patch has been selected. Periodic tilings always fulfil these requirements. Non-periodic tilings that adhere to these properties provide models for quasicrystals. These tilings and the corresponding crystal structures will be called aperiodic to distinguish them from the general lack of translational symmetry.

An aperiodic tiling of the plane can be obtained with two different prototiles.

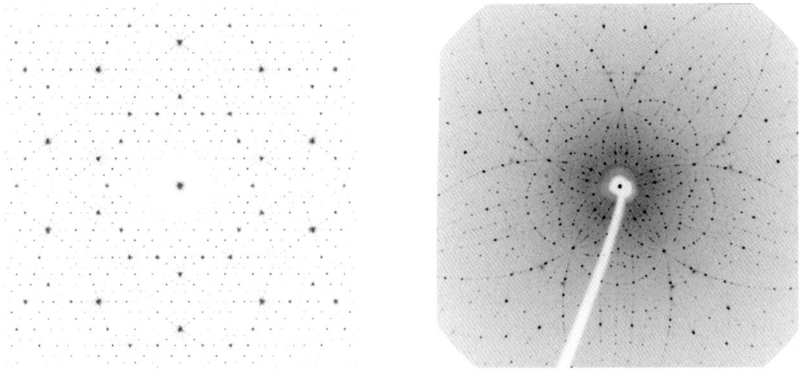
(a) (b)

FIG. 1.7. (a) Optical diffraction of the Penrose tiling. Reprinted with permission from Welberry (2004). (b) Laue diffraction along a fivefold axis of the icosahedral quasicrystal *i*-Zn–Mg–Ho. Courtesy of H. Takakura (Osaka University, Japan).

In the simplest form they are rhombuses with equal edges and related angles that are multiples of $(360/10)^\circ$. The skinny rhombus is based on an angle of 36° between the edges, while the fat rhombus has an angle of 72° [Fig. 1.6(a)]. Aperiodic order is enforced by so-called matching rules. Single arrows and double arrows are assigned to the edges of the two rhombuses according to Fig 1.6(a), and only those tilings are admitted in which each edge obtains the same marking when considered by either one of the two neighbouring rhombuses. Both the type and the direction of the arrow need to match. With disregard of the matching rules, the two prototiles can be used to devise an infinite number of different periodic and non-periodic tilings. One example is given in Fig. 1.6(b), in which a periodic tiling contains skinny and fat rhombuses in the ratio 2 : 1. When the matching rules are fulfilled the resulting tiling is the Penrose tiling (Fig. 1.6(c); Penrose 1974).

The Penrose tiling shows crystalline properties in a number of ways (Gardner, 1976; Senechal, 1995). The edges occur in five different orientations only, thus representing hidden fivefold rotational symmetry. Translational symmetry is replaced by local isomorphism: any finite patch of the Penrose tiling is repeated within a distance of two times its diameter. With 'atoms' of unit scattering power placed at the vertices of the Penrose tiling, the Fourier transform appears to consist of Bragg reflections (Fig. 1.7(a); Mackay 1982). The sharpness of the reflections is an indication for long-range order of the tiling, while the tenfold point symmetry of the diffraction pattern (symmetry of the tiling plus inversion symmetry) represents the hidden fivefold symmetry of the tiling. Other properties that indicate order in the Penrose tiling make use of the golden ratio $\tau = \frac{1}{2} + \frac{1}{2}\sqrt{5}$. The area of the fat rhombus is τ times the area of the skinny rhombus and the number of fat rhombuses is τ times the number of skinny rhombuses in the tiling. Quasicrystalline structures, including the Penrose tiling, can be obtained as intersections of periodic structures in five-dimensional or six-dimensional superspace. The latter is deterministic, and therefore the Penrose tiling must be an ordered structure too. The superspace description of quasicrystals is essentially different from the superspace description of modulated and composite crystals, and will not be pursued in this book.

Atomic structures of quasicrystals can be constructed by decorating the Penrose tiling with atoms. Both unit cells are independently filled with atoms, where care should be taken that all eight vertices must have the same atom, and that all edges with single arrows must be decorated in the same way, and similarly for the edges with double arrow markings. A three-dimensional structure is obtained as a stack of quasiperiodic planes, thus leading to a crystal that is quasiperiodic in two dimensions and periodic in the third direction. Most known structures with planes of atoms with fivefold symmetry posses three-dimensional structures based on a point group with tenfold symmetry. This is brought about by the alternate stacking of a fivefold plane and the same plane rotated over 36°. In this way the periodic direction encompasses at least two layers of atoms in these decagonal quasicrystals (Steurer, 2004). Tilings of the plane are possible with any n-fold rotational symmetry (n is an integer), and corresponding axial quasicrystals can be made as stacks of quasiperiodic planes. They will have translational symmetry in the stack direction. Experimentally, axial quasicrystals have been found that are based on eightfold symmetry (octagonal quasicrystals), tenfold symmetry (decagonal quasicrystals) or twelvefold symmetry (dodecagonal quasicrystals) (Steurer, 2004).

Many compounds have been found that are quasicrystals based on the icosahedral point group, *i.e.* their diffraction patterns have icosahedral point symmetry. The icosahedral point group contains six fivefold rotation axes that point towards the vertices of an icosahedron. Therefore, a single unique axis of rotation does not exist, and icosahedral quasicrystals are quasicrystalline in all directions of space. An icosahedral tiling of space can been obtained as a generalization of the Penrose tiling, now employing a prolate and an oblate rhombohedron as the two prototiles (Janot, 1994).

Further understanding can be obtained from the notion that the principal building blocks of quasicrystals are clusters of atoms. All known quasicrystals are intermetallic compounds. In periodic crystals of the same elements and with similar compositions as quasicrystals, large clusters of atoms (*e.g.* of diameter $\sim$20 Å) have been identified that are are arranged on a lattice. The atoms near the boundaries are part of two clusters, which can be considered to overlap. The mode of overlap is specific to each compound. In quasicrystals similar clusters again form overlaps, but now several modes of overlap are possible for a single compound, and the clusters are arranged in an aperiodic way (Fig. 1.8). It was then shown by Gummelt (1996) that a covering of the plane by overlapping decagons produces the Penrose tiling, if the allowed modes of overlap of decagons

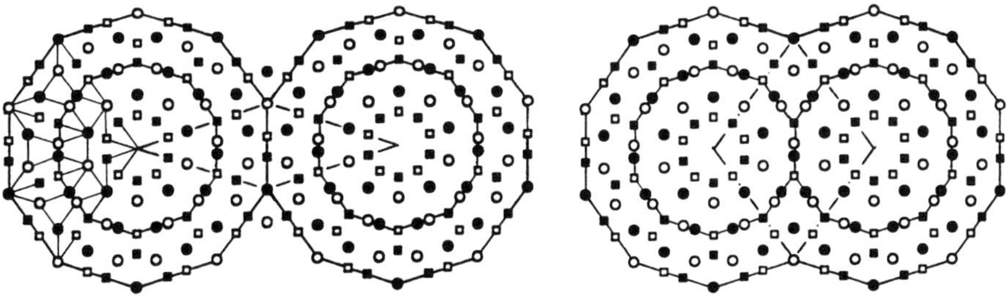

FIG. 1.8. Cluster model for the atomic structure of *decagonal*-Al–Cu–Co. Reprinted with permission from Burkov (1991). Copyright (1991) by the American Physical Society.

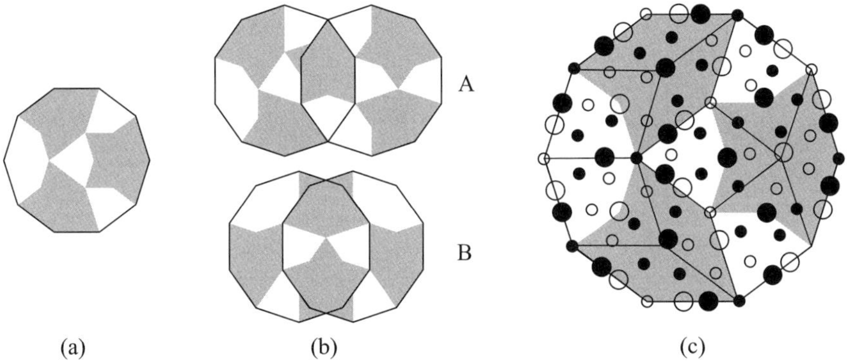

FIG. 1.9. (a) Gummelt decagon. (b) One of the four type A overlaps and the type B overlap. (c) Decoration of the Gummelt decagon with atoms in a model for the crystal structure of *decagonal*-Al–Co–Ni after Yan and Pennycook (2001). Small circles represent Al atoms and large circles denote Co/Ni atoms. Filled and open circles represent atoms from layers at $x_3 = \frac{1}{2}$ and $x_3 = 0$, respectively. A possible subdivision into Penrose rhombuses is indicated.

are enforced by matching rules, as provided by a suitable shading of the decagon [Fig. 1.9(a)]. Overlap of decagon clusters is only allowed if white and shaded areas exactly match, and if the area of overlap is at least the size of the type A overlap in [Fig. 1.9(b)]. Accordingly, four small (type A) overlaps and one large (type B) overlap were defined. In a further step, high-resolution electron microscopy was used to determine the atomic structures of the decagon clusters, and it was found that these atomic decorations provide a chemical coding for the matching rules between overlapping clusters (Steinhardt *et al.*, 1998).

The cluster model explains that quasicrystals have been found for special chemical compositions only, while periodic crystals of intermetallic compounds can be stable in wide ranges of compositions. Quasicrystals are only stable if the

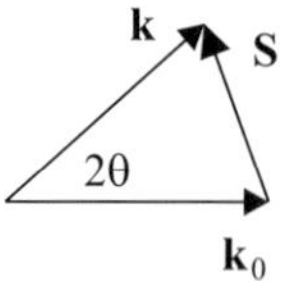

FIG. 1.10. The scattering vector **S** is defined as the difference of the wave vectors of the primary ($\mathbf{k}_0$) and scattered (**k**) X-ray beams.

cluster is a particularly stable configuration of atoms, that is reached for very specific mixtures of elements only. The model of overlapping clusters explains that quasicrystals do not form in a greater variety, because the atomic structures of clusters must be such as to allow overlap. Even then, overlap is probably not perfect, and remaining disorder can be expected in the overlap regions. Intrinsic disorder might be one of the features that prevents structural analysis of quasicrystals with the same accuracy, as has been achieved for periodic crystals (Steurer, 2004).

1.2 Diffraction by periodic crystals

Aperiodic crystals scatter X-rays in the form of Bragg reflections. The shape and sharpness of Bragg reflections is the same in periodic and aperiodic crystals. They reflect the perfect long-range order that can be found in any type of crystalline material. However, both periodic and aperiodic crystals may suffer from imperfections, and broad reflections and diffuse scattering occur frequently for real materials.

Translational symmetry of periodic crystals is expressed in their diffraction by the fact that the scattering vector **S** of any Bragg reflection is given by a reciprocal lattice vector **G**,

$$\mathbf{G} = h_1\,\mathbf{a}_1^* + h_2\,\mathbf{a}_2^* + h_3\,\mathbf{a}_3^*\,, \tag{1.14}$$

where h_1, h_2 and h_3 are integers. As usual, the scattering vector is defined as the difference between the wave vector of the incoming beam ($\mathbf{k}_0$) and the wave vector of the scattered beam (**k**) (Fig. 1.10), and it is related to the wavelength (λ) and scattering angle (2θ) by

$$|\mathbf{S}| = S = 2\frac{\sin(\theta)}{\lambda}\,, \tag{1.15}$$

where $S = |\mathbf{S}|$ denotes the length of the vector **S**. Reciprocal lattice vectors **G** give the scattering vectors of the maxima of the Bragg reflections of periodic crystals. Scattered radiation with scattering vectors $\mathbf{S} \neq \mathbf{G}$ occurs because of the finite width of Bragg reflections, imperfections of the crystal structures and thermal vibrations, and in cases where the crystal does not have translational symmetry.

In first approximation, the amplitude, $E(\mathbf{S})$, of the scattered wave is proportional to the Fourier transform of the electron density

$$E(\mathbf{S}) = \int \rho(\mathbf{x}) \exp[2\pi i \mathbf{S}\cdot\mathbf{x}]\, d\mathbf{x}\,, \tag{1.16}$$

where the integration extends over all of space. The electron density distribution $\rho(\mathbf{x})$ is measured in number of electrons per unit volume, and it is defined by $\rho(\mathbf{x})\,\mathrm{d}\mathbf{x}$ being the number of electrons in an infinitesimal volume $\mathrm{d}\mathbf{x}$ at position $\mathbf{x}$ in space. In this way eqn (1.16) defines the scattered wave relative to the scattering of a single free electron (Thomson scattering). All matter, whether solid, liquid or gaseous, is composed of atoms. A good approximation to eqn (1.16) is then obtained, when the electron density is written as the sum of electron densities of free atoms, $\rho_j(\mathbf{x})$, according to

$$\rho(\mathbf{x}) = \sum_{j=1}^{N_{\mathrm{vol}}} \rho_j[\mathbf{x} - \mathbf{x}(j)]\,, \tag{1.17}$$

where $\mathbf{x}(j)$ is the position of atom j and the summation extends over all (N_{vol}) atoms in the diffracting matter. Substitution of eqn (1.17) into eqn (1.16) gives, after mathematical manipulations employing properties of the Fourier transform,

$$E(\mathbf{S}) = \sum_{j=1}^{N_{\mathrm{vol}}} f_j(\mathbf{S}) \exp[2\pi i \mathbf{S}\cdot\mathbf{x}(j)]\,, \tag{1.18}$$

where $f_j(\mathbf{S})$ is the atomic scattering factor, that is the Fourier transform of the electron density of a single atom,

$$f_j(\mathbf{S}) = \int \rho_j(\mathbf{x}) \exp[2\pi i \mathbf{S}\cdot\mathbf{x}]\,\mathrm{d}\mathbf{x}\,. \tag{1.19}$$

Equations (1.17) and (1.18) are known as the independent atom approximation.

The atomic electron density $\rho_j(\mathbf{x})$ consists of a single peak centred on the origin of the coordinate system. This peak has a Full Width at Half Maximum (FWHM) of the order of 1 Å, and $\rho_j(\mathbf{x})$ exponentially approaches zero in the limit of large $|\mathbf{x}|$. The argument in eqn (1.17) shifts this peak to the position $\mathbf{x}(j)$ of atom j. While different atoms have different positions, the number of different functions $\rho_j(\mathbf{x})$ is limited to the number of chemical elements. $\rho_j(\mathbf{x})$ is spherically symmetric for a free atom, and the atomic scattering factor becomes a function of the length of the scattering vector only [eqn (1.15)]. The expression for the atomic scattering factor [eqn (1.19)] can be simplified by a transformation to spherical coordinates, and by noting that a spherical symmetric function depends on $r = |\mathbf{x}|$ only, resulting in

$$f_j(S) = \int_0^{\infty} \rho_j(r)\, \frac{\sin[2\pi S r]}{S}\, 2r\,\mathrm{d}r\,. \tag{1.20}$$

Values of $f(S)$ have been tabulated for all elements in the *International Tables for Crystallography Vol. C* (Wilson, 1995).

Equation (1.18) applies to all kinds of matter, including gases, liquids, liquid crystals, amorphous materials, periodic crystals and aperiodic crystals (Guinier,

1994). The amplitude of the scattered wave can be computed from the positions of all atoms contained in a volume that is coherently illuminated by the radiation. With a coherence length of 0.8 μm for radiation generated by X-ray tubes, at least $\sim 10^{10}$ atoms contribute to the coherent scattering of solid matter. Other X-ray sources, like synchrotrons, deliver radiation with much larger coherence lengths. The fundamental quantity to be considered is the intensity, $I(\mathbf{S})$, rather than the amplitude of the scattered radiation, because intensities can be measured, while amplitudes cannot be measured. The intensity and amplitude of radiation are related by

$$I(\mathbf{S}) = |E(\mathbf{S})|^2 = E(\mathbf{S})E(\mathbf{S})^* \,, \tag{1.21}$$

where E^* is the complex conjugate of E. Substitution of the expression for the scattered wave [eqn (1.16) or eqn (1.18)] gives

$$I(\mathbf{S}) = \int \left(\int \rho(\mathbf{y})\rho(\mathbf{x}+\mathbf{y})\mathrm{d}\mathbf{y} \right) \exp[2\pi i \mathbf{S}\cdot\mathbf{x}]\,\mathrm{d}\mathbf{x} \tag{1.22a}$$

$$= \sum_{j=1}^{N_{\mathrm{vol}}} \sum_{j'=1}^{N_{\mathrm{vol}}} f_j(\mathbf{S}) f_{j'}(\mathbf{S}) \exp[2\pi i \mathbf{S}\cdot(\mathbf{x}[j] - \mathbf{x}[j'])] \,. \tag{1.22b}$$

The integral over $\mathbf{y}$ defines the Patterson function or pair correlation function. If the positions of the atoms are exactly known, like for perfect periodic and aperiodic structures, then the intensity can be obtained by first computing the scattered amplitude [eqn (1.18)], followed by application of eqn (1.21). In real crystals, however, atoms have time dependent positions, $\mathbf{x}(j;t)$, as governed by thermal vibrations. The intensity follows as the average over time, $\langle \cdots \rangle_t$, of the instantaneous scattered intensity [eqn (1.22)]:

$$I(\mathbf{S}) = \sum_{j=1}^{N_{\mathrm{vol}}} \sum_{j'=1}^{N_{\mathrm{vol}}} f_j(\mathbf{S}) f_{j'}(\mathbf{S}) \left\langle \exp[2\pi i \mathbf{S}\cdot(\mathbf{x}[j;t] - \mathbf{x}[j';t])] \right\rangle_t \,. \tag{1.23}$$

Evaluation of the average over time for the vibrations in periodic crystals results in the Debye–Waller factor. Modifications to the Debye–Waller factor as they might be required for aperiodic crystals are discussed in Chapter 6. Because thermal vibrations of different atoms are independent from each other, the scattered intensities of Bragg reflections can be computed as the squares of scattered amplitudes [eqn (1.21)]. The latter are obtained by Fourier transform of the time-averaged electron density, *i.e.* from atoms with thermally smeared densities centred on positions $\mathbf{x}(j)$ [eqn (1.19)]. In less than perfectly ordered matter, correlations between the positions of different atoms are essential to compute the scattered intensity.

Disregarding thermal vibrations and disorder, the scattering of a periodic crystal can be computed by equating the positions of the atoms with the lattice periodic positions of eqn (1.3):

$$\mathbf{x}(j) = \mathbf{L} + \mathbf{x}^0(\mu) \,. \tag{1.24}$$

The N_{vol} atoms j are described by a relatively small number of N atoms in the unit cell, that are repeated by the lattice translations $\mathbf{L}$ in $N_{\text{cell}} = N_{\text{vol}}/N$ different unit cells. With eqn (1.24) the scattered amplitude becomes [eqn (1.18)]

$$E(\mathbf{S}) = F(\mathbf{S}) \sum_{\mathbf{L}}^{N_{\text{cell}}} \exp[2\pi i \mathbf{S}\cdot\mathbf{L}] \,, \tag{1.25}$$

with the structure factor $F(\mathbf{S})$ defined as the scattered amplitude of a single unit cell,

$$F(\mathbf{S}) = \sum_{\mu=1}^{N} f_{\mu}(\mathbf{S}) \exp\left[2\pi i \mathbf{S}\cdot\mathbf{x}^0(\mu)\right] . \tag{1.26}$$

Following Guinier (1994) the form function $\sigma_V(\mathbf{x})$ of a crystal of volume V is defined by

$$\sigma_V(\mathbf{x}) = \begin{cases} 1 & \text{for } \mathbf{x} \text{ within the crystal} \\ 0 & \text{for } \mathbf{x} \text{ outside the crystal.} \end{cases} \tag{1.27}$$

The electron density of the crystal can then be written as the product of the electron density of an infinite crystal with the form function,

$$\rho(\mathbf{x}) = \rho_{\infty}(\mathbf{x})\, \sigma_V(\mathbf{x}) \,. \tag{1.28}$$

Employing the convolution theorem of Fourier integrals, the scattered amplitude is obtained as [eqn (1.16)]

$$E(\mathbf{S}) = \int \hat{\rho}_{\infty}(\mathbf{S}')\, \hat{\sigma}_V(\mathbf{S} - \mathbf{S}')\, \mathrm{d}\mathbf{S}' \,. \tag{1.29}$$

The Fourier transform $\hat{\rho}_{\infty}(\mathbf{S})$ of the electron density of the infinite crystal is obtained by extending the summation over the lattice points towards infinity, resulting in [eqn (1.25)]

$$\hat{\rho}_{\infty}(\mathbf{S}) = F(\mathbf{S}) \sum_{\mathbf{L}} \exp[2\pi i \mathbf{S}\cdot\mathbf{L}] = F(\mathbf{S}) \sum_{\mathbf{G}} \delta(\mathbf{S} - \mathbf{G}) \,, \tag{1.30}$$

where the second summation is over all reciprocal lattice vectors $\mathbf{G}$. The Dirac delta function $\delta(\mathbf{S})$ is defined by $\delta(x - a) = 0$ except at $x = a$ and

$$\int_{-\infty}^{\infty} f(x)\, \delta(x - a)\, \mathrm{d}x = f(a) \,, \tag{1.31}$$

for well-behaved functions $f(x)$. In deriving eqn (1.30) the property

$$\sum_{m=-\infty}^{\infty} \exp[2\pi i s m] = \sum_{m=-\infty}^{\infty} \delta(s - m) \tag{1.32}$$

has been used. The scattered amplitude of a finite crystal follows from eqn (1.29) and eqn (1.30) as

$$E(\mathbf{S}) = F(\mathbf{S}) \sum_{\mathbf{G}} \hat{\sigma}_V(\mathbf{S} - \mathbf{G})\,. \tag{1.33}$$

The Fourier transform of the form function is denoted by $\hat{\sigma}_V(\mathbf{S})$. Because $\sigma_V(\mathbf{x})$ is a broad function, its Fourier transform is sharply peaked. The maximum value is $\hat{\sigma}_V(0) = V$ and its width is $1/V^{1/3}$. The precise shape of $\hat{\sigma}_V(\mathbf{S})$ depends on the shape of the crystal, but for typical crystal sizes $\hat{\sigma}_V(\mathbf{S})$ is much narrower than the distance between reciprocal lattice points. In the limit of large scattering volumes, $\hat{\sigma}_V(\mathbf{S})$ is related to the Dirac δ function according to

$$\delta(\mathbf{S}) = \lim_{V\to\infty} \hat{\sigma}_V(\mathbf{S}) = \lim_{V\to\infty} \frac{1}{V} \left[\hat{\sigma}_V(\mathbf{S})\right]^2\,. \tag{1.34}$$

For a sufficiently large crystal, the functions $\hat{\sigma}_V(\mathbf{S}-\mathbf{G})$ and $\hat{\sigma}_V(\mathbf{S}-\mathbf{G}')$ will never be non-zero for the same $\mathbf{S}$ if $\mathbf{G}$ and $\mathbf{G}'$ are different reciprocal lattice vectors. Employing this property, the scattered intensity of a finite crystal follows as

$$I(\mathbf{S}) = \sum_{\mathbf{G}} |\, F(\mathbf{G})\,|^2 \, \left[\hat{\sigma}_V(\mathbf{S} - \mathbf{G})\right]^2\,. \tag{1.35}$$

Equation (1.35) describes scattering in the form of Bragg reflections centred on the nodes of the reciprocal lattice, and with widths of the order of $1/V^{1/3}$. From eqn (1.34) it follows that the scattered intensity is proportional to the scattering volume V.

1.3 Diffraction by modulated crystals

Scattering of X-rays by aperiodic crystals is in the form of Bragg reflections (Fig. 1.11). Unlike for periodic crystals, the scattering vectors are not given by reciprocal lattice vectors, but they can be indexed with integers. For a modulated crystal with modulation wave vector $\mathbf{q}$ [eqn (1.4)], scattering vectors of Bragg reflections are

$$\mathbf{H} = h_1\,\mathbf{a}_1^* + h_2\,\mathbf{a}_2^* + h_3\,\mathbf{a}_3^* + h_4\,\mathbf{a}_4^*\,, \tag{1.36}$$

where $\mathbf{a}_4^* = \mathbf{q}$ and $h_1, \cdots, h_4$ are integers. Especially in the case of modulated crystals, the four reflection indices are often denoted by

$$(h_1\, h_2\, h_3\, h_4) = (h\, k\, l\, m)\,. \tag{1.37}$$

Scattering vectors of Bragg reflections of modulated structures thus become

$$\mathbf{H} = \mathbf{G} + m\,\mathbf{q}\,. \tag{1.38}$$

Equation (1.38) shows that each Bragg reflection of the basic structure ($\mathbf{G}$) is surrounded by a series of equally spaced satellite reflections at distances $\pm m\,\mathbf{q}$ from the main Bragg reflections (Fig. 1.11). Satellite reflections ($m \neq 0$) are

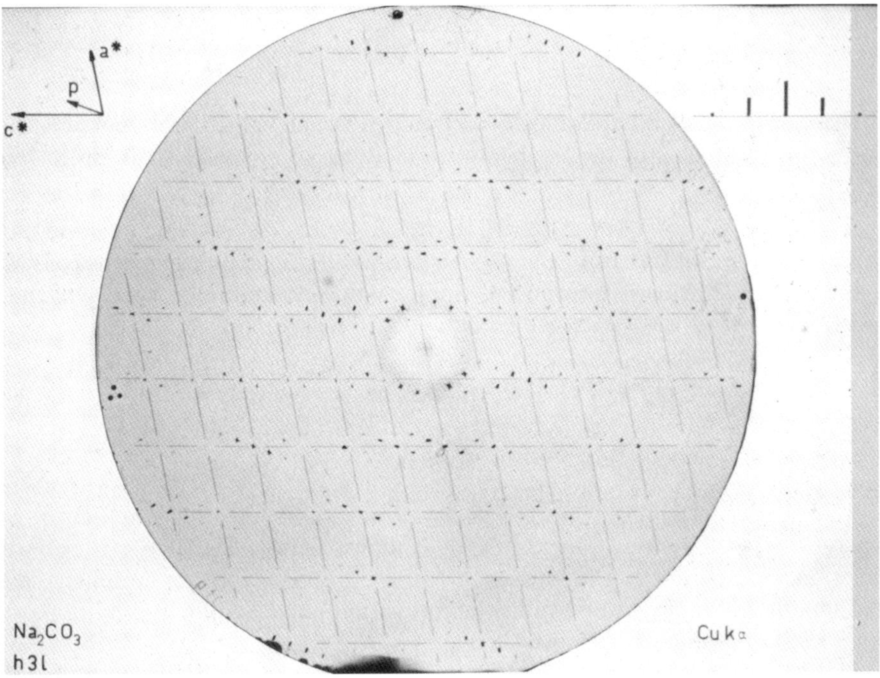

FIG. 1.11. Diffraction pattern of Na_2CO_3 (retigram; similar to a precession photograph). The measured X-ray diffraction has been overlayed with thin lines, highlighting the reciprocal lattice of the basic structure. p indicates the modulation wave vector. Reprinted from Tuinstra and Fraase Storm (1972) by courtesy of F. Tuinstra (Delft, The Netherlands).

much weaker than main reflections ($m = 0$) if the amplitude of the modulation is small—as is the case in most known modulated crystals. The intensities of satellite reflections decrease with increasing satellite order $|m|$, such that often only first- and second-order satellites can be observed. Exceptions do exist, and satellite reflections up to order $|m| = 9$ have been found in the incommensurately modulated phase of α-U (Marmeggi *et al.*, 1990).

The starting point for the computation of the elastic scattering of X-rays by modulated crystal structures is the expression for the scattered wave in the independent-atom approximation [eqn (1.18)]. The difference with periodic crystals enters through the positions of the atoms, that are now given by [eqn (1.9)]

$$\mathbf{x}(j) = \mathbf{L} + \mathbf{x}^0(\mu) + \mathbf{u}^\mu(\bar{x}_4)\,, \tag{1.39}$$

where $\mu = 1, \cdots, N$ enumerates the atoms in the unit cell of the basic structure, and $\mathbf{L}$ are lattice vectors of the basic structure. Substitution into eqn (1.18) gives for the scattered wave,

$$E(\mathbf{S}) = \sum_{\mathbf{L}}^{N_{\text{cell}}} \sum_{\mu=1}^{N} f_\mu(\mathbf{S}) \exp\left[2\pi i \mathbf{S} \cdot \left(\mathbf{L} + \mathbf{x}^0(\mu) + \mathbf{u}^\mu[\bar{x}_4]\right)\right]. \tag{1.40}$$

A separation of the summations over $\mathbf{L}$ and μ is not immediately possible, because the arguments of the modulation functions also depend on $\mathbf{L}$ and μ [eqn (1.5)]:

$$\bar{x}_4 = t + \mathbf{q} \cdot \left[\mathbf{L} + \mathbf{x}^0(\mu)\right]. \tag{1.41}$$

In order to remove $\mathbf{L}$ and μ from the arguments of the modulation functions, an additional integration is introduced, employing the following properties of the δ function,

$$\begin{aligned} \mathcal{F}(x) &= \int_{-\infty}^{\infty} \delta(x-\tau)\, \mathcal{F}(\tau)\, \mathrm{d}\tau \\ &= \int_0^1 \sum_{m=-\infty}^{\infty} \delta(x-\tau-m)\, \mathcal{F}(\tau+m)\, \mathrm{d}\tau\,, \end{aligned}$$

where $\mathcal{F}(x)$ can be any function with complex values. If $\mathcal{F}(x)$ is a periodic function with period 1, the integer m can be dropped from its argument, and the summation index m occurs in the δ function only. Employing the property eqn (1.32) of δ functions leads to

$$\begin{aligned} \mathcal{F}(x) &= \int_0^1 \left(\sum_{m=-\infty}^{\infty} \exp[-2\pi i m (x-\tau)] \right) \mathcal{F}(\tau)\, \mathrm{d}\tau \\ &= \sum_{m=-\infty}^{\infty} \left(\int_0^1 \exp[2\pi i m \tau]\; \mathcal{F}(\tau)\, \mathrm{d}\tau \right) \exp[-2\pi i m x]\,. \end{aligned} \tag{1.42}$$

With $\mathcal{F}(x) = \exp\left[2\pi i \mathbf{S} \cdot \mathbf{u}^\mu(\bar{x}_4)\right]$ and $x = \bar{x}_4 - t = \mathbf{q} \cdot \left[\mathbf{L} + \mathbf{x}^0(\mu)\right]$ the scattered amplitude becomes [eqn (1.40)]

$$\begin{aligned} E(\mathbf{S}) = \sum_{\mathbf{L}}^{N_{\text{cell}}} \sum_{\mu=1}^{N} \sum_{m=-\infty}^{\infty} \int_0^1 & f_\mu(\mathbf{S}) \exp\left[2\pi i \mathbf{S} \cdot \left(\mathbf{L} + \mathbf{x}^0[\mu]\right)\right] \exp[2\pi i m \tau] \\ & \times \exp[2\pi i \mathbf{S} \cdot \mathbf{u}^\mu(t+\tau)] \exp\left[-2\pi i m \mathbf{q} \cdot \left(\mathbf{L} + \mathbf{x}^0[\mu]\right)\right] \mathrm{d}\tau\,. \end{aligned}$$

Because the integral over τ encompasses exactly one period of the modulation functions, t can be dropped from their arguments. Rearranging terms and factorising the triple sum finally leads to

$$\begin{aligned} E(\mathbf{S}) = \sum_{m=-\infty}^{\infty} \sum_{\mu=1}^{N} & f_\mu(\mathbf{S}) \int_0^1 \exp[2\pi i \mathbf{S} \cdot \mathbf{u}^\mu(\tau)] \exp[2\pi i m \tau]\, \mathrm{d}\tau \\ & \times \exp\left[2\pi i (\mathbf{S} - m\mathbf{q}) \cdot \mathbf{x}^0(\mu)\right] \left(\sum_{\mathbf{L}}^{N_{\text{cell}}} \exp[2\pi i (\mathbf{S} - m\mathbf{q}) \cdot \mathbf{L}] \right). \end{aligned}$$

Assuming an infinite crystal, the summation over $\mathbf{L}$ can be transformed into a summation over reciprocal lattice vectors $\mathbf{G}$ of the basic structure, in a procedure similar to eqn (1.30). Together with the form function of the crystal, $\sigma_V(\mathbf{x})$, this gives for the scattered amplitude of a finite modulated crystal [eqn (1.29)]

$$E(\mathbf{S}) = \sum_{\mathbf{G}} \sum_{m=-\infty}^{\infty} F(\mathbf{S};m)\,\hat{\sigma}_V[\mathbf{S}-(\mathbf{G}+m\mathbf{q})]\,, \tag{1.43}$$

where $F(\mathbf{S};m)$ is the structure factor for a modulated crystal, that involves a summation over atoms in the unit cell of the basic structure only:

$$F(\mathbf{S};m) = \sum_{\mu=1}^{N} f_\mu(\mathbf{S}) g_\mu(\mathbf{S};m) \exp\big[2\pi i(\mathbf{S}-m\mathbf{q})\cdot\mathbf{x}^0(\mu)\big] \tag{1.44a}$$

and

$$g_\mu(\mathbf{S};m) = \int_0^1 \exp[2\pi i\mathbf{S}\cdot\mathbf{u}^\mu(\tau)]\exp[2\pi i m\tau]\,\mathrm{d}\tau\,. \tag{1.44b}$$

The dependence of the diffracted intensity on the modulation is concentrated in the function $g_\mu(\mathbf{S};m)$. This function can be considered as an additional factor that modifies the atomic scattering factor. $|g_\mu(\mathbf{S};m)|$ is always less than one, as can easily be derived from eqn (1.44b). In the limit of zero amplitude of the modulation functions, $g_\mu(\mathbf{S};m) = 1$ for main reflections ($m = 0$) and it is equal to 0 for satellite reflections ($m \neq 0$). An important property is that $g_\mu(\mathbf{S};m) \to 0$ in the limit of large $|m|$, as it follows from the fact that the magnitudes of the harmonics n of $\mathbf{u}^\mu(\bar{x}_4)$ go to zero in the limit of large n (Section 1.1.2) and from the presence of the oscillatory factor $\exp[2\pi i m\tau]$ in the integrand of eqn (1.44b). Both the structure factor and the additional atomic scattering factor g_μ depend independently on the scattering vector $\mathbf{S}$ and the satellite index m.

The scattered intensity [eqn (1.21)] is defined as the square of the scattered amplitude [eqn (1.43)], resulting in

$$I(\mathbf{S}) = \sum_{\mathbf{G}} \sum_{m=-\infty}^{\infty} |F(\mathbf{S};m)|^2\, |\hat{\sigma}_V[\mathbf{S}-(\mathbf{G}+m\mathbf{q})]|^2\,. \tag{1.45}$$

Equation (1.45) describes Bragg reflections with scattering vectors indexed with four integers according to eqns (1.36) and (1.38). The widths of main reflections and satellite reflections are the same as the widths of the Bragg reflections of a periodic crystal of the same dimensions, and they are entirely determined by the form function of the crystal. Because of the incommensurability between $\mathbf{G}$ and $\mathbf{q}$, eqn (1.45) predicts that satellite reflections exist with scattering vectors that are arbitrary close to scattering vectors of main reflections. This property questions the assumption that functions $\hat{\sigma}_V[\mathbf{S}-(\mathbf{G}+m\mathbf{q})]$ do not overlap when they are centred on different vectors $(\mathbf{G} + m\mathbf{q})$; an assumption that was used in the computation of the scattered intensity. Furthermore, this property would imply that individual satellite reflections can never be observed, because each reflection would always have other reflections that are arbitrary close to it in reciprocal space. However, it was shown above that non-zero values of $|F(\mathbf{S}; m)|$ are only obtained for reflections up to some maximum value of $|m|$. Because two reflections with scattering vectors that differ less than a small amount ε will differ in their satellite orders $|m|$ by a number that is proportional to $1/\varepsilon$, at least one of the two reflections has a very large satellite index $|m|$, and it will have zero intensity. Thus, the natural requirement on modulations, that atomic shifts may not exceed a maximum value, eventually leads to a diffraction pattern that is composed of isolated Bragg peaks, in accordance with experiment (Fig. 1.11).

1.4 More than one modulation wave vector

A limit to the number of independent modulation waves does not exist. Each of the waves is characterized by its own modulation wave vector $\mathbf{q}^j$ $(j = 1, \cdots, d)$ with components,

$$\mathbf{q}^j = \sigma_{j1}\, \mathbf{a}_1^* + \sigma_{j2}\, \mathbf{a}_2^* + \sigma_{j3}\, \mathbf{a}_3^* \,. \tag{1.46}$$

Here σ is a $d \times 3$ matrix of components of the modulation wave vectors. The number (d) of independent modulation waves can be obtained as the minimum number of modulation wave vectors that is required for an integer indexing of the diffraction pattern. Accordingly, the scattering vectors of Bragg reflections are given by

$$\begin{aligned} \mathbf{H} &= h_1\, \mathbf{a}_1^* + \cdots + h_{3+d}\, \mathbf{a}_{3+d}^* \\ &= h\, \mathbf{a}^* + k\, \mathbf{b}^* + l\, \mathbf{c}^* + m_1\, \mathbf{q}^1 + \cdots + m_d\, \mathbf{q}^d \,, \end{aligned} \tag{1.47}$$

where reflection indices $(h_1 \;\cdots\; h_{3+d})$ are identified with $(h\,k\,l\,m_1 \;\cdots\; m_d)$ and $\mathbf{a}_{3+j}^* = \mathbf{q}^j$ $(j = 1, \cdots, d)$. A modulated crystal with a diffraction pattern that can be indexed with d modulation wave vectors is called a d-dimensionally modulated crystal.

Modulation functions of a d-dimensionally modulated crystal are periodic functions of arguments $\bar{x}_{3+j}$ $(j = 1, \cdots, d)$. The latter are defined in a way similar to the definition of $\bar{x}_4$ as [eqn (1.5)]

$$\bar{x}_{3+j} = t_j + \mathbf{q}^j \cdot \bar{\mathbf{x}} \,. \tag{1.48}$$

Because $\mathbf{q}^j$ are incommensurate with respect to the lattice of the basic structure but also with respect to each other, the initial phases t_j of the different waves $j = 1, \cdots, d$ can independently be given arbitrary values. For truly independent waves, modulation functions are obtained by adding the d functions $\mathbf{u}^{\mu j}(\bar{x}_{3+j})$ [eqn (1.8)]. However, inter-modulations between the waves often occur (see below), and mixed terms involving two or more arguments $\bar{x}_{3+j}$ will be present.

This feature is apparent from the Fourier expansion of the modulation functions, that is not simply the sum of d functions of the form of eqn (1.10), but that involves mixed higher-harmonic coefficients too,

$$u_i^\mu(\bar{x}_4, \cdots, \bar{x}_{3+d}) = \sum_{n_1=0}^{\infty} \cdots \sum_{n_d=0}^{\infty} A_i^{n_1 \cdots n_d}(\mu) \sin[2\pi(n_1\bar{x}_4 + \cdots + n_d\bar{x}_{3+d})] + B_i^{n \cdots n_d}(\mu) \cos[2\pi(n_1\bar{x}_4 + \cdots + n_d\bar{x}_{3+d})], \quad (1.49)$$

for $i = 1, 2, 3$. The term with all n_j equal to zero is excluded from the summation, because it represents a shift of the basic structure position $x_i^0(\mu)$. Fourier terms $A_i^{n_1 \cdots n_d}(\mu)$ and $B_i^{n_1 \cdots n_d}(\mu)$ $(i = 1, 2, 3)$ define the modulation functions of atom μ.

Compounds with independent modulation waves do exist. For example, in the low-temperature phase of $NbSe_3$ two independent modulations reside on different but collinear atomic chains of the basic structure (van Smaalen *et al.*, 1992). A second example is provided by the incommensurate composite crystal $Eu_{1-p}Cr_2Se_{4-p}$ $(p = 0.284)$ (Fig. 1.5). The three subsystems have independent periodicities along the channel axis, thus requiring an indexing of the diffraction pattern by five indices [eqn (1.47)].

In many crystals with two-dimensional modulations the two modulation wave vectors are not independent, but they are related by symmetry instead. Yet, both wave vectors are necessary for an integer indexing of the Bragg reflections. A typical example is Co-Åkermanite $Ca_2CoSi_2O_7$ with tetragonal symmetry (Hagiya *et al.*, 1993). The modulation wave vector $\mathbf{q}^1 = (0.2913, 0.2913, 0)$ is mapped onto the vector $\mathbf{q}^2 = (-0.2913, 0.2913, 0)$ by the fourfold rotation [Fig. 1.12(a)], but both vectors are required for an indexing of the Bragg reflections by five integers $(h\,k\,l\,m_1\,m_2)$. Next to satellite reflections with $(h\,k\,l\,1\,0)$ and $(h\,k\,l\,0\,1)$ also the mixed higher-order satellite reflections $(hkl11)$ and $(hkl{-11})$ have been observed in the X-ray diffraction. Accordingly, the structure model involves the Fourier components A_i^{10}, B_i^{10}, A_i^{01}, B_i^{01}, A_i^{11}, B_i^{11}, $A_i^{\bar{1}1}$ and $B_i^{\bar{1}1}$ for $i = 1, 2, 3$ [eqn (1.49)]. In this book modulations like the one in Åkermanite are called two-dimensional modulations, because two rationally independent wave vectors suffice for an integer indexing of the Bragg reflections. On the other hand, an interpretation is possible in which the modulation in Åkermanite is considered to be composed of four waves, represented by the wave vectors $\mathbf{q}^1$, $\mathbf{q}^2$, $\mathbf{q}^1 + \mathbf{q}^2$ and $-\mathbf{q}^1 + \mathbf{q}^2$, and with magnitudes as given by the corresponding Fourier amplitudes of the modulation functions.

1.5 Definition of satellite order

In the case of two-dimensional modulations in triclinic, monoclinic, orthorhombic and tetragonal symmetries, the satellite order of reflections is defined as the sum of the absolute values of the satellite indices:

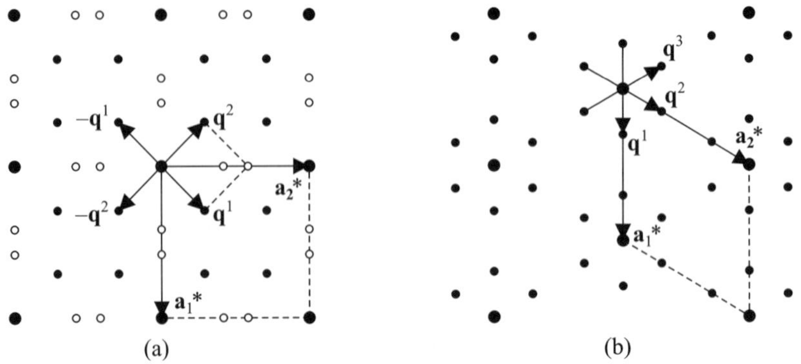

FIG. 1.12. Diffraction by two-dimensionally modulated crystals. (a) $(h\,k\,0)$ section of the diffraction pattern of $Ca_2CoSi_2O_7$. (b) $(hk0)$ section of the diffraction pattern of $2H$-$TaSe_2$. Large discs denote main reflections, small discs represent satellite reflections. Open circles indicate second-order satellites of the type $\mathbf{q}^1 \pm \mathbf{q}^2$ in $Ca_2CoSi_2O_7$.

$$m = \sum_{j=1}^{d} |m_j| = \sum_{j=1}^{d} |h_{3+j}| \,. \tag{1.50}$$

In crystals with a hexagonal lattice this formula is inadequate, as is illustrated by the example of $2H$-$TaSe_2$ (Moncton *et al.*, 1977). $2H$-$TaSe_2$ is one representative of the class of transition metal dichalcogenides with layered structures on a hexagonal lattice. An incommensurate modulation develops at low temperatures due to the charge-density-wave (CDW) mechanism. The diffraction pattern can be indexed with integers on the basis of the two modulation wave vectors, $\mathbf{q}^1 = (0.327, 0, 0)$ and $\mathbf{q}^2 = (0, 0.327, 0)$. Together with $\mathbf{q}^3 = -\mathbf{q}^1 + \mathbf{q}^2 = (-0.327, 0.327, 0)$, they form a set of three symmetry equivalent wave vectors [Fig. 1.12(b)]. Any pair of them can be used for an integer indexing of the Bragg reflections, and the third vector can always be expressed as the sum or difference of the other two. From a physical point of view the modulated crystal contains three equivalent modulation waves, but for crystallographic purposes an indexing of the Bragg reflections with two wave vectors is most fruitful ($d = 2$) and $2H$-$TaSe_2$ will be denoted as a two-dimensionally modulated crystal.

The equivalence of $\mathbf{q}^1$, $\mathbf{q}^2$ and $(-\mathbf{q}^1 + \mathbf{q}^2)$ makes the proper definition of satellite order of a reflection a problem. Because $(h\;k\;l\;1\;0)$, $(\bar{k}\;h{+}k\;l\;0\;1)$ and $(\bar{h}{+}\bar{k}\;h\;l\;\bar{1}\;1)$ are equivalent by symmetry, it seems awkward to denote the last reflection as a mixed higher-order satellite, as it was the correct designation in the case of tetragonal symmetry [eqn (1.50)]. Employing the analogy with the four-integer indexing of the diffraction of periodic crystals with a hexagonal lattice, the satellite order of reflections of crystals with a two-dimensional modulation can be defined to properly reflect the symmetry of the hexagonal lattice. Consider the case that the two modulation wave vectors have been chosen such that the angle between them is 60° [Fig. 1.12(b)]. Then, for each reflection $(h\,k\,l\,m_1\,m_2)$ the order m is defined by

$$m = \tfrac{1}{2}\left(|m_1| + |m_2| + |m_1 + m_2|\right). \tag{1.51}$$

It is easily checked that all reflections with either $m_1 = 0$ or $m_2 = 0$ do have reflection orders $|m_2|$ and $|m_1|$, respectively, as common sense requires. However, all reflections in each group of reflections that are equivalent by symmetry attain the same order by this definition. Of course, the order m is not a unique designation, and second-order reflections $(h\,k\,l\,2\,0)$ still need to be distinguished from the mixed second-order satellite reflections $(h\ k\ l\ 1\ 1)$. This definition is easily extended towards other modulations within hexagonal and trigonal symmetries, employing the mathematical theory of eutactic stars (Coxeter, 1973).

Modulations of dimensions higher than two also exist. An example is provided by wustite $Fe_{1-x}O$ with $x = 0.098$, that has a cubic crystal structure (Yamamoto, 1982*b*). Vacancy ordering corresponds to an incommensurately modulated crystal with a three-dimensional modulation. The three modulation wave vectors, $(0.398, 0, 0)$, $(0, 0.398, 0)$ and $(0, 0, 0.398)$, are equivalent by the cubic symmetry. Again, with the exception of compounds with hexagonal lattices, satellite orders of reflections in crystals with a d-dimensional modulation are defined as the sum of the absolute values of the satellite reflection indices [eqn (1.50)].

2

MODULATED CRYSTALS IN SUPERSPACE

2.1 Introduction

The building principles of aperiodic crystals have been introduced in Chapter 1. A mathematically exact description was given of the positions of the atoms in incommensurately modulated structures and composite crystals. It was shown that crystal structures of incommensurately modulated crystals are completely defined by the positions of the atoms in the unit cell of the periodic basic structure, combined with modulation functions for each of these atoms (Section 1.1). The mathematics of this description is straightforward, but a series of problems is not solved by this simple approach.

The first issue is that of symmetry. Most periodic crystals possess point symmetry in addition to their translational symmetry. It is not immediately clear how point symmetry of the basic structure can be extended towards symmetry of modulation functions, or whether incommensurately modulated structures have symmetry at all. Knowledge of the symmetry relations between Fourier amplitudes of modulation functions [eqn (1.10)] is imperative for a successful structural analysis. Any missed symmetry relation introduces parameters that are perfectly correlated and structure refinements will fail. Symmetry is also a central issue for the understanding of physical properties of materials. This includes the anisotropy of elasticity and other tensor properties, degeneracies between quantum states, as they are of profound consequences in spectroscopy, and the question about the presence of inversion symmetry in relation to non-linear properties.

The second issue is that of structure solution. Classical crystallographic methods fail for aperiodic crystals, because direct methods, Patterson Methods and more recent developments rely on translational symmetry of the crystal structures.

Thirdly, the crystal-chemical analysis of incommensurately modulated compounds is difficult, because the incommensurability dictates that coordination polyhedra assume infinitely many different shapes. Related to this issue is the quantitative interpretation of crystal structures, *e.g.* by calculations of the lattice energy, that in its classical form relies on three-dimensional periodicity.

An elegant solution to all these problems is provided by the superspace approach. Superspace is a four-dimensional space, in which the first three coordinate axes represent the three-dimensional space in which we live (3D space or physical space). The fourth coordinate axis is identified with the argument of the modulation functions [eqn (1.9)]. Originally, superspace was introduced to describe the symmetry of incommensurately modulated crystals (de Wolff, 1974). However, the concept of superspace is simpler than that of superspace symmetry. Furthermore, superspace has applications that go beyond a mere consideration of the symmetry of aperiodic crystals. Therefore, in this chapter the superspace approach to the atomic structures of incommensurately modulated structures is presented without reference to symmetry, while superspace groups are introduced in Chapter 3. Two applications are given: the analysis of interatomic distances and the computation of the structure factor.

2.2 Reciprocal superspace

The superspace description should be considered as an aid to the quantitative analysis and visualization of incommensurate crystal structures. A physical meaning need not be given to it. Nevertheless, the concept of superspace is firmly rooted in observations on aperiodic crystals. In particular, the development of superspace relies on the observation that diffraction patterns of incommensurately modulated crystals consist of Bragg reflections that can be indexed by four integers $(h_1\, h_2\, h_3\, h_4)$ according to [eqn (1.36)],

$$\mathbf{H} = h_1\, \mathbf{a}_1^* + h_2\, \mathbf{a}_2^* + h_3\, \mathbf{a}_3^* + h_4\, \mathbf{a}_4^* \,, \tag{2.1}$$

where $\mathbf{H}$ are scattering vectors corresponding to the maxima of the Bragg reflections. The first three basis vectors define a reciprocal lattice,

$$\Lambda^* = \{\mathbf{a}_1^*,\, \mathbf{a}_2^*,\, \mathbf{a}_3^*\} = \{\mathbf{a}^*,\, \mathbf{b}^*,\, \mathbf{c}^*\}\,. \tag{2.2}$$

The maximum number of independent vectors in three-dimensional space is three, and the fourth basis vector can be expressed in Λ^* according to [eqn (1.4)],

$$\mathbf{a}_4^* = \sigma_1\, \mathbf{a}_1^* + \sigma_2\, \mathbf{a}_2^* + \sigma_3\, \mathbf{a}_3^*\,. \tag{2.3}$$

Incommensurability is mathematically defined by at least one of the components σ_i $(i = 1, 2, 3)$ being an irrational number. In that case, the set

$$M = \{\mathbf{a}_1^*,\, \mathbf{a}_2^*,\, \mathbf{a}_3^*,\, \mathbf{a}_4^*\} \tag{2.4}$$

is rationally independent. This implies that all points $(h_1\, h_2\, h_3\, h_4)$ are different from each other, and that the indexing $(h_1\, h_2\, h_3\, h_4)$ is unique for a given set of basis vectors M.

The set of points in eqn (2.1) form a dense set, if $\mathbf{a}_4^*$ is incommensurate with the reciprocal lattice Λ^*. For example, if $\mathbf{a}_4^*$ has coordinates $\sigma = (\sigma_1, 0, 0)$ with $\sigma_1 = 1/\sqrt{7}$ being an irrational number, any reciprocal lattice line parallel to $\mathbf{a}_1^*$ is densely filled with points of the form of eqn (2.1). This is easily verified by computing the three-dimensional indices of the scattering vectors [eqns (2.1) and (2.3)]):

$$\mathbf{H} = (h_1 + \sigma_1\, h_4)\, \mathbf{a}_1^* + h_2\, \mathbf{a}_2^* + h_3\, \mathbf{a}_3^*\,. \tag{2.5}$$

Any point of eqn (2.5) can be approximated to arbitrarily good accuracy by other points $(h_1\, h_2\, h_3\, h_4)$, if arbitrary large numbers h_1 and h_4 are allowed. Although all combinations of indices $(h_1 h_2 h_3 h_4)$ define different points in reciprocal space, the finite resolution of any experiment would prevent distinguishing one point from the infinite number of neighbouring points within the resolution boundary. However, intensities of Bragg reflections are zero for $|h_4|$ exceeding some upper bound $h_4^{\max}$. Usually $h_4^{\max}$ is a small number in the range $1 < h_4^{\max} < 10$, and reflections with finite intensities do not form a dense set. Consequently, a unique indexing can be retrieved from the diffraction experiment. The uniqueness of the indexing of eqn (2.1) is the essential property that allows superspace to be defined. Whether the components of modulation wave vectors are rational or irrational numbers is only of secondary importance, because an experiment can never distinguish between an irrational number and some rational approximant.

The concept of superspace is based on the notion that four independent vectors in four-dimensional space define a lattice. The four reciprocal basis vectors $\mathbf{a}_k^*$ $(k = 1, \cdots, 4)$ in three-dimensional space [eqn (2.4)] are considered to be the projections of four reciprocal basis vectors, $\mathbf{a}_{sk}^*$, in four-dimensional space. The rational independence of the four vectors in M then allows the inverse of the projection operation to be performed for all Bragg reflections. In this way a Bragg reflection $(h_1\ h_2\ h_3\ h_4)$ is identified with the reciprocal lattice point in four-dimensional space given by the same indices [eqn (2.1)],

$$\mathbf{H}_s = h_1\,\mathbf{a}_{s1}^* + h_2\,\mathbf{a}_{s2}^* + h_3\,\mathbf{a}_{s3}^* + h_4\,\mathbf{a}_{s4}^* \,. \tag{2.6}$$

The embedding of three-dimensional space in four-dimensional space is obtained by identifying $\mathbf{a}_i^*$ with $\mathbf{a}_{si}^*$ for $i = 1, 2, 3$, while a fourth coordinate axis, $\mathbf{b}$, is introduced perpendicular to physical space. The basis vector $\mathbf{b}$ is dimensionless and of arbitrary length, but here we will make the choice $b = 1$. The reciprocal basis vector $\mathbf{b}^*$ is parallel to $\mathbf{b}$, and it has a length defined by $b^*\,b = 1$. With $\mathbf{b}^*$ perpendicular to physical space, the vectors $\mathbf{a}_{si}^*$ $(i = 1, 2, 3)$ have components zero along $\mathbf{b}^*$, while $\mathbf{a}_{s4}^* = (\mathbf{a}_4^*, \mathbf{b}^*)$. A reciprocal lattice in superspace, Σ^*, is then defined by (Fig. 2.1)

$$\Sigma^* : \begin{cases} \mathbf{a}_{si}^* = (\mathbf{a}_i^*, 0) & i = 1, 2, 3 \\ \mathbf{a}_{s4}^* = (\mathbf{a}_4^*, \mathbf{b}^*) \,. \end{cases} \tag{2.7}$$

Independent of the dimension of space, the relation between direct and reciprocal lattice vectors is given by

$$\mathbf{a}_{sk}^* \cdot \mathbf{a}_{sk'} = \delta_{kk'} \,, \tag{2.8}$$

with $\delta_{kk'} = 1$ for $k = k'$ and otherwise zero. With this definition the direct lattice corresponding to the reciprocal lattice Σ^* is (Fig. 2.2)

$$\Sigma : \begin{cases} \mathbf{a}_{si} = (\mathbf{a}_i, -\sigma_i \mathbf{b}) & i = 1, 2, 3 \\ \mathbf{a}_{s4} = (0, \mathbf{b}) \,, \end{cases} \tag{2.9}$$

where

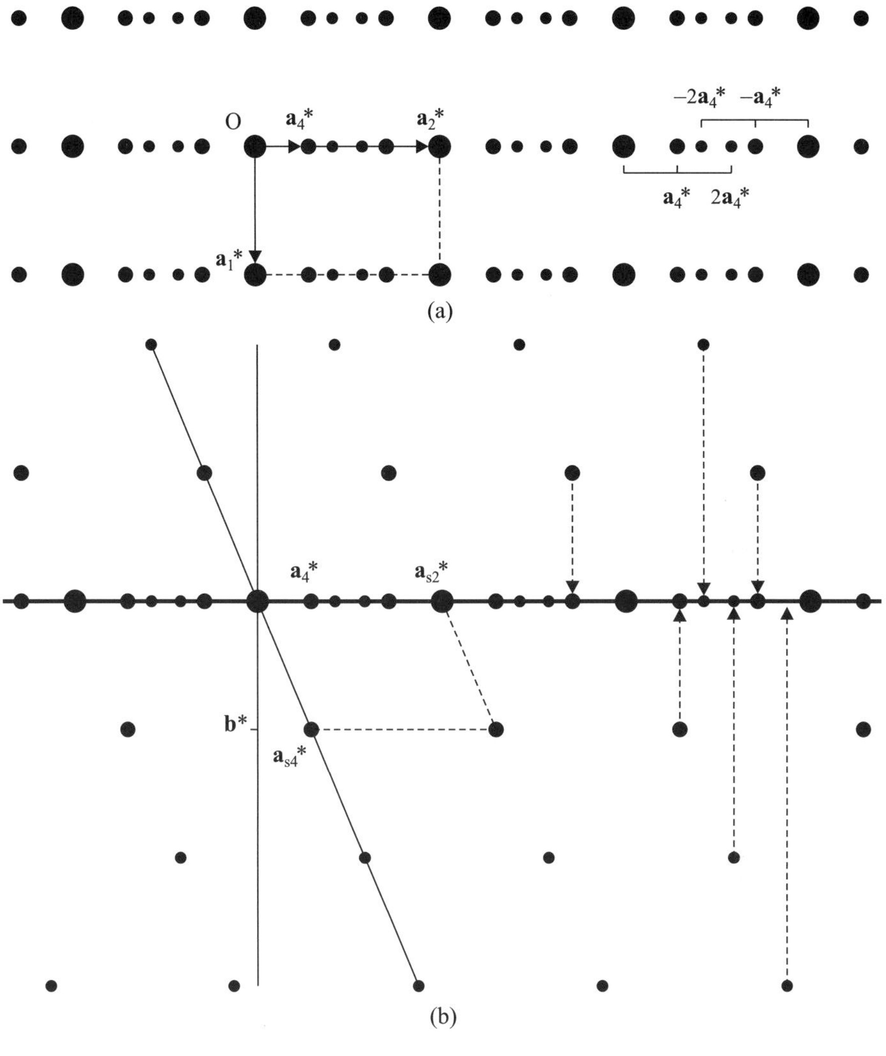

FIG. 2.1. (a) $(h_1\ h_2\ 0)$ Section of reciprocal space of a crystal with a one-dimensional incommensurate modulation. (b) $(0\ h_2\ 0\ h_4)$ Section of reciprocal superspace, illustrating the projection of reciprocal lattice points of superspace onto Bragg reflections in physical space (horizontal thick line).

$$\Lambda = \{\mathbf{a}_1, \mathbf{a}_2, \mathbf{a}_3\} = \{\mathbf{a}, \mathbf{b}, \mathbf{c}\} \tag{2.10}$$

is the direct lattice corresponding to Λ^*.

Vectors in direct superspace with coordinates relative to Σ are defined by an additional subscript s,

$$\mathbf{x}_s = x_{s1}\,\mathbf{a}_{s1} + x_{s2}\,\mathbf{a}_{s1} + x_{s3}\,\mathbf{a}_{s3} + x_{s4}\,\mathbf{a}_{s4}\,. \tag{2.11}$$

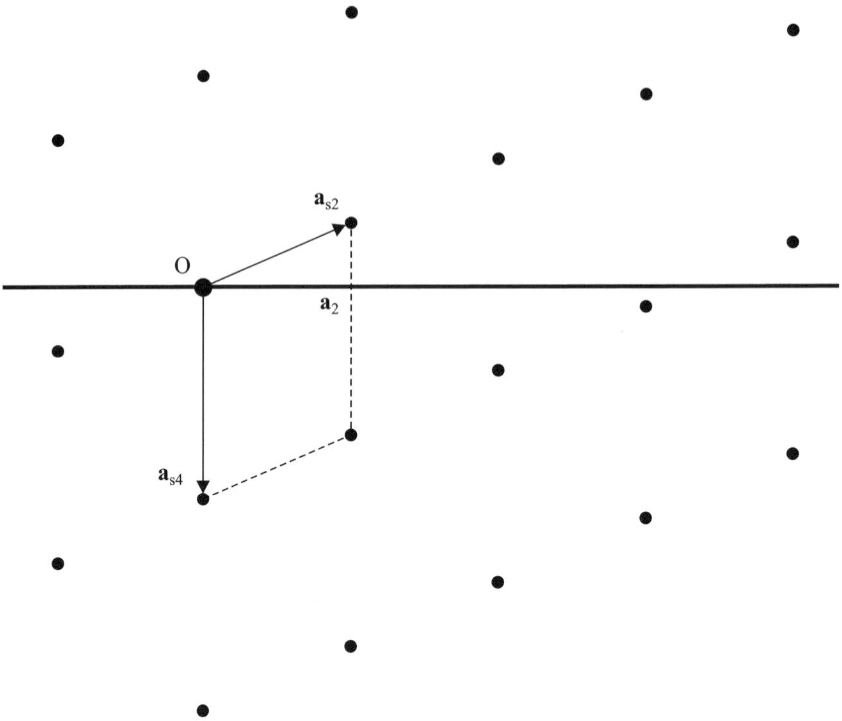

FIG. 2.2. The direct lattice Σ corresponding to the reciprocal lattice Σ^* of Fig. 2.1. Physical space is represented by the horizonal thick line.

Because Σ is defined for a three-dimensional periodic basic structure with a one-dimensional modulation, this superspace will be denoted as (3+1)-dimensional superspace. The importance of the distinction between (3+1)-dimensional superspace and general four-dimensional space will be encountered at many places in this book.

Up to this point, the definition of direct and reciprocal lattices in (3+1)-dimensional superspace is a purely geometrical one. The third and final step in the development of the superspace description is the assignment of the integrated intensities of Bragg reflections to the reciprocal lattice points in superspace. Similarly, each structure factor $F(h_1\, h_2\, h_3\, h_4)$ [eqn (1.44a)] is assigned to a reciprocal lattice point in superspace that has the same indices $(h_1\, h_2\, h_3\, h_4)$. This allows the definition of a generalized electron density, $\rho_s(\mathbf{x}_s)$, as the inverse Fourier transform of the structure factors in superspace,

$$\rho_s(\mathbf{x}_s) = \frac{1}{V} \sum_{\mathbf{H}_s} F(h_1\, h_2\, h_3\, h_4) \exp[-2\pi i\, \mathbf{H}_s \cdot \mathbf{x}_s] \,, \tag{2.12}$$

where the summation extends over all reciprocal lattice vectors in superspace.

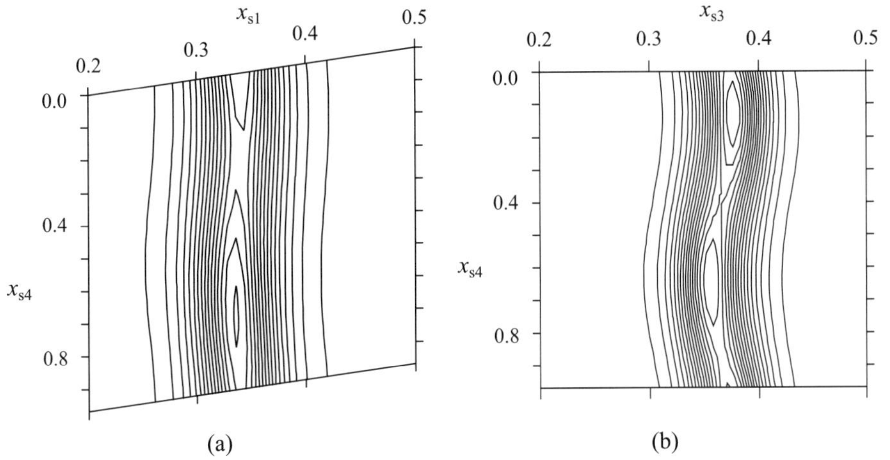

FIG. 2.3. The generalized electron density of $(NH_4)_2BeF_4$ in (3+1)-dimensional superspace. The modulation wave vector is $\mathbf{q} = (0.4796, 0, 0)$. (a) (x_{s1}, x_{s4})-Section centred on the atom F3 at $\mathbf{x}^0 = (0.336, 0.034, 0.364)$, showing a small longitudinal modulation amplitude. (b) (x_{s3}, x_{s4})-Section, showing a larger transversal modulation amplitude along $\mathbf{a}_3$. Contours of constant electron density are drawn at intervals of 2.0 electrons/Å^3. Coordinates from Palatinus *et al.* (2004).

The generalized electron density is a periodic function of the superspace coordinates x_{sk} $(k = 1, 2, 3, 4)$, with a periodicity provided by the lattice Σ. Figure 2.3 shows that $\rho_s(\mathbf{x}_s)$ is composed of local maxima in the form of strings, that are—on the average—parallel to the fourth superspace coordinate.

Equation (2.12) can be compared to the electron density in physical space, $\rho(\mathbf{x})$, that is obtained by the inverse Fourier transform of the same structure factors in physical space,

$$\rho(\mathbf{x}) = \frac{1}{V} \sum_{\mathbf{H}} F(h_1\, h_2\, h_3\, h_4) \exp[-2\pi i\, \mathbf{H} \cdot \mathbf{x}]\,, \tag{2.13}$$

where the summation is over all reciprocal vectors in physical space that describe Bragg reflections [eqn (2.1)], and $\mathbf{H}$ is the projection of $\mathbf{H}_s$ onto physical space [Fig. 2.1(b)]. Employing a property of the Fourier transform, $\rho(\mathbf{x})$ then is obtained as the physical-space section of $\rho_s(\mathbf{x}_s)$. Figure 2.2 shows that this section lacks translational symmetry, in accordance with eqn (2.13). The relation between physical space and superspace is summarized in Fig. 2.4.

2.3 Direct superspace

2.3.1 *Construction of superspace*

The relation between crystal structures in superspace and physical space can be obtained without reference to reciprocal space. A heuristic derivation starts

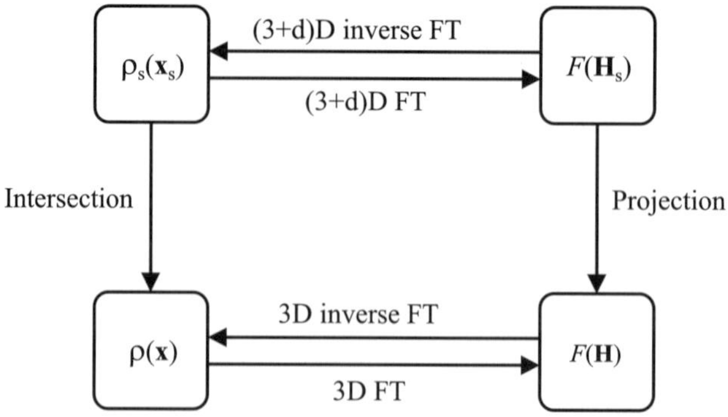

FIG. 2.4. The relation between the electron density and structure factors is given by a Fourier transform (FT) in three-dimensional (3D) or (3+d)D space.

with the description of the atomic structure of an incommensurately modulated crystal in terms of a periodic basic structure and modulation functions (Section 1.1.2). Consider a crystal with orthorhombic symmetry and a modulation wave vector $\sigma = (0, \sigma_2, 0)$ parallel to the lattice line $\mathbf{a}_2$. Then, the row of atoms along $\mathbf{a}_2$ and the modulation function can be superimposed onto each other, such that the values of the wave at the basic structure positions ($\bar{x}_2$) give the longitudinal components of the displacements of the atoms $[u_2(\bar{x}_4)]$ towards their positions (x_2) in the modulated structure [Fig. 2.5(a),(b)]. The superspace equivalent of a point atom in physical space is obtained by rotation of the modulation wave over 90°, employing the basic structure position of the atoms as pivot. Applying this procedure to the atom at the origin of the coordinate system (atom number 0) results in a string that is on the average perpendicular to physical space, and that has the shape of the modulation function [Fig. 2.5(c)]. The rotated string defines the fourth coordinate axis in (3+1)-dimensional superspace, with the fundamental translation $\mathbf{b}$ equal to one period of the modulation. For a structure of parallel strings, this definition guarantees translational symmetry in the direction perpendicular to physical space, because the modulation is periodic.

Rotation of the modulation wave about atom 1 results in a string that, on the average, is again perpendicular to physical space [Fig. 2.5(d)]. The intersection of this string with physical space is a point that is exactly the position of the atom in physical space. This property of the rotated string follows directly from Fig. 2.5(b), where it is shown that the displacement of an atom is equal in size to the value of the modulation wave at the basic structure position of this atom. Pivoting copies of the modulation wave about the basic structure positions of all atoms results in a collection of congruent strings that, on the average, are collinear with the fourth superspace axis [Fig. 2.5(e)]. Each string intersects physical space at the position (x_2) of the atom that was used as pivot for rotation of this string. The collection of strings forms the superspace equivalent of the

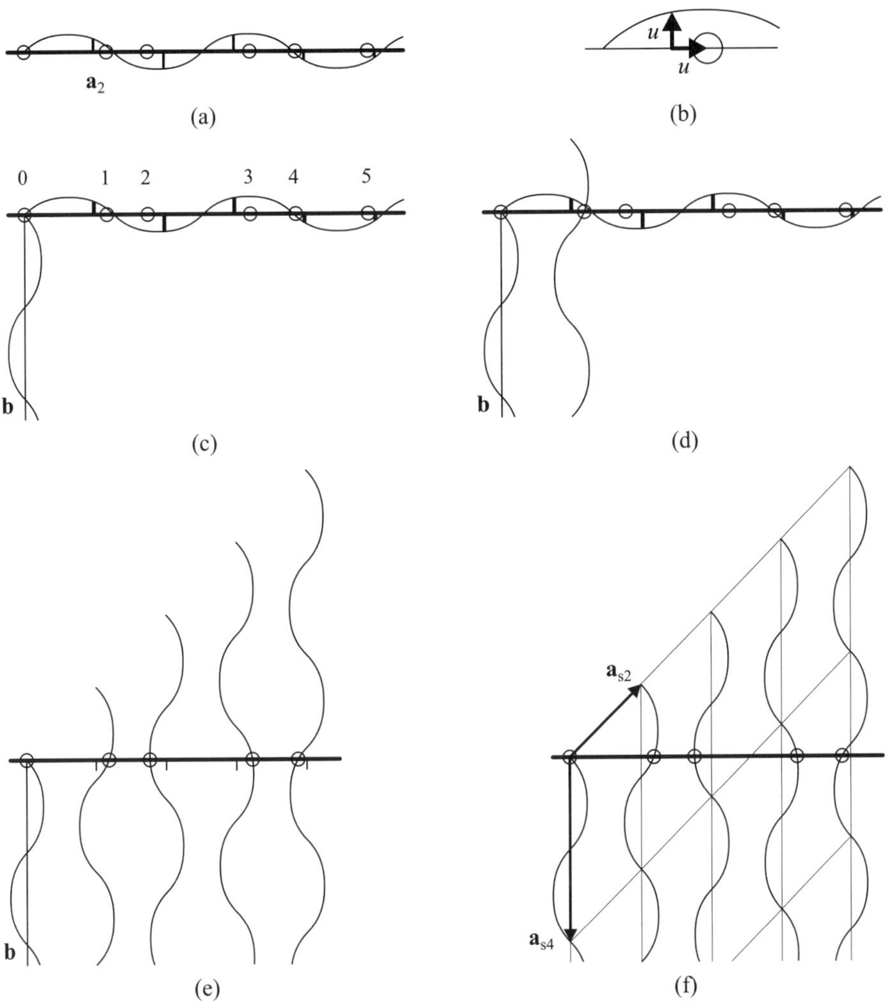

FIG. 2.5. (a) Modulation wave superimposed on a row of atoms. (b) Displacement of an atom derived from the value of the modulation wave at the basic structure position. (c) Rotation about atom 0. (d) Rotation about atom 1. (e) Rotated waves for all atoms. Vertical bars indicate the basic structure positions of the atoms. (f) Same as Fig. 2.5(e) with grid lines highlighting the translational symmetry in superspace.

crystal structure, because intersection with physical space recovers exactly the modulated structure.

Translational symmetry along the fourth axis in superspace follows from the periodicity of the modulation:

$$\mathbf{a}_{s4} = (0, \mathbf{b}) . \tag{2.14}$$

The fundamental translations of the basic structure do not provide translational symmetry in superspace. However, it is noticed that subsequent atoms j, with $\bar{\mathbf{x}} = (0, j, 0)$, have values for $\bar{x}_4$ given by [eqn (1.5)]

$$\bar{x}_4(j) = \sigma_2\, \bar{x}_2(j) = \sigma_2\, j . \tag{2.15}$$

Each translation $\mathbf{a}_2$ corresponds to an increment of the phase of the modulation wave by σ_2. It follows that the collection of strings possesses translational symmetry in superspace, whereby the translation $\mathbf{a}_2$ is combined with a translation of $-\sigma_2$ along $\mathbf{b}$, thus defining one basis vector in superspace as

$$\mathbf{a}_{s2} = (\mathbf{a}_2, -\sigma_2 \mathbf{b}) . \tag{2.16}$$

The lattice plane ($\mathbf{a}_1$,$\mathbf{a}_3$) is perpendicular to the direction of the modulation wave. Therefore this plane represents translational symmetry of the modulated structure, and the corresponding basis vectors in superspace are

$$\begin{aligned} \mathbf{a}_{s1} &= (\mathbf{a}_1, 0) , \\ \mathbf{a}_{s3} &= (\mathbf{a}_3, 0) . \end{aligned} \tag{2.17}$$

Noticing that $\sigma_1 = \sigma_3 = 0$ in the present example, it is found that eqns (2.14)–(2.17) represent the four basis vectors of the direct lattice in superspace, as they were previously derived by consideration of reciprocal superspace [eqn (2.9)]. It is a straightforward procedure to include additional atoms and transversal components of the modulation waves into the theory.

A collection of strings has been constructed that has lattice symmetry in superspace, and that recovers the positions of the atoms of the modulated crystal on intersection with physical space. Accordingly, all information about the structure of the modulated crystal is contained in a single unit cell of the superspace lattice. While the position of an atom in physical space is a point [eqn (1.9)], the equivalent object in superspace is a string that extends for exactly one period along the fourth superspace axis $\mathbf{a}_{s4}$. The superspace atom then is repeated in all unit cells by the superspace translations Σ [eqn (2.9)].

Real atoms consist of a point-like nucleus together with an electron density of finite width of the order of 1 Å. The position of the nucleus of a spherical atom coincides with the maximum of its electron density, and thus provides a unique measure for the position of this atom. The superspace equivalent of a real atom is obtained by the juxtaposition of real atoms centred on different parts of the strings, such that the intersection of the decorated string with physical space results in a real atom. This generalization of atoms to superspace atoms follows more easily from the second method of construction of superspace models for crystal structures, as is presented below.

The construction of the superspace equivalent of real atoms starts with the superspace lattice. The latter is completely defined by the translational symmetry of the basic structure in physical space together with the modulation wave

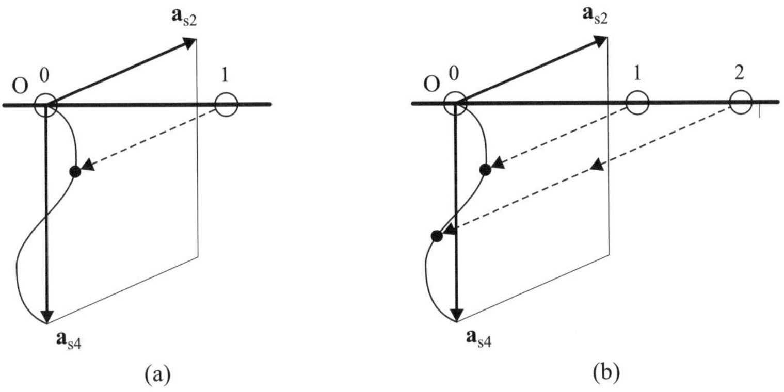

FIG. 2.6. Continued on next page.

vector [eqn (2.9)]. Application of the superspace translation $-\mathbf{a}_{s2}$ to atom 1 in Fig. 2.6(a) results in an atom within the first unit cell of superspace, but at a position that is not in physical space. The distance of this atom to the fourth axis is equal to the displacement u of atom 1 out of its basic structure position, because the basic structure position would translate to a point on the fourth axis, as is easily shown by consideration of the geometry of parallel lines. The translated atom thus is centred on a point that belongs to the string defining the superspace atom. Application of $-2\mathbf{a}_{s2}$ to atom 2 produces a different point in the first unit cell of superspace, that is part of the string describing the superspace atom for the same reasons as given above [Fig. 2.6(b)]. The procedure can be repeated for all atoms in physical space, each time employing a suitable lattice translation in superspace so as to arrive in the first unit cell [Fig. 2.6(c)]. For an incommensurate modulation each atom in physical space is translated to a different position in the first unit cell. Together, the positions of the translated atoms densily fill a string, and the string of atoms of length a_{s4} is the superspace equivalent of an atom in physical space [Fig. 2.6(d)]. Subsequently, lattice translations in superspace are applied to this string, resulting in the generalized electron density of the modulated crystal [Fig. 2.6(e)]. Because each translated string of atoms intersects physical space in the form of an atomic electron density centred on the position of the atom in the modulated structure, the string of atoms defines a string of points (the positions of the atoms) that is identical to the string obtained by the first construction. Comparison of the generalized electron density constructed in this way with a Fourier map in superspace (Fig. 2.3) shows the equivalence of the reciprocal space and direct space approaches.

2.3.2 *Coordinates in superspace*

The coordinates $(x_{s1}, x_{s2}, x_{s3}, x_{s4})$ of a point in superspace are relative coordinates with respect to the basis vectors Σ in superspace [eqn (2.11)]. The orientation of physical space with respect to Σ is completely determined by the lattice

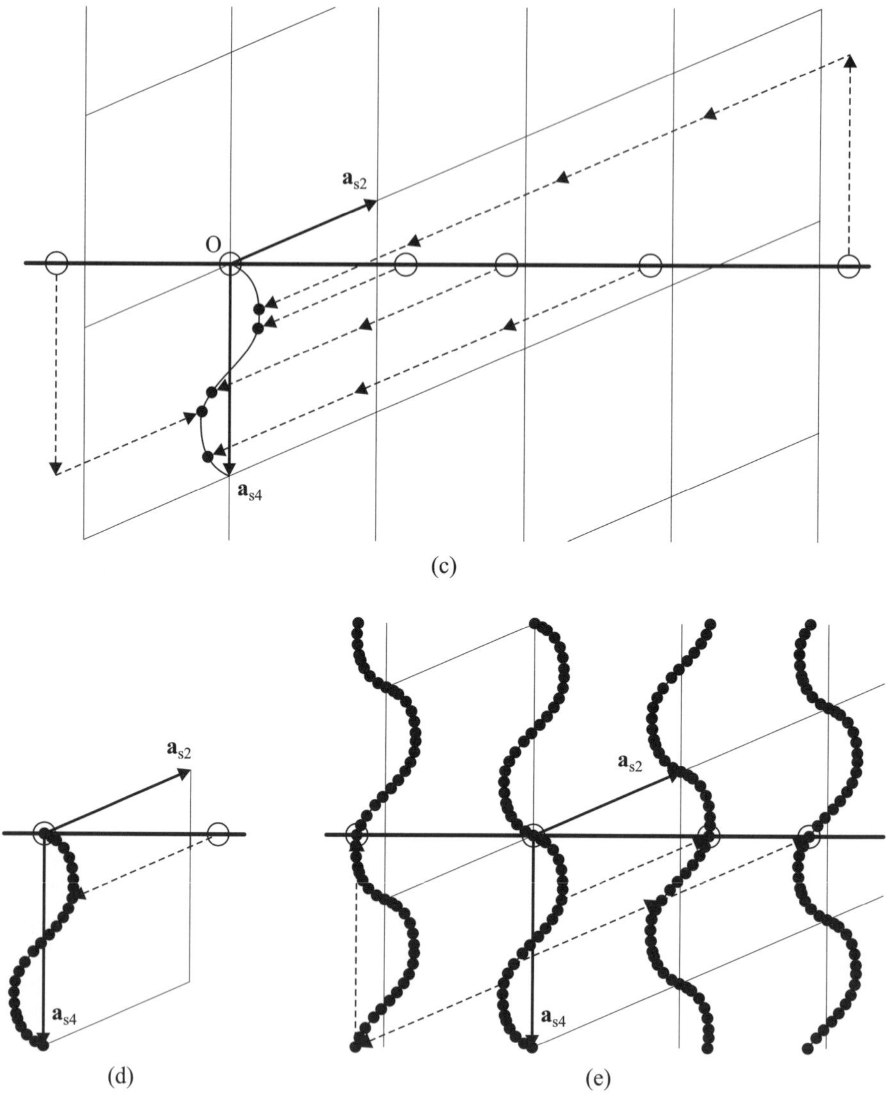

FIG. 2.6. Generalized electron density derived by superspace translations applied to the atoms in physical space. (a) Translation of atom 1 towards the first unit cell. (b) Translation of atom 2. (c) Translation of several atoms. (d) The result of translations applied to many atoms. (e) The string of atoms in the first unit cell translated to the other unit cells in superspace. The superspace lattice is highlighted by grid lines. Circles represent atoms in physical space. Dots are atoms in superspace obtained by translation from atoms in physical space.

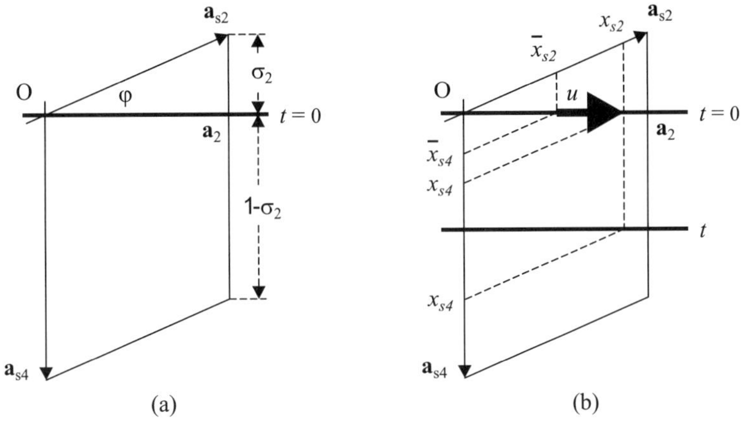

FIG. 2.7. (a) Geometry of the unit cell of the direct lattice in superspace. (b) Superspace coordinates for the basic and modulated structures.

of the basic structure and the modulation wave vector, both defined in physical space [eqn (2.9) and Fig. 2.7(a)]. Employing this simple geometrical relation, the superspace coordinates of a point in physical space are obtained as [Fig. 2.7(b)],

$$\begin{aligned} x_{si} &= x_i \qquad\qquad i = 1,2,3 \\ x_{s4} &= \mathbf{q}\cdot\mathbf{x}\,, \end{aligned} \tag{2.18}$$

where $\mathbf{x} = (x_1, x_2, x_3)$ are the physical-space coordinates with respect to the lattice, Λ, of the basic structure. In the present example, with one non-zero component of the modulation wave vector, the fourth superspace coordinate is $x_{s4} = \sigma_2\, x_2 = \sigma_2\, x_{s2}$, in agreement with Fig. 2.7. The coordinates of a point outside physical space are obtained by the addition to $\mathbf{q}\cdot\mathbf{x}$ of the distance between this point and physical space. This distance will be indicated by the parameter t [Fig. 2.7(b)]:

$$x_{s4} = t + \mathbf{q}\cdot\mathbf{x}\,. \tag{2.19}$$

All points with the same distance to physical space form a three-dimensional space that intersects the fourth coordinate axis in the point $\mathbf{x}_s = (0,0,0,t)$. Physical space is recovered for $t = 0$.

With these definitions, the superspace coordinates of the basic-structure position of atom j follow as [eqn (1.3)]

$$\begin{aligned} \bar{x}_{si}(j) &= \bar{x}_i(j) = l_i + x_i^0(\mu) \qquad i = 1,2,3 \\ \bar{x}_{s4}(j) &= \bar{x}_4(j) = t + \mathbf{q}\cdot\bar{\mathbf{x}}(j)\,. \end{aligned} \tag{2.20}$$

As defined in eqn (1.24), the number j stands for atom μ in unit cell $\mathbf{L}$ of the basic structure. Notice that the fourth superspace coordinate is equal to the argument $(\bar{x}_4)$ of the modulation functions, as it was previously defined in eqn

(1.5). The superspace coordinates for the position of atom j in the modulated structure are obtained as [eqn (1.9) and Fig. 2.7(b)]

$$\begin{aligned} x_{si}(j) &= x_i(j) = \bar{x}_{si}(j) + u_i^{\mu}(\bar{x}_{s4}) \qquad i = 1, 2, 3 \\ x_{s4}(j) &= \bar{x}_{s4}(j) + \mathbf{q}\cdot\mathbf{u}^{\mu}(\bar{x}_{s4}), \end{aligned} \tag{2.21}$$

where $\bar{x}_{s4}$ has been substituted for $\bar{x}_4$ in the argument of the modulation functions.

A superspace atom is defined by the set of points $\mathbf{x}_s(j)$ in eqn (2.21) for $0 \leqslant t < 1$. The generalized electron density in a point $\mathbf{x}_s = (x_{s1}, x_{s2}, x_{s3}, t+\mathbf{q}\cdot\bar{\mathbf{x}})$ is obtained as the atomic electron density of a physical-space atom centred on the point $\mathbf{x}_s(j)$ in section t of superspace (Section 1.2),

$$\rho_{sj}(\mathbf{x}_s) = \rho_j\left[x_{s1} - x_{s1}(j), x_{s2} - x_{s2}(j), x_{s3} - x_{s3}(j)\right], \tag{2.22}$$

where the property has been used that $\mathbf{a}_i$ is the projection of $\mathbf{a}_{si}$ onto physical space with $x_i = x_{si}$ $(i = 1, 2, 3)$. It is emphasized again that $\mathbf{x}_s$ and $\mathbf{x}_s(j)$ must be chosen in a single section t, in order to obtain the electron density in $\mathbf{x}_s$ by eqn (2.22).

2.4 *t*-Plots

2.4.1 *Interatomic distances*

A section of superspace perpendicular to the fourth axis $\mathbf{a}_{s4}$ is a three-dimensional subspace of superspace. Decomposition of the fourth superspace coordinate into the sum of $\mathbf{q}\cdot\mathbf{x}$ and a remaining amount t [eqn (2.19)] allows a section to be defined by the relation

$$t = \text{constant}. \tag{2.23}$$

Subspaces intersecting the first unit cell in different points on the fourth axis are obtained for values of t in the interval $0 \leqslant t < 1$. Translational symmetry of the crystal structure in superspace determines that different sections t are equivalent to each other, if they differ by a lattice translation in superspace. For example, application of the translation $\mathbf{a}_{s2}$ to the section t in Fig. 2.8(a) results in a section that must be equivalent to the original one, and that is defined by

$$t' = t - \mathbf{q}\cdot\mathbf{L}, \tag{2.24}$$

where $\mathbf{L}$ is a superspace lattice translation, equal to $\mathbf{a}_{s2}$ in the example. For incommensurate crystals, a lattice translation in superspace can always be found, such that the section t is translated to the section $t = 0$ (compare Fig. 2.6). It follows that any section t is equivalent to physical space. Each section of superspace perpendicular to $\mathbf{a}_{s4}$ provides an alternative description of the modulated crystal structure, all of which are equivalent.

The equivalence of different sections t allows the interpretation of the generalized crystal structure and generalized electron density as the juxtaposition

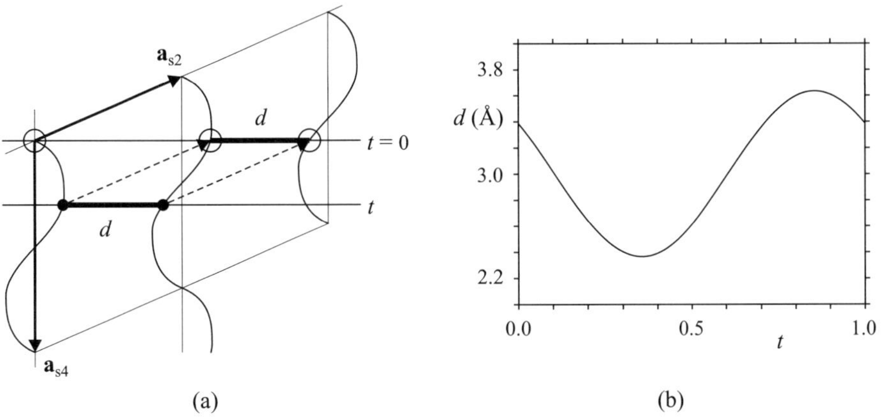

FIG. 2.8. (a) An interatomic distance in the first unit cell of section t is equal to an interatomic distance in section $t = 0$. (b) t-plot of the distance between atoms with $\bar{x}_2 = 0$ and $\bar{x}_2 = 1$ in Fig. 2.8(a). The basic-structure distance is equal to $a_2 = 3.0$ Å.

of an infinite number of copies of the physical-space structure or physical-space electron density. The phase of the modulation wave in copy t of physical space is shifted by an amount t as compared to the section $t = 0$. Alternatively, this phase shift is brought about by a lattice translation $\mathbf{L}$, such that $t = \mathbf{q} \cdot \mathbf{L}$ (modulo 1). The required translation might be short, as for $t = \sigma_2$, or it can be impossibly long, as for values of t close to σ_2.

The second method of construction of superspace has already shown that the displacement of an atom in section t out of its basic structure position is equal to the displacement of this atom somewhere in section $t = 0$. Here, this property is found to be a property of the entire crystal structure. Thus, relative displacements of neighbouring atoms in section t are equal to relative displacements of neighbouring atoms somewhere in physical space [Fig. 2.8(a)]. Incommensurability determines that distances between neighbouring atoms assume different values for pairs of atoms in different unit cells of the basic structure. Instead of studying the table of infinitely many values for the distance, the dependence of this distance on the physical space section t can be analysed. Because of the periodicity in superspace, all information is contained in one period along the fourth coordinate, i. e. for $0 \leqslant t < 1$ [Fig. 2.8(b)].

The distance between two atoms of the modulated structure can be computed from the physical-space coordinates of these atoms [eqn (1.9)]. The vector connecting the atoms 0 and 1 is

$$\begin{aligned} \Delta\mathbf{x} &= \mathbf{x}(1) - \mathbf{x}(0) \\ &= [\bar{\mathbf{x}}(1) - \bar{\mathbf{x}}(0)] + \left(\mathbf{u}^1[t + \mathbf{q}\cdot\bar{\mathbf{x}}(1)] - \mathbf{u}^0[t + \mathbf{q}\cdot\bar{\mathbf{x}}(0)]\right) . \end{aligned} \tag{2.25}$$

The distance follows as the length of the vector $\Delta\mathbf{x}$, and it is given in Fig. 2.8(b) as a function of t. Equation (2.25) shows that the distance between two atoms involves the distance in the basic structure, $|\bar{\mathbf{x}}(1) - \bar{\mathbf{x}}(0)|$, and a deviation from it. The variation of the distance in the modulated structure, $(\,|\Delta\mathbf{x}| - |\bar{\mathbf{x}}(1) - \bar{\mathbf{x}}(0)|\,)$, is not equal to the magnitude of $\big(\mathbf{u}^1[\bar{x}_{s4}(1)] - \mathbf{u}^0[\bar{x}_{s4}(0)]\big)$, because $\big(\mathbf{u}^1[\bar{x}_{s4}(1)] - \mathbf{u}^0[\bar{x}_{s4}(0)]\big)$ is not parallel to $[\bar{\mathbf{x}}(1) - \bar{\mathbf{x}}(0)]$. It follows that the average distance (average of $|\Delta\mathbf{x}|$ over t) is not equal to the distance in the basic structure, although in many compounds both values are nearly equal.

The distance between two points in superspace is obtained from the superspace coordinates. Substitution of atomic coordinates [eqn (2.18)] into the general expression for the distance in superspace shows that eqn (2.25) is not easily recovered. However, the distance between arbitrary points in superspace is a meaningless quantity, because the result will depend on the length of $\mathbf{a}_{s4}$, that can be chosen arbitrarily. Properties of crystals can only be obtained if atoms from which they are calculated are in a single physical-space section. Distances within a physical-space section are most easily calculated after a coordinate transformation in superspace towards a coordinate system defined by [compare eqn (2.9)]

$$\Sigma_r : \begin{cases} \mathbf{a}_{rsi} = (\mathbf{a}_i, 0) & i = 1, 2, 3 \\ \mathbf{a}_{rs4} = (0, \mathbf{b})\,. \end{cases} \tag{2.26}$$

Coordinates with respect to Σ_r follow from the coordinates with respect to Σ by arguments similar to arguments in Section 2.3.2 that gave the relation between superspace and physical-space coordinates. The result is

$$\begin{aligned} x_{rsi}(j) &= x_{si}(j) = x_i(j) \\ x_{rs4}(j) &\phantom{= x_{si}(j)} = t\,. \end{aligned} \tag{2.27}$$

Unlike vectors $\mathbf{a}_{si}$, the vectors $\mathbf{a}_{rsi}$ $(i = 1, 2, 3)$ do not describe translational symmetry of the crystal, but they are perpendicular to $\mathbf{a}_{rs4}$. The latter property facilitates the calculation of distances between points in superspace. The distance d between points on the strings representing atoms 0 and 1 is

$$d = \Big(\,|\Delta\mathbf{x}|^2 + ([x_{rs4}(1) - x_{rs4}(0)]a_{rs4})^2\Big)^{1/2}\,. \tag{2.28}$$

The case of interest is that both points are in a single physical-space section. Then $x_{rs4}(1) = x_{rs4}(0) = t$, and the second term in eqn (2.28) is zero. The expression for the distance in superspace reduces to that for the distance in physical space [eqn (2.25)]. Different physical-space sections of superspace are represented by different values of t. This variation of t directly corresponds to the variation of the initial phase of the modulation wave in physical space, which was denoted by the parameter t too.

A plot like Fig. 2.8(b) is called a t-plot. t-Plots can be made for many quantities derived from the modulated structure, including the components of modulation functions, magnitudes of displacements, occupation probabilities of atomic

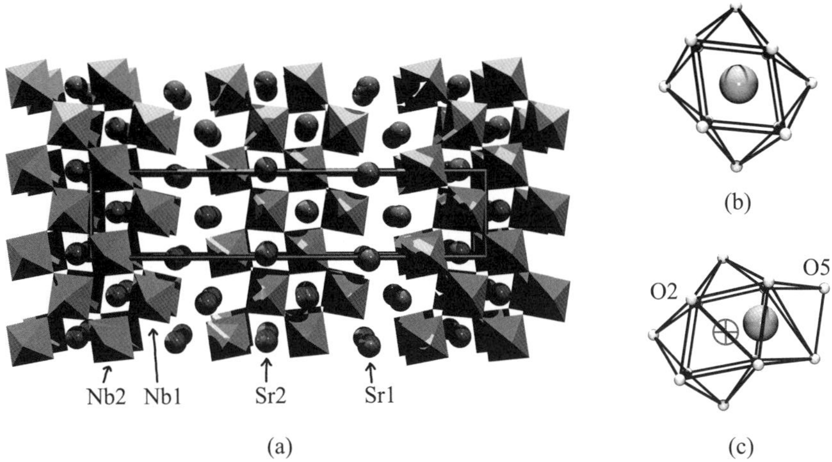

FIG. 2.9. (a) Perspective view of the basic structure of $Sr_2Nb_2O_7$. Spheres represent Sr atoms. NbO_6 groups are shown as shaded octahedra. $\mathbf{a}_2$ is from left to right and $\mathbf{a}_3$ is pointing upwards. (b) Coordination of Sr2 by oxygen atoms. (c) Coordination of Sr1 by oxygen atoms. Reprinted with permission from Daniels *et al.* (2002), copyright (2002) IUCr.

sites, directions of magnetic moments, distances between atoms, bond angles, torsion angles and orientations of atomic groups. t-Plots provide a concise representation of the variation of a single quantity in the modulated structure, as the variation of this quantity over the phase of the modulation wave. However, the most important property of t-plots is that values of different quantities can directly be compared. Distances and angles within a cluster of atoms at a single value of t define the geometry of this cluster somewhere in physical space. Different values of t then provide the variation of the geometry of the cluster, as it occurs in different unit cells of the basic structure. These properties of t-plots will be illustrated by an analysis of the incommensurately modulated structure of $Sr_2Nb_2O_7$ (Daniels *et al.*, 2002).

2.4.2 *Modulated structure of $Sr_2Nb_2O_7$*

$Sr_2Nb_2O_7$ is a layered compound with a crystal structure derived from the perovskite structure type. Layers of corner-sharing NbO_6 octahedra are separated by planes of additional oxygen. Within the layers, Sr atoms are in distorted cuboctahedral coordination by oxygen. The sizes of the cavities are adapted to the size of Sr by cooperative rotations of the NbO_6 octahedra [Fig. 2.9(a)]. At the borders of the layers, one corner of the cuboctahedron is replaced by a pair of oxygen atoms of the next layer, with the result that the Sr atom is displaced from the centre of the cuboctahedron [encircled cross in Fig. 2.9(c)] towards the neighbouring layer. The incommensurate modulation is the result of internal

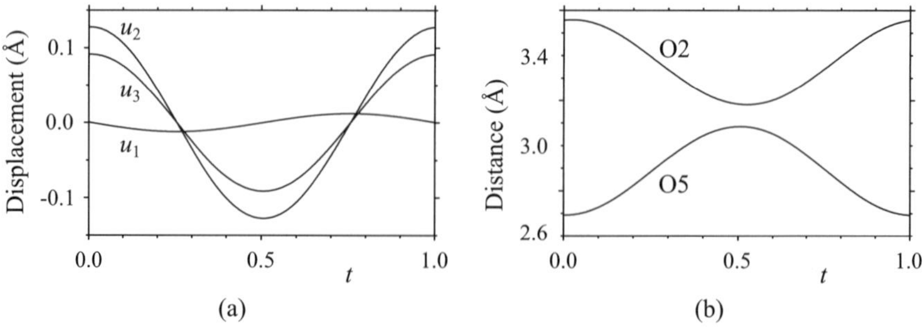

FIG. 2.10. (a) The three components of the modulation function of Sr1 in $Sr_2Nb_2O_7$. (b) Distances between Sr1 and O2 and between Sr1 and O5. O2 and O5 are defined in Fig. 2.9(c). Coordinates from Daniels *et al.* (2002).

strain between the layers about the Sr1 atoms (Daniels *et al.*, 2002).

The basic structure is orthorhombic $Cmc2_1$, with lattice parameters $a_1 = 3.9544$, $a_2 = 26.767$ and $a_3 = 5.6961$ Å. The primitive unit cell of the C-centred lattice of the basic structure contains $N = 22$ atoms, of which 11 atoms are independent by symmetry. The one-dimensional, incommensurate modulation is a displacive modulation characterized by a modulation wave vector $\mathbf{q} = (0.488, 0, 0)$. The refined structure model involves single-harmonic modulation functions [$n = 1$ in eqn (1.10)] for each atom in the unit cell of the basic structure.

t-Plots of the three components of the modulation function of Sr1 show that the modulation of this atom is almost a transverse wave [Fig. 2.10(a)]. Values of the displacements along $\mathbf{a}_1$, $\mathbf{a}_2$ and $\mathbf{a}_3$ at a single value of t represent displacements of a single Sr atom. Accordingly, t-plots allow the property to be derived, that maximum values of displacements along $\mathbf{a}_2$ and $\mathbf{a}_3$ occur simultaneously, both at $t = 0$ and at $t = 0.5$.

Coordination polyhedra are the most important structural feature from a chemical point of view. In a simplified analysis, they are characterized by the distances between a central atom and its neighbours. Here, the distances between Sr1 and two of its oxygen neighbours have been selected for display [Fig. 2.9(c)]. Rather large variations are found for both Sr1–O distances [Fig. 2.10(b)]; large in the sense that $d_{\mathrm{max}} - d_{\mathrm{min}}$ is of similar size as the amplitudes of the modulation functions. Inspection of the t-plot shows that a minimum of d(Sr1–O2) coincides with a maximum of d(Sr1–O5), and the other way around. This feature is easily understood from the geometry of the coordination of Sr1, where displacements of Sr1 towards O2 correspond to an elongation of the distance between Sr1 and O5, and displacements of Sr1 towards O5 correspond to an elongation of the distance between Sr1 and O2. This interpretation is supported by the t-plot of the modulation function of Sr1: maximum displacements in Fig. 2.10(a) are found at the same values of t at which the distances Sr1–O2 and Sr1–O5 assume

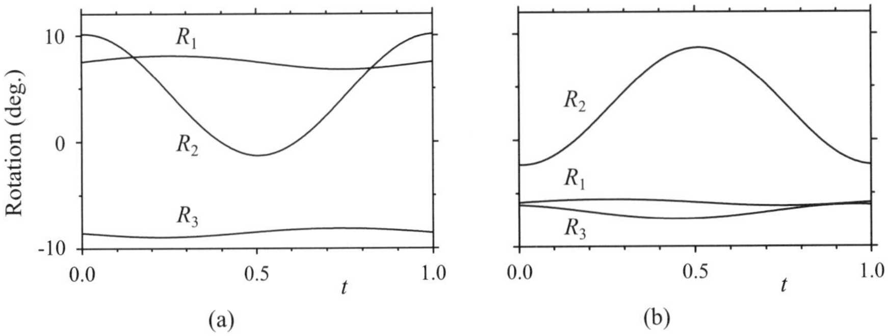

FIG. 2.11. Orientations of NbO_6 octahedral groups in $Sr_2Nb_2O_7$. (a) For Nb1. (b) For Nb2. Rotations (R_i) about the coordinate axis $i = 1, 2, 3$ are relative to an orientation aligned with the coordinate axes. Coordinates from Daniels *et al.* (2002).

extremal values in Fig. 2.10(b).

Modulation amplitudes of Nb are almost zero, while those of O are comparable in size to the displacements of Sr1. With Nb–O distances nearly independent of t, this suggests that a major part of the modulation can be described as variations of the orientations of rigid NbO_6 octahedra. Orientations of rigid groups can be described in many different ways. Here, the angles of rotation about the three crystallographic axes are given as a function of t for the octahedral groups surrounding Nb1 and Nb2. It appears that the modulations of the oxygen atoms can be described as rotations of NbO_6 groups about $\mathbf{a}_2$ (Fig. 2.11). The t-plots show that rotations are in anti-phase for two octahedra sharing an oxygen atom [Nb1 and Nb2 in Fig. 2.9(a)], in accordance with the absence of distortions of the octahedra. Taking correlations between t-plots one step further, Fig. 2.10 and Fig. 2.11 show that maximum rotations of NbO_6 groups coincide with maximum displacements of Sr1 atoms. This correlation suggest that NbO_6 groups rotate so as to avoid too short Sr1–O contacts.

2.5 Structure factor

2.5.1 *Fourier transform in superspace*

In Section 1.3 it has been shown that modulated crystals diffract in the form of Bragg reflections. The structure factor has been calculated from the positions of the atoms in a modulated structure, without reference to superspace [eqn (1.44)]. The generalized electron density, $\rho_s(\mathbf{x}_s)$, is defined as the inverse Fourier transform of the structure factors in superspace [eqn (2.12)]. Therefore, the structure factors of a modulated crystal are the Fourier transform of the generalized electron density in one unit cell of superspace. The latter can be expressed as a sum of atomic electron densities [compare eqn (2.22)],

$$\rho_s(\mathbf{x}_s) = \sum_{\mu=1}^{N} \rho_\mu \left[x_{s1} - x_{s1}(\mu),\, x_{s2} - x_{s2}(\mu),\, x_{s3} - x_{s3}(\mu)\right], \tag{2.29}$$

where $\mathbf{x}_s(\mu)$ corresponds to the set of points $\mathbf{x}_s(j)$ in eqn (2.21) with $0 \leqslant t < 1$. Equation (2.29) is only valid if both $\mathbf{x}_s$ and $\mathbf{x}_s(\mu)$ refer to the same section t.

The Fourier transform of the generalized electron density is in the independent atom approximation,

$$\begin{aligned} F(\mathbf{S}_s) &= \int_{\text{cell}} \rho_s(\mathbf{x}_s) \exp[2\pi i \mathbf{S}_s \cdot \mathbf{x}_s]\, \mathrm{d}\mathbf{x}_s \\ &= \sum_{\mu=1}^{N} \int \rho_\mu \left[x_{s1} - x_{s1}(\mu),\, x_{s2} - x_{s2}(\mu),\, x_{s3} - x_{s3}(\mu)\right] \\ &\qquad \times \exp[2\pi i \mathbf{S}_s \cdot \mathbf{x}_s]\, \mathrm{d}\mathbf{x}_s\,. \end{aligned} \tag{2.30}$$

The integral can be evaluated after transformation to the coordinate system Σ_r [eqn (2.26)]. The fourth coordinates of points in section t then are equal to $x_{rs4} = x_{rs4}(\mu) = t$. Scattering vectors $\mathbf{S}_s = (S_1, S_2, S_3, S_4)$ are defined by components with respect to the reciprocal lattice Σ^*, and the scalar product follows as

$$\mathbf{S}_s \cdot [\mathbf{x}_{rs} - \mathbf{x}_{rs}(\mu)] = \sum_{i=1}^{3} (S_i + S_4\sigma_i)[x_{rsi} - x_{rsi}(\mu)] = \mathbf{S} \cdot [\mathbf{x} - \mathbf{x}(\mu)] \tag{2.31}$$

$$\mathbf{S}_s \cdot \mathbf{x}_{rs}(\mu) = \mathbf{S} \cdot \mathbf{x}(\mu) + S_4 t\,, \tag{2.32}$$

where $\mathbf{S}$ and $\mathbf{x}$ are vectors in physical space. Recognizing that $\bar{\mathbf{x}}(\mu) = \mathbf{x}^0(\mu)$ for atoms in the first unit cell, the structure factor becomes

$$\begin{aligned} F(\mathbf{S}_s) = \sum_{\mu=1}^{N} f_\mu(\mathbf{S}) \exp\left[2\pi i \mathbf{S} \cdot \mathbf{x}^0(\mu)\right] \\ \times \int_0^1 \exp\left[2\pi i \left(\mathbf{S} \cdot \mathbf{u}^\mu \left[t + \mathbf{q} \cdot \mathbf{x}^0(\mu)\right] + S_4\, t\right)\right] \mathrm{d}t\,. \end{aligned} \tag{2.33}$$

The basic-structure positions can be removed from the remaining integral by choosing the new integration variable $\tau = t + \mathbf{q} \cdot \mathbf{x}^0(\mu)$. Employing the periodicity of modulation functions this gives

$$F(\mathbf{S}_s) = \sum_{\mu=1}^{N} f_\mu(\mathbf{S}) g_\mu(\mathbf{S}_s) \exp\left[2\pi i (\mathbf{S} - S_4 \mathbf{q}) \cdot \mathbf{x}^0(\mu)\right] \tag{2.34a}$$

with

$$g_\mu(\mathbf{S}_s) = \int_0^1 \exp[2\pi i \left(\mathbf{S} \cdot \mathbf{u}^\mu[\tau] + S_4\, \tau\right)]\, \mathrm{d}\tau\,. \tag{2.34b}$$

The result is equal to the expression for the structure factor that was derived from the coordinates of the atoms in physical space [eqn (1.44)]. It shows that the diffraction by incommensurately modulated crystals can be obtained as the Fourier transform of the superspace electron density. Bragg reflections follow from the lattice periodicity in superspace, by arguments that are a simple extension of the derivation of Bragg reflections of periodic crystals (Section 1.2).

2.5.2 *Displacive modulations*

The structure factor of eqn (2.34) can be computed for modulation functions of arbitrary shapes by numerical evaluation of the integral in eqn (2.34b). The sample frequency of the integrand limits the maximum order of harmonics in the Fourier expansions of the modulation functions, that can be taken into account [eqn (1.10)]. Therefore, it is sometimes advisable to obtain analytical solutions of the integral for highly anharmonic functions, like block waves or saw tooth functions (Böhm, 1977; Petricek *et al.*, 1990). On the other hand, many incommensurate crystals have smooth modulation functions that can be described by a few harmonics. The integral can then be evaluated towards expressions involving Bessel functions.

As an example a simple modulation is considered, with $\mathbf{q} = (0, \sigma_2, 0)$ and $\mathbf{u}(\bar{x}_{s4}) = (A_1, A_2, A_3)\sin(2\pi\bar{x}_{s4})$ [eqn (1.10)]. With the Jacobi–Auger expansion of $\exp[2\pi i\mathbf{S}\cdot\mathbf{A}\sin(2\pi\tau)]$ into an infinite series of Bessel functions, the modulation scattering factor of Bragg reflection $\mathbf{H}_s$ becomes,

$$\begin{aligned} g_\mu(\mathbf{H}_s) &= \int_0^1 \exp[2\pi i\,\mathbf{H}\cdot\mathbf{A}\sin(2\pi\tau) + h_4\,\tau]\,\mathrm{d}\tau \\ &= \sum_{m=-\infty}^{\infty} J_{-m}(2\pi\,\mathbf{H}\cdot\mathbf{A}) \int_0^1 \exp[2\pi i(-m\tau + h_4\tau)]\,\mathrm{d}\tau \\ &= J_{-h_4}(2\pi\,\mathbf{H}\cdot\mathbf{A})\,. \end{aligned} \tag{2.35}$$

The structure factor follows as

$$F(\mathbf{H}_s) = \sum_{\mu=1}^{N} f_\mu(\mathbf{H})\,J_{-h_4}[2\pi\,\mathbf{H}\cdot\mathbf{A}(\mu)]\,\exp\left[2\pi i(\mathbf{H} - h_4\mathbf{q})\cdot\mathbf{x}^0(\mu)\right]. \tag{2.36}$$

$J_m(x)$ is a Bessel function of the first kind of order m with $J_{-m}(x) = (-1)^m J_m(x)$. The scattered intensity of Bragg reflection $\mathbf{H}_s$ is proportional to $|F(\mathbf{H}_s)|^2$. Its magnitude for satellite reflections of order h_4 is proportional to $[J_{h_4}(2\pi\,\mathbf{H}\cdot\mathbf{A})]^2$, where $\mathbf{A}$ is some average modulation amplitude of the N atoms in the unit cell.

Equation (2.36) allows some general properties of the diffraction of modulated crystals to be derived. The first result is that a single-harmonic displacive modulation gives non-zero intensities for—in principle—all orders of satellite reflections. The intensities rapidly decrease with increasing satellite order, as is shown by the values of $[J_m(x)]^2$ in Fig. 2.12, because the arguments $2\pi\,\mathbf{H}\cdot\mathbf{A}$ are less than 1 for small modulation amplitudes and accessible reflections.

Compared to the basic structure, the intensities of main reflections are reduced by a factor $[J_0(2\pi\,\mathbf{H}\cdot\mathbf{A})]^2$. The amount of intensity diffracted into the

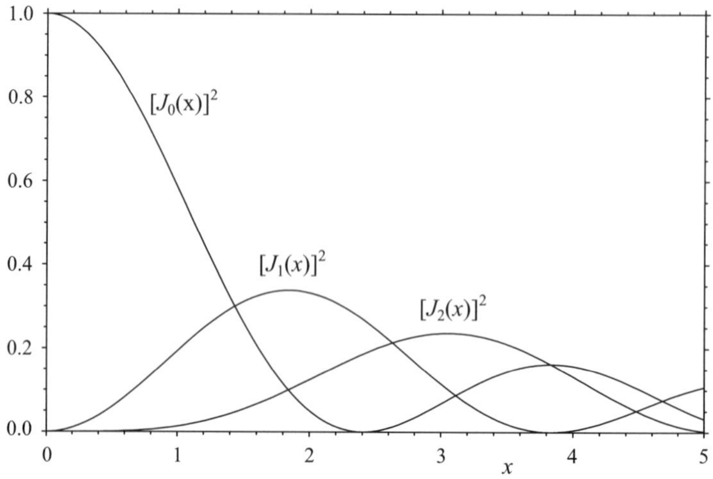

FIG. 2.12. Square of the Bessel functions, $[J_m(x)]^2$ for $m = 0, 1, 2$.

satellite reflections is of the same order of magnitude as the reduction of the intensities of main reflections. This is easily checked by comparing $[J_0(x)]^2$ and $[J_1(x)]^2$ in the limit of small x. The failure to determine modulation amplitudes from the intensities of main reflections is explained by the fact that the Debye–Waller factor provides a reasonable approximation to $[J_0(x)]^2$ in the limit of small x, and increased values of the temperature parameters can mimic the modulations as far as they affect the intensities of main reflections (see Section 10.2).

2.5.3 *Occupational modulations*

The second type of structural modulations is occupational modulation. A specific atom site in the unit cell of the basic structure might have a probability of occupation, p, that varies from one unit cell to the next, according to a modulation wave (Fig. 1.3). Similarly, different chemical elements might concurrently occupy one site or one atom might occupy different sites with complementary occupation probabilities.

As an example an occupational modulation is considered with $\mathbf{q} = (0, \sigma_2, 0)$ and with an occupational probability of site μ given by

$$p^{\mu}(\bar{x}_{s4}) = P^0(\mu)\left[1 + P^1(\mu)\sin(2\pi\bar{x}_{s4})\right]. \tag{2.37}$$

$P^0(\mu)$ is the average occupancy, that must be less than one. $P^1(\mu)$ is the single modulation parameter for atom μ. Because the probability of occupation must be less than or equal to one for every site in the structure, the modulation parameter must fulfil the relation $0 \leqslant P^0(\mu)\left[1 \pm P^1(\mu)\right] \leqslant 1$. $p^{\mu}(\bar{x}_{s4})$ can be introduced into the expression for the structure factor as an additional factor to the atomic scattering factor, resulting in

$$F(\mathbf{H}_s) = \sum_{\mu=1}^{N} f_\mu(\mathbf{H})\, P^0(\mu)\, g_\mu(\mathbf{H}_s)\, \exp\bigl[2\pi i(\mathbf{H} - h_4\mathbf{q})\cdot\mathbf{x}^0(\mu)\bigr]\,, \tag{2.38a}$$

with

$$\begin{aligned} g_\mu(\mathbf{H}_s) &= \int_0^1 \bigl[1 + P^1(\mu)\, \sin(2\pi\tau)\bigr]\, \exp[2\pi i h_4\, \tau]\; \mathrm{d}\tau \\ &= \delta_{h_4 0} + \frac{1}{2i} P^1(\mu)\, \bigl(\delta_{h_4 1} + \delta_{h_4 \bar{1}}\bigr)\,, \end{aligned} \tag{2.38b}$$

where δ_{nm} is the Kronecker delta defined by $\delta_{nm} = 1$ if $n = m$ and otherwise $\delta_{nm} = 0$.

A major difference to structures with displacive modulations is that each harmonic wave $\sin(2\pi n\bar{x}_{s4})$ of occupational modulation leads to only one pair of satellite reflections around each main reflection, with satellite order $|h_4| = n$. The intensities of main reflections correspond to intensities of a periodic basic structure with occupational probabilities equal to the average occupational probability, and they are independent of the modulation. Disorder, expressed by occupational probabilities smaller than one, gives rise to diffuse scattering. The intensities of the satellite reflections are obtained by a reduction of the diffuse scattering, rather than borrowed from the main reflections.

Occupational modulations are usually accompanied by displacement modulations, and both parts of the modulation will contribute to the intensities of Bragg reflections.

3

SYMMETRY OF MODULATED CRYSTALS

3.1 Introduction

Incommensurately modulated crystals do not posses translational symmetry according to a three-dimensional lattice. Translational symmetry of the basic structure is lost in the modulated structure, because the modulation wave determines shifts of the atoms that are different in different unit cells of the basic structure. However, lattice translations perpendicular to the modulation wave vector survive as symmetry of the modulated structure, because the phase of a wave is constant in a plane perpendicular to the wave vector. Modulated crystals with modulation wave vectors along reciprocal lattice vectors of the basic structure possess two-dimensional translational symmetry according to the lattice plane perpendicular to the modulation wave vector. Modulation wave vectors in general directions [three irrational components of $\mathbf{q}$ in eqn (1.4)] lead to complete loss of translational symmetry, while for modulation wave vectors in reciprocal lattice planes of the basic structure (two irrational components of $\mathbf{q}$), only a lattice line survives as translational symmetry of the modulated structure.

Lost translational symmetry re-appears as hidden symmetry of the modulated structure. The essential step towards uncovering this hidden symmetry is the notion that the atomic positions can be obtained as the sum of lattice periodic positions $\bar{\mathbf{x}}$ and values of modulation functions $\mathbf{u}^{\mu}(\bar{x}_{s4})$ (Section 1.1.2). Subsequently it is recognized that a lattice translation $\mathbf{L}$ of the basic structure, combined with a shift of the phase of the modulation wave by $(-\mathbf{q}\cdot\mathbf{L})$ represents symmetry of modulated crystals. The superspace approach has formalized this hidden symmetry in the form of translational symmetry of the generalized electron density according to a lattice Σ in a four-dimensional space: in (3+1)-dimensional superspace (Chapter 2).

The symmetry of periodic crystals is given by space groups. Space groups are based on a three-dimensional lattice of translations (Λ) that give the translational symmetry. Other symmetry operators that may be part of a space group are proper and improper rotations as well as screw axes and glide planes. The counterpart of translational symmetry of periodic crystals is the lattice Σ for aperiodic crystals. Σ gives symmetry of modulated structures through true translational symmetry of the generalized electron density in superspace. The question emerges about rotational symmetry of aperiodic crystals. First, it is noticed that modulated crystals do not possess rotational symmetry, even if a two-dimensional lattice were to survive as translational symmetry. Suppose that an axis or plane through the origin of the lattice of translations of the basic structure gives rotational or mirror symmetry of the modulated crystal. This symmetry element is then repeated by lattice translations in physical space, that form at most a lattice plane. These points of symmetry are an exception rather than a property of the modulated crystal, because they are not homogeneously distributed throughout space. It is easy to select a finite crystal of any size, that does not contain these symmetry points. Homogeneity is one of the fundamental properties that apply to both periodic and aperiodic crystals.

Hidden rotational symmetry is of importance to incommensurate crystallography, because it provides relations between structural parameters. Symmetries of physical properties depend on the symmetry of the crystal structure. For example, optical activity and other non-linear properties are absent in crystals with inversion symmetry. Extensive experimental studies have been made towards establishing the presence of optical activity in incommensurately modulated crystals that have but hidden inversion symmetry (Kobayashi, 1990). However, more recent studies have shown that optical activity is zero in incommensurately modulated crystals with hidden inversion symmetry, despite the absence of true inversion symmetry in physical space (Folcia *et al.*, 1993). These results suggest that hidden rotational symmetry exists, and that it is to be considered as real symmetry of incommensurate crystals. Uncovering the hidden rotational symmetries is best done through an analysis of the diffraction of aperiodic crystals.

3.2 Diffraction symmetry

Diffraction is one kind of physical property to which Neumann's Principle applies:

> Symmetries of a physical property of a material include the crystal point group, but may include more symmetry.

Experimental observations of the diffraction of aperiodic crystals have shown that diffraction possesses true rotational symmetry, in complete analogy to rotational symmetry of the diffraction of periodic crystals. Because diffraction takes place in physical space, its symmetry is given by a three-dimensional point group. By Neumann's Principle any rotational symmetry or hidden rotational symmetry of the atomic structures of aperiodic crystals then is based on a three-dimensional point group. Quasicrystals are aperiodic crystals with non-crystallographic point symmetries of their diffraction patterns (Fig. 1.7 in Section 1.1.4). The non-crystallographic point groups are easily enumerated as the two icosahedral point groups and the uniaxial groups n/mmm with $n =$ integer, that—together with their subgroups—are defined in complete analogy to $4/mmm$ and $6/mmm$ crystallographic point groups. Quasicrystals have been found with icosahedral, $5/mmm$, $8/mmm$, $10/mmm$ and $12/mmm$ symmetries (Steurer, 2004). They are not further analysed in this book.

Diffraction patterns of incommensurately modulated crystals contain main reflections that can be indexed on the basis of the three-dimensional lattice of the basic structure (Janner and Janssen, 1979). Point symmetry of the diffraction must be point symmetry of the main reflections. The latter define the reciprocal lattice of the basic structure, that has point symmetry according to one of the 32

TABLE 3.1. Admissible incommensurate wave vectors for one-dimensional modulations. Unique axis is $\mathbf{a}_3$. Hexagonal setting for the rhombohedral lattice.

Crystal system	Modulation wave vector	Crystal system	Modulation wave vector
Triclinic	$(\sigma_1, \sigma_2, \sigma_3)$	Tetragonal	$(0, 0, \sigma_3)$
Monoclinic	$(\sigma_1, \sigma_2, 0)$	Trigonal	$(0, 0, \sigma_3)$
	$(0, 0, \sigma_3)$	Hexagonal	$(0, 0, \sigma_3)$
Orthorhombic	$(\sigma_1, 0, 0)$	Cubic	none
	$(0, \sigma_2, 0)$		
	$(0, 0, \sigma_3)$		

crystallographic point groups. The diffraction symmetry of an incommensurate crystal thus is given by one of the 32 crystallographic point groups. An immediate consequence is that the lattice of the basic structure belongs to one of the 14 Bravais classes in three-dimensional space, with corresponding restrictions on the lattice parameters.

A symmetry operator R transforms main reflections amongst each other. Consequently, satellite reflections are imaged onto satellite reflections by the operator R too. For a one-dimensional modulation this implies that $R : \mathbf{q} \rightarrow \pm\mathbf{q}$. With the parameter $\epsilon = \pm 1$ the following relation is obtained,

$$\sigma R^{-1} - \epsilon^{-1}\sigma = \mathbf{0} \,. \tag{3.1}$$

The 1×3 matrix σ contains the components $(\sigma_1, \sigma_2, \sigma_3)$ of the modulation wave vector $\mathbf{q}$ [eqn (1.4)]. $\mathbf{0}$ is the 1×3 null matrix. R denotes a rotation in physical space as well as its representation by a 3×3 matrix. R^{-1} is the inverse of R. Juxtaposition of the symbols of two matrices implies the matrix product. Matrices R are defined by active transformation of the atomic coordinates according to

$$\begin{pmatrix} \bar{x}'_1 \\ \bar{x}'_2 \\ \bar{x}'_3 \end{pmatrix} = R \begin{pmatrix} \bar{x}_1 \\ \bar{x}_2 \\ \bar{x}_3 \end{pmatrix} . \tag{3.2}$$

Equation (3.1) defines admissible modulation wave vectors for one-dimensional modulations in each of the seven Bravais systems in three-dimensional space (Table 3.1).

As an example consider a primitive orthorhombic lattice of the basic structure. The lattice symmetry is generated by three mirror planes perpendicular to the coordinate axes. Application of eqn (3.1) for R equal to the mirror perpendicular to $\mathbf{a}_1$ (m_x) and with $\epsilon = 1$ gives,

$$(-\sigma_1, \sigma_2, \sigma_3) - (\sigma_1, \sigma_2, \sigma_3) = (0, 0, 0) \implies (-2\sigma_1, 0, 0) = (0, 0, 0) \,,$$

from which follows $\sigma_1 = 0$. Employing $\epsilon = 1$ for all three mirrors, m_x, m_y and m_z, then leads to $\sigma_1 = \sigma_2 = \sigma_3 = 0$, *i.e.* a modulated structure is not allowed for this choice of symmetry. With $\epsilon = -1$ for $R = m_z$ follows,

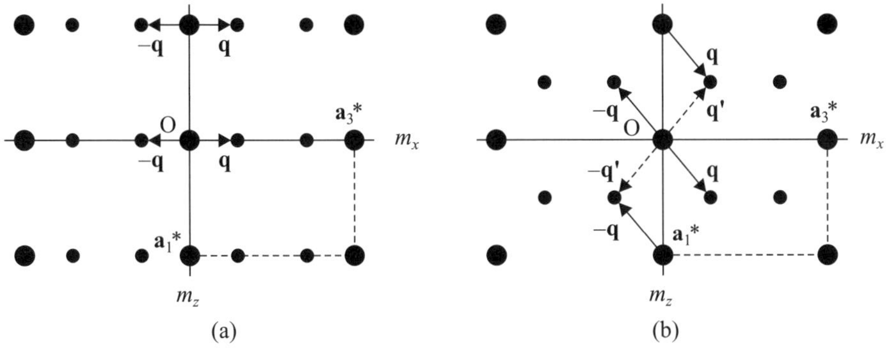

FIG. 3.1. Diffraction patterns of orthorhombic crystals with one-dimensional modulations. (a) $\mathbf{q} = (0,\, 0,\, \sigma_3)$. (b) $\mathbf{q} = (\frac{1}{2},\, 0,\, \sigma_3)$. m_x and m_z are mirror planes perpendicular to $\mathbf{a}_1$ and $\mathbf{a}_3$, respectively.

$$(\sigma_1, \sigma_2, -\sigma_3) - (-\sigma_1, -\sigma_2, -\sigma_3) = (0,0,0) \implies (2\sigma_1, 2\sigma_2, 0) = (0,0,0)\,,$$

from which follows $\sigma_1 = \sigma_2 = 0$. Combination of operators $(R, \epsilon) = (m_x, 1)$, $(m_y, 1)$ and $(m_z, \bar{1})$ leads to the condition $\sigma_1 = \sigma_2 = 0$, and the admissible incommensurate modulation wave vector in orthorhombic symmetry is $(0, 0, \sigma_3)$. Modulation wave vectors $(\sigma_1, 0, 0)$ and $(0, \sigma_2, 0)$ are obtained for the alternative choices $\epsilon = -1$ for m_x and $\epsilon = -1$ for m_y, respectively. The admissible incommensurate modulation wave vectors, *i.e.* solutions to eqn (3.1), depend on the point group. A solution does not exist for cubic point groups, and one-dimensional modulations are incompatible with cubic symmetry (Table 3.1).

The value ± 1 for ϵ is obtained in accordance with the action of R on the modulation wave vector leading to $\pm\mathbf{q}$. The values of ϵ for each three-dimensional rotation R are thus fixed by the choice of σ. In Section 3.4 it will be shown that (R, ϵ) define the rotational symmetry operators in superspace.

The requirement that satellite reflections are imaged onto satellite reflections can alternatively be fulfilled by allowing Umklapp terms. For a one-dimensional modulation this implies that $R : \mathbf{q} \to \mathbf{G} \pm \mathbf{q}$, where $\mathbf{G}$ is a reciprocal lattice vector of the basic structure. Equation (3.1) is replaced by the more general relation between R, ϵ and $\mathbf{q}$ given by

$$\sigma R^{-1} - \epsilon^{-1}\sigma = \mathbf{m}^*\,. \tag{3.3}$$

$\mathbf{m}^*$ is the 1×3 matrix of components (m_1^*, m_2^*, m_3^*) of a reciprocal lattice vector of the basic structure. Modulations with non-zero $\mathbf{m}^*$ and with $\mathbf{m}^*$ equal to the null vector are illustrated in Fig. 3.1 for a one-dimensional modulation in an orthorhombic lattice. Application of mirror symmetry to $\mathbf{q} = (0,\, 0,\, \sigma_3)$ gives $m_x : \mathbf{q} \to \mathbf{q}$ $(\epsilon = 1)$ and $m_z : \mathbf{q} \to -\mathbf{q}$ $(\epsilon = -1)$, in accordance with eqn (3.1). However, if $\mathbf{q} = (\sigma_1,\, 0,\, \sigma_3)$ has two non-zero components, $m_x : \mathbf{q} \to \mathbf{q}'$ and $m_z : \mathbf{q} \to -\mathbf{q}'$ [Fig. 3.1(b)]. The condition of a one-dimensional modulation

TABLE 3.2. Admissible incommensurate wave vectors with a non-zero rational part $\mathbf{q}_r$, for one-dimensional modulations. Unique axis is $\mathbf{a}_3$. Symbols P, A, B, C, F, I and R indicate the type of centring of the three-dimensional lattice.

Three-dimensional Bravais class	Modulation wave vectors		
Monoclinic – P	$(\sigma_1,\, \sigma_2,\, \frac{1}{2})$	$(\frac{1}{2},\, 0,\, \sigma_3)$	
Monoclinic – B	$(0,\, \frac{1}{2},\, \sigma_3)$		
Monoclinic – A	$(\frac{1}{2},\, 0,\, \sigma_3)$		
Orthorhombic – P	$(\sigma_1,\, 0,\, \frac{1}{2})$	$(\sigma_1,\, \frac{1}{2},\, 0)$	$(\sigma_1,\, \frac{1}{2},\, \frac{1}{2})$
	$(\frac{1}{2},\, \sigma_2,\, 0)$	$(0,\, \sigma_2,\, \frac{1}{2})$	$(\frac{1}{2},\, \sigma_2,\, \frac{1}{2})$
	$(0,\, \frac{1}{2},\, \sigma_3)$	$(\frac{1}{2},\, 0,\, \sigma_3)$	$(\frac{1}{2},\, \frac{1}{2},\, \sigma_3)$
Orthorhombic – C	$(1,\, 0,\, \sigma_3)$	$(0,\, 1,\, \sigma_3)$	
Orthorhombic – A	$(\frac{1}{2},\, 0,\, \sigma_3)$		
Orthorhombic – B	$(0,\, \frac{1}{2},\, \sigma_3)$		
Orthorhombic – F	$(1,\, 0,\, \sigma_3)$	$(0,\, 1,\, \sigma_3)$	
Tetragonal – P	$(\frac{1}{2},\, \frac{1}{2},\, \sigma_3)$		
Trigonal – P	$(\frac{1}{3},\, \frac{1}{3},\, \sigma_3)$		

is violated, because $\mathbf{q}' \neq \pm\mathbf{q}$. Umklapp terms can be introduced in the special case of $\sigma_1 = \frac{1}{2}$, resulting in $\mathbf{q}' = (-\mathbf{a}_1^* + \mathbf{q})$. Equation (3.3) is now fulfilled by $\mathbf{q} = (\frac{1}{2},\, 0,\, \sigma_3)$ with $\mathbf{m}^* = (\bar{1},\, 0,\, 0)$ for $R = m_x$ and with $\mathbf{m}^* = (1,\, 0,\, 0)$ for $R = m_z$.

Modulation wave vectors can be decomposed into a sum of a wave vector with rational components ($\mathbf{q}_r$) and a wave vector with irrational (unrestricted) components and components equal to zero ($\mathbf{q}_i$),

$$\mathbf{q} = \mathbf{q}_r + \mathbf{q}_i\,. \tag{3.4}$$

For example, for $\mathbf{q} = (\frac{1}{2},\, 0,\, \sigma_3)$ the rational part is $\mathbf{q}_r = (\frac{1}{2},\, 0,\, 0)$ and the irrational part is $\mathbf{q}_i = (0,\, 0,\, \sigma_3)$ [Fig. 3.1(b)]. If $\mathbf{q}$ is a solution to eqn (3.3) for a given combination of (R, ϵ), then $\mathbf{q}_i$ is a solution to eqn (3.1) for the same (R, ϵ). Unrestricted (irrational) components of $\mathbf{q}$ are thus given by eqn (3.1). Non-zero reciprocal lattice vectors in eqn (3.3) correspond to rational components of $\mathbf{q}$, that replace zero components as they occur in $\mathbf{q}_i$ (Table 3.2).

Additional complications arise for centred lattices. Employing the conventional centred unit cell, certain reciprocal lattice points describe forbidden reflections, and they are not allowed to act as vectors $\mathbf{m}^*$ in eqn (3.3). For example, in a C-centred lattice, reflections $(h_1\, h_2\, h_3)$ with $h_1 + h_2 =$ odd are extinct, and the reciprocal lattice vector $\mathbf{a}_1^*$ is forbidden. Accordingly, $\mathbf{q} = (\frac{1}{2},\, 0,\, \sigma_3)$ is not compatible with the C-centred orthorhombic lattice. Instead, the occurrence of a modulation with this wave vector implies the loss of the C-centre of the basic-structure lattice. On the other hand, centred lattices allow additional solutions

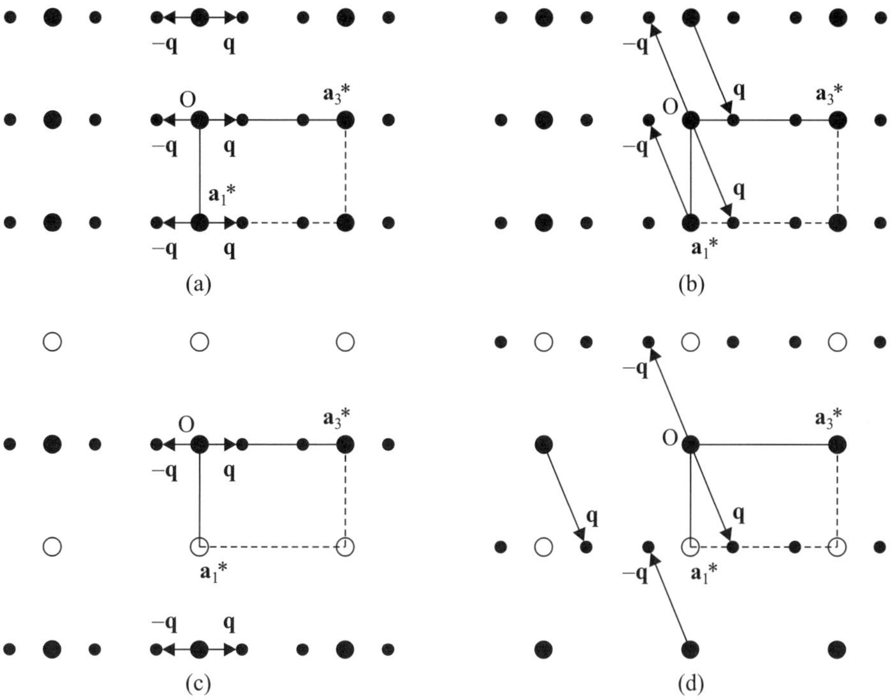

FIG. 3.2. Sections $h_2 = 0$ of reciprocal planes in physical space of orthorhombic crystals with one-dimensional modulations. (a) Primitive lattice and $\mathbf{q} = (0,\, 0,\, \sigma_3)$. (b) Same primitive lattice with an alternative indexing of the satellite reflections by $\mathbf{q} = (1,\, 0,\, \sigma_3)$. (c) C-centred lattice and $\mathbf{q} = (0,\, 0,\, \sigma_3)$. (d) C-centred lattice and $\mathbf{q} = (1,\, 0,\, \sigma_3)$. Filled circles represent scattering vectors of main reflections and satellite reflections. Open circles represent scattering vectors of extinct reflections.

to eqn (3.3). With $\mathbf{m}^* = -2\mathbf{a}_1^*$ for $R = m_x$ and $\mathbf{m}^* = 2\mathbf{a}_1^*$ for $R = m_z$, $\mathbf{q} = (1,\, 0,\, \sigma_3)$ is obtained as possible wave vector for a one-dimensional modulation in the C-centred orthorhombic lattice. This modulation wave vector is not a new solution in a primitive lattice, because the alternative choice of modulation wave vector, $(\mathbf{q} - \mathbf{a}_1^*) = (0,\, 0,\, \sigma_3)$, is equally good for indexing the satellite reflections (Fig. 3.2). But the reciprocal vector $\mathbf{a}_1^*$ is forbidden in a C-centred lattice, and $(\mathbf{q} - \mathbf{a}_1^*)$ is not equivalent to $\mathbf{q}$ in this case. For $\mathbf{q} = (0,\, 0,\, \sigma_3)$, rows of satellite reflections are found on reciprocal lattice lines containing allowed reciprocal lattice points of the basic structure [Fig. 3.2(c)], while for $\mathbf{q} = (1,\, 0,\, \sigma_3)$ rows of satellites reflections coincide with reciprocal lattice lines containing forbidden reciprocal lattice points of the basic structure [Fig. 3.2(d)].

Admissible modulation wave vectors with $\mathbf{q}_r = (0,\, 0,\, 0)$ are given in Table 3.1. They are valid for both primitive and centred lattices. A complete list of

TABLE 3.3. Admissible rational wave vectors for one-dimensional modulations [$\mathbf{q}_i = (0,0,0)$]. Unique axis is $\mathbf{a}_3$. Symbols P and I indicate the type of centring of the three-dimensional lattice.

Three-dimensional Bravais class	Modulation wave vector	Three-dimensional Bravais class	Modulation wave vector
Orthorhombic – I	$(\frac{1}{2}, \frac{1}{2}, \frac{1}{2})$	Cubic – I	$(\frac{1}{2}, \frac{1}{2}, \frac{1}{2})$
Tetragonal – I	$(\frac{1}{2}, \frac{1}{2}, \frac{1}{2})$	Cubic – I	$(1, 0, 0)$
Hexagonal – P	$(\frac{1}{3}, \frac{1}{3}, \frac{1}{2})$	Cubic – I	$(0, 1, 0)$
Hexagonal – P	$(\frac{1}{3}, \frac{1}{3}, 0)$	Cubic – I	$(0, 0, 1)$
Cubic – P	$(\frac{1}{2}, \frac{1}{2}, \frac{1}{2})$		

modulation wave vectors with non-zero rational components is given in Table 3.2 for the 14 Bravais classes of three-dimensional space groups. This list is longer than that published in the *International Tables for Crystallography Vol. C* (Janssen *et al.*, 1995), because different solutions $\mathbf{q}$ have been included, that are equivalent by a suitable coordinate transformation.

Shortly after the introduction of the superspace concept, Janner, Janssen and de Wolff noticed that the superspace theory can be applied to crystals with commensurate modulations as well (Janner and Janssen, 1979). Modulation wave vectors of superstructures are obtained by making the unrestricted components of $\mathbf{q}$ equal to the appropriate rational values. However, other wave vectors for one-dimensional commensurate modulations exist, that cannot be obtained in this way (van Smaalen, 1987). They follow as solutions of eqn (3.3) with $\mathbf{q}_i = (0, 0, 0)$ and $\mathbf{q}_r \neq (0, 0, 0)$. Because $\mathbf{q}_i$ is the null vector, $R\mathbf{q}_i = \pm\mathbf{q}_i$, and both choices $\epsilon = \pm 1$ are valid for all symmetry operators. In particular, the inversion operator (that represents symmetry of any lattice) combined with $\epsilon = 1$ gives the condition,

$$(-\sigma_1, -\sigma_2, -\sigma_3) - (\sigma_1, \sigma_2, \sigma_3) = \mathbf{m}^* \implies (2\sigma_1, 2\sigma_2, 2\sigma_3) = \mathbf{m}^* .$$

Each component of $\mathbf{q}$ is restricted and $\mathbf{q}_i$ is the null vector. Components of $\mathbf{m}^*$ equal to zero result in $\sigma_i = 0$, while components $m_i^* = 1$ lead to $\sigma_i = \frac{1}{2}$. Most wave vectors obtained in this way are not new, because they follow from the set of incommensurate wave vectors by making the unrestricted components equal to zero or $\frac{1}{2}$. However, for the primitive cubic lattice, $\mathbf{q} = (\frac{1}{2}, \frac{1}{2}, \frac{1}{2})$ is a solution to eqn (3.3) for all symmetry operators of $m\bar{3}m$, that is not contained in Table 3.1 or Table 3.2. A complete list of wave vectors for one-dimensional commensurately modulated crystals is provided in Table 3.3.

Several materials are known, that are modulated according to a symmetry-restricted commensurate wave vector. They include $Ag_{0.35}TiS_2$ with a trigonal space group of the basic structure and $\mathbf{q} = \mathbf{q}_r = (\frac{1}{3}, \frac{1}{3}, \frac{1}{2})$, and $PrPu_4P_{12}$ with I-centred cubic symmetry of the basic structure and $\mathbf{q} = \mathbf{q}_r = (1, 0, 0)$ (van Smaalen, 1987; Lee *et al.*, 2004). A comprehensive discussion of the application of the superspace theory to commensurately modulated structures is given in Chapter 5.

3.3 Space group symmetry of periodic crystals

Symmetry of periodic crystals is given by three-dimensional space groups. Elements of space groups are symmetry operators that consist of a rotation (R) followed by a translation ($\mathbf{v}$). The two parts of a symmetry operator are combined into the Seitz symbol as $\{R|\mathbf{v}\}$, where the vertical bar separates the rotation on the left from the translation on the right. Rotations include the crystallographic rotations, represented by the symbols 1, 2, 3, 4 and 6. Improper rotations are based on the inversion operator (symbol: $i = \bar{1}$), and they include combinations of i with any of the proper rotations, resulting in rotoinversion operators $i \circ 3 = \bar{3}$, $i \circ 4 = \bar{4}$ and $i \circ 6 = \bar{6}$. The symbol $\circ$ indicates that the operator to the right should be applied first, followed by the application of the operator to the left of this symbol. The rotoinversion $i \circ 2$ is equal to the mirror operator m perpendicular to the twofold axis. $i \circ 1 = i = \bar{1}$, and 1 is the identity operator defined by the absence of any transformation. The alternative symbol E will often be used to denote this operator.

Translations $\mathbf{v}$ are three-dimensional vectors, that—in general—depend on the relative locations of the origin of the coordinate system and the symmetry element R. Translations parallel to a rotation axis n or parallel to a mirror m are independent of the choice of the origin of the coordinate system. They are called intrinsic translations belonging to the symmetry element. Non-zero intrinsic translations define screw axes and glide planes, while an intrinsic translation equal to the null vector indicates pure rotational symmetry of the crystal structure.

Translations $\mathbf{v}$ are often defined through their coordinates (v_1, v_2, v_3) with respect to the basis $\Lambda = \{\mathbf{a}_1, \mathbf{a}_2, \mathbf{a}_3\}$ (Section 1.1.1). In general, vectors in direct space are represented by 3×1 matrices of components with respect to Λ. Vectors in reciprocal space are represented by 1×3 matrices of components on Λ^*. The same variable R is used to denote a rotational operator as well as its representation by a 3×3 matrix. The matrix R is defined such that it transforms the coordinates of a direct vector as specified in eqn (3.2). These definitions give for the action of $\{R|\mathbf{v}\}$ on a point $\mathbf{x}$ in direct space,

$$\{R|\mathbf{v}\} : \mathbf{x} \rightarrow \mathbf{x}' = R\mathbf{x} + \mathbf{v}\,. \tag{3.5}$$

The action of the same matrix R on a reciprocal vector $\mathbf{S}$ with components (S_1, S_2, S_3) then is,

$$\{R|\mathbf{v}\} : \mathbf{S} \rightarrow \mathbf{S}' = \mathbf{S}R^{-1}\,, \tag{3.6}$$

where R^{-1} is the inverse of R. The inverse of $\{R|\mathbf{v}\}$ is

$$\{R|\mathbf{v}\}^{-1} = \{R^{-1}| -R^{-1}\mathbf{v}\}\,. \tag{3.7}$$

The operator $\{E|\mathbf{0}\}$ is the identity, defined by zero rotation and null translation. Lattice translations $\{E|l_1,\, l_2,\, l_3\}$ with l_i = integer ($i = 1, 2, 3$) belong to all space groups (G). Together they define the translation subgroup ($\mathbb{T}$) of G. Centred lattices contain centring translations of the form $\{E|0,\, \frac{1}{2},\, \frac{1}{2}\}$ (A-centre), $\{E|\frac{1}{2},\, 0,\, \frac{1}{2}\}$ (B-centre), $\{E|\frac{1}{2},\, \frac{1}{2},\, 0\}$ (C-centre) or $\{E|\frac{1}{2},\, \frac{1}{2},\, \frac{1}{2}\}$ (I-centre). A four-fold rotation axis parallel to $\mathbf{a}_3$ is denoted by 4^z. In a space group the corresponding operator is $\{4^z|v_1,\, v_2,\, 0\}$, where $(v_1,\, v_2,\, 0)$ is the origin-dependent translation. A screw axis 4_1^z becomes in the Seitz notation $\{4_1^z|v_1,\, v_2,\, \frac{1}{4}\}$, where $(0,\, 0,\, \frac{1}{4})$ is the intrinsic translation, and $(v_1,\, v_2,\, 0)$ is the origin-dependent part of the translation.

A notation of symmetry elements of superspace groups will be used, that is a straightforward extension of the definitions in the present section.

3.4 Symmetry operators in superspace

Hidden translational symmetry of modulated crystals has been introduced in Chapter 2 as true translational symmetry of the generalized electron density in superspace. The superspace theory is based on the experimental observation of Bragg reflections in the diffraction of modulated crystals, that can be indexed by integers on the basis of four reciprocal vectors [eqn (1.36)]. The reciprocal lattice, Σ^*, in (3+1)-dimensional superspace is obtained by identifying the four reciprocal vectors M, used for indexing in physical space, with projections onto physical space of the four reciprocal basis vectors in superspace (Section 2.2). In this way, the single set of indices $(h_1\, h_2\, h_3\, h_4)$ may indicate a reciprocal vector in physical space [eqn (2.1)] and a reciprocal lattice vector in superspace [eqn (2.6)]. The one-to-one relation between $\mathbf{a}_i^*$ and $\mathbf{a}_{si}^*$ [eqn (2.7)] determines that rotational symmetry of the set M must be rotational symmetry of Σ^* as well. Rotational symmetry of M is given by a three-dimensional point group. The three basis vectors of the reciprocal lattice of the basic structure transform under the action of a rotation R as,

$$\begin{pmatrix} \mathbf{a}_1'^* \\ \mathbf{a}_2'^* \\ \mathbf{a}_3'^* \end{pmatrix} = R \begin{pmatrix} \mathbf{a}_1^* \\ \mathbf{a}_2^* \\ \mathbf{a}_3^* \end{pmatrix}, \tag{3.8}$$

where R is the 3×3 matrix equal to the matrix for transformation of direct-space coordinates given in eqn (3.2). The action of R on the fourth basis vector $\mathbf{q} = \mathbf{a}_4^*$ has been defined in Section 3.2 as,

$$R : \mathbf{q} \rightarrow \mathbf{n}^* + \epsilon\mathbf{q}\,, \tag{3.9}$$

with $\epsilon = \pm1$. This allows the construction of a 4×4 matrix, R_s, that defines the transformation of the four reciprocal basis vectors M under the action of the operator R according to,

$$\begin{pmatrix} \mathbf{a}_1'^* \\ \mathbf{a}_2'^* \\ \mathbf{a}_3'^* \\ \mathbf{a}_4'^* \end{pmatrix} = R_s \begin{pmatrix} \mathbf{a}_1^* \\ \mathbf{a}_2^* \\ \mathbf{a}_3^* \\ \mathbf{a}_4^* \end{pmatrix}, \tag{3.10}$$

with

$$R_s = \begin{pmatrix} & & 0 \\ & R & 0 \\ & & 0 \\ & \mathbf{n}^* & \epsilon \end{pmatrix}, \tag{3.11}$$

and with $\mathbf{n}^*$ the 1×3 matrix of integer components (n_1^*, n_2^*, n_3^*) of a reciprocal lattice vector. The inverse of R_s is

$$(R_s)^{-1} = \begin{pmatrix} & & 0 \\ & R^{-1} & 0 \\ & & 0 \\ & -\left(\epsilon^{-1}\,\mathbf{n}^*\,R^{-1}\right) & \epsilon^{-1} \end{pmatrix} = \begin{pmatrix} & & 0 \\ & R^{-1} & 0 \\ & & 0 \\ & \mathbf{m}^* & \epsilon^{-1} \end{pmatrix}. \tag{3.12}$$

The relation between $\mathbf{q}$, R, ϵ and $\mathbf{n}^*$ is defined in eqn (3.3) through a relation between R^{-1}, ϵ^{-1}, $\mathbf{m}^*$ and the components of $\mathbf{q}$ on Λ^*. Equation (3.12) provides the relation between $\mathbf{n}^*$ and $\mathbf{m}^*$ as,

$$\mathbf{m}^* = -\left(\epsilon^{-1}\,\mathbf{n}^*\,R^{-1}\right) \Longleftrightarrow \mathbf{n}^* = -\left(\epsilon\,\mathbf{m}^*\,R\right). \tag{3.13}$$

Combination with eqn (3.3) leads to a condition for $\mathbf{n}^*$ that is equally simple as the condition for $\mathbf{m}^*$,

$$\sigma R - \epsilon\sigma = \mathbf{n}^*. \tag{3.14}$$

The one-to-one correspondence between M and Σ^* implies that the 4×4 matrix R_s defines a transformation of the basis vectors of the reciprocal lattice in superspace,

$$\begin{pmatrix} \mathbf{a}_{s1}'^* \\ \mathbf{a}_{s2}'^* \\ \mathbf{a}_{s3}'^* \\ \mathbf{a}_{s4}'^* \end{pmatrix} = R_s \begin{pmatrix} \mathbf{a}_{s1}^* \\ \mathbf{a}_{s2}^* \\ \mathbf{a}_{s3}^* \\ \mathbf{a}_{s4}^* \end{pmatrix}. \tag{3.15}$$

The matrix R_s can be considered as a matrix representation of a point group operator

$$R_s = (R,\,\epsilon) \tag{3.16}$$

in superspace. In this notation it is understood that, for a given lattice in superspace, the pair of operators R and ϵ uniquely define the vector $\mathbf{n}^*$ [eqns (3.3) and (3.14)].

The 'diffraction' pattern in superspace has been defined by assigning the scattered intensity $I(h_1, \cdots, h_{3+d})$ of Bragg reflection $\mathbf{H}$ to the reciprocal lattice vector $\mathbf{H}_s$ in superspace, that has the same components has $\mathbf{H}$. From this construction it follows immediately that point symmetry of the diffraction pattern of a modulated crystal in physical space must define point symmetry of the 'diffraction' pattern in superspace, through a one-to-one correspondence between rotational operators R and R_s. In complete analogy to the analysis of the diffraction and symmetry of periodic crystals in physical space, point symmetry of the scattered intensity in superspace implies that the symmetry of the generalized electron density is given by a four-dimensional space group. Similar to periodic crystals, the diffracted intensities may exhibit inversion symmetry, $R_s = (i, \bar{1})$, while the structure is acentric. Apart from this ambiguity, each point symmetry element R_s of the superspace diffraction pattern corresponds to an operator $\{R_s|\mathbf{v}_s\}$ in superspace, that may or may not have a non-zero translational part,

$$\mathbf{v}_s = (v_{s1}, v_{s2}, v_{s3}, v_{s4})\,. \tag{3.17}$$

$\{R_s|\mathbf{v}_s\}$ transforms coordinates $\mathbf{x}_s$ in direct superspace towards $\mathbf{x}'_s$ according to,

$$\begin{pmatrix} x'_{s1} \\ x'_{s2} \\ x'_{s3} \\ x'_{s4} \end{pmatrix} = R_s \begin{pmatrix} x_{s1} \\ x_{s2} \\ x_{s3} \\ x_{s4} \end{pmatrix} + \begin{pmatrix} v_{s1} \\ v_{s2} \\ v_{s3} \\ v_{s4} \end{pmatrix}, \tag{3.18}$$

where R_s is identical to the matrix R_s for transformation of the reciprocal lattice vectors [eqn (3.15)]. In an alternative notation for $\{R_s|\mathbf{v}_s\}$, the result of the action of the operator on the set of coordinates $(x_{s1}, x_{s2}, x_{s3}, x_{s4})$ can be given, as it follows from eqn (3.18).

Symmetry $\{R_s|\mathbf{v}_s\}$ of the generalized electron density implies symmetry $\{R|\mathbf{v}\}$ of the basic structure in physical space, with $\mathbf{v} = (v_1, v_2, v_3) = (v_{s1}, v_{s2}, v_{s3})$.

The intrinsic translation contained in $\{R_s|\mathbf{v}_s\} = \{R, \epsilon|\mathbf{v}_s\}$ may include screw or glide components along the three physical-space directions v_{s1}, v_{s2} and v_{s3}. For $\epsilon = 1$ it may also include a non-zero fourth component, while the value of v_{s4} depends on the choice of the origin of Σ for operators with $\epsilon = -1$. Possible intrinsic translations can be derived from the condition,

$$\{R_s|\mathbf{v}_s\}^n = \{E_s|\mathbf{L}_s\}\,, \tag{3.19}$$

where n is the smallest positive integer for which $(R_s)^n = E_s$; n is the order of the symmetry element R_s. $\mathbf{L}_s = (l_{s1}, l_{s2}, l_{s3}, l_{s4})$ is a lattice translation in superspace, defined by,

$$\mathbf{L}_s = l_{s1}\,\mathbf{a}_{s1} + l_{s2}\,\mathbf{a}_{s2} + l_{s3}\,\mathbf{a}_{s3} + l_{s4}\,\mathbf{a}_{s4}\,, \tag{3.20}$$

with l_{si} = integer $(i = 1, 2, 3, 4)$. For a symmetry element of order two it follows that,

$$R_s\,\mathbf{v}_s + \mathbf{v}_s = \mathbf{L}_s\,. \tag{3.21}$$

In case of a twofold axis along $\mathbf{a}_2$ in a modulated crystal with $\mathbf{q} = (0,\, \sigma_2,\, 0)$, the condition on $\mathbf{v}_s$ becomes,

$$(0,\, 2v_{s2},\, 0,\, 2v_{s4}) = (l_{s1},\, l_{s2},\, l_{s3},\, l_{s4})\,. \tag{3.22}$$

v_{s1} and v_{s3} are not restricted, and their values will depend on the choice of the origin. Equation (3.22) has solutions with $v_{s2} = 0$ or $\frac{1}{2}$ and $v_{s4} = 0$ or $\frac{1}{2}$. Four different superspace operators are obtained from the point symmetry operator defined by $R = 2$ and $\epsilon = 1$:

TABLE 3.4. Symbols for the fourth component of intrinsic translation vectors of symmetry operators $\{R_s|\mathbf{v}_s\}$ in (3+1)-dimensional superspace.

v_{s4}	0	$\frac{1}{2}$	$\frac{1}{3}$	$-\frac{1}{3}$	$\frac{1}{4}$	$-\frac{1}{4}$	$\frac{1}{6}$	$-\frac{1}{6}$
Symbol	0	s	t	$\bar{t}$	q	$\bar{q}$	h	$\bar{h}$

1. A twofold rotation in superspace with the symbol $(2, 0)$.
2. A twofold screw axis with zero translational component along $\mathbf{a}_{s4}$, and with the symbol $(2_1, 0)$.
3. A twofold rotation with an intrinsic translation of $\frac{1}{2}$ along $\mathbf{a}_{s4}$, and with the symbol $(2, s)$.
4. A twofold screw axis with translational component $\frac{1}{2}$ along $\mathbf{a}_{s4}$, and with the symbol $(2_1, s)$.

The previously introduced notation for point symmetry operators, $R_s = (R, \epsilon)$, is extended to include the intrinsic translation. The point symmetry operator R is replaced by the appropriate symbol indicating possible screw or glide components in physical space. For operators with $\epsilon = -1$, the second part of the symbol (R, ϵ) is kept as $\bar{1}$, because the fourth component of the intrinsic translation is zero. For operators with $\epsilon = 1$, the value of ϵ in (R, ϵ) is replaced by a symbol that indicates the value of v_{s4} in the intrinsic translation vector. Thus, point symmetry operator $(2, 1)$ corresponds to space-group operator $(2, 0)$, where the intrinsic translation is zero. Values of v_{s4} of the intrinsic translation vector are restricted to the same set of values that appear as possible screw components in physical space. Symbols as introduced by de Wolff *et al.* (1981) will be used to indicate these values (Table 3.4).

As second example a modulated structure is considered with $\mathbf{q} = (\frac{1}{2}, \sigma_2, 0)$. The twofold rotation along $\mathbf{a}_2$ is represented by the 4×4 matrix [eqn (3.11)]

$$R_s = \begin{pmatrix} \bar{1} & 0 & 0 & 0 \\ 0 & 1 & 0 & 0 \\ 0 & 0 & \bar{1} & 0 \\ \bar{1} & 0 & 0 & 1 \end{pmatrix}. \tag{3.23}$$

The condition on intrinsic translation vectors of the operator $\{2^y, 1\,|\mathbf{v}_s\}$ follows as [eqn (3.21)],

$$(0, 2v_{s2}, 0, v_{s1} + 2v_{s4}) = (l_{s1}, l_{s2}, l_{s3}, l_{s4})\,. \tag{3.24}$$

Solutions are obtained for $v_{s2} = 0$ and $\frac{1}{2}$. Despite the relation $v_{s1} + 2v_{s4} =$ integer, neither v_{s1} nor v_{s4} are restricted to particular values. The only possible non-zero component of the intrinsic translation vector is v_{s2}, and two different operators are obtained: $(2, 0)$ and $(2_1, 0)$.

Centring translations may replace $\mathbf{L}_s$ in the condition on $\mathbf{v}_s$ of eqn (3.19). They give further possibilities for intrinsic translational components, like $\frac{1}{4}$ for a mirror operator. An easier understanding of possible and impossible intrinsic translations is obtained from the consideration of symmetry in reciprocal space in conjunction with the derivations of extinction conditions on Bragg reflections. This discussion is deferred to Section 3.8.

3.5 Superspace groups

3.5.1 *Crystal classes, Bravais classes and space group types*

Symmetry of the generalized electron density of a d-dimensionally modulated crystal ($d \geqslant 1$) contains a lattice of translations (Chapter 2). Furthermore, symmetry of $\rho_s(\mathbf{x}_s)$ may include rotations, improper rotations, screw and glide operators (Section 3.4). This implies that the collection of all symmetry operators of $\rho_s(\mathbf{x}_s)$ forms a space group in (3+d)-dimensional space. The (3+d)-dimensional space group also gives symmetry of the modulated structure itself, because the generalized electron density in superspace is uniquely determinded by the atomic structure of the modulated crystal in physical space.

The theory of space groups is based on the notion of equivalence, that is employed at different levels, and that leads to a classification of space groups according to crystal classes, Bravais classes and crystal systems (Hahn, 2002). It is not the aim of this book to give a complete account of the theory of space groups in arbitrary dimensions, and the reader is referred to the extensive literature on this subject (Brown *et al.*, 1978; Engel, 1986; Hahn, 2002). Nevertheless, for an understanding of superspace groups it is necessary to consider the concept of equivalence in more detail. The starting point is the choice of a conventional basis for the lattice Σ, such that all lattice translations are integer linear combinations of the $3 + d$ basis vectors [eqn (3.20)]. An infinite number of different sets of basis vectors is possible, providing equivalent coordinate systems that are related by coordinate transformations Q. The $(3{+}d) \times (3{+}d)$ matrix Q is defined by the transformation of basis vectors,

$$\left(\mathbf{a}'_{s1} \cdots \mathbf{a}'_{s(3+d)}\right) = \left(\mathbf{a}_{s1} \cdots \mathbf{a}_{s(3+d)}\right) Q^{-1} \,, \tag{3.25}$$

where both Q^{-1} and Q have integer components only. The determinant of Q^{-1} is equal to 1 or -1,

$$|det(Q)| = |det(Q^{-1})| = 1 \,, \tag{3.26}$$

because the volume of the unit cell is independent of the choice of the basis vectors. Q is called a unimodular matrix. The matrices R_s of point symmetry operators that leave the lattice invariant are unimodular too.

Two space groups are said to be equivalent if a unimodular matrix Q can be found, such that the matrix representation of $\{R_s{}^{(1)}|\mathbf{v}_s{}^{(1)}\}$ of space group 1 is identical to the matrix representation $\{R_s{}^{(2)'}|\mathbf{v}_s{}^{(2)'}\}$, that is obtained by application of the coordinate transformation Q to the matrix representation $\{R_s{}^{(2)}|\mathbf{v}_s{}^{(2)}\}$ of space group 2. In order to obtain equal matrices for pairs of operators in space groups 1 and 2, allowance must be made for a shift of the origin of the coordinate system. The latter is described by a vector $\mathbf{v}_s^{\rm O}$ with respect to the coordinate system of space group 2. The condition for equivalence of space groups can then be expressed mathematically as,

$$R_s{}^{(1)} = QR_s{}^{(2)}Q^{-1} \tag{3.27a}$$

$$\mathbf{v}_s{}^{(1)} = Q\mathbf{v}_s{}^{(2)} + Q\left(1 - R_s{}^{(2)}\right)\mathbf{v}_s{}^{\rm O}, \tag{3.27b}$$

where all pairs of elements $\{R_s{}^{(1)}|\mathbf{v}_s{}^{(1)}\} \sim \{R_s{}^{(2)}|\mathbf{v}_s{}^{(2)}\}$ are related by a single matrix Q and a single vector $\mathbf{v}_s^{\rm O}$. All equivalent space groups together define a space group type. In this book 'space group' will often be used when 'space group type' would have been the more correct designation.

Equivalence of space groups is not always obvious from a simple comparison of the lists of symmetry elements of the two groups. For example, the three-dimensional space groups $A2/a$ and $B2/n$ are equivalent (monoclinic symmetry, unique axis $\mathbf{a}_3$), but the required coordinate transformation,

$$Q = \begin{pmatrix} 0 & \bar{1} & 0 \\ 1 & \bar{1} & 0 \\ 0 & 0 & 1 \end{pmatrix},$$

is not immediately clear to everyone. This is one reason that the *International Tables for Crystallography Vol. A* provide different settings for several space groups. Equivalence of space groups in higher dimensions is even more evasive; it is discussed in Section 3.9.

A second classification of space groups is according to their point symmetry. The point group of a space group is obtained by removal of the translational part from each operator $\{R_s|\mathbf{v}_s\}$. The resulting collection of operators R_s form a finite group, K_s, of the $(3+d)$-dimensional space group. Two space groups belong to the same crystal class (or geometric crystal class) if their point groups are equivalent by a transformation of coordinates according to eqn (3.27a). The difference from the classification of space group types is, that any matrix Q with $|\det(Q)| = 1$ is allowed for defining equivalence in the sense of crystal classes.

The third classification discussed here is that of Bravais classes. The holohedry is the point group that gives the point symmetry of the lattice Σ. The holohedry contains all elements of K_s, but may contain more symmetry. Point symmetry of the lattice implies restrictions on the lattice parameters, but a primitive lattice and lattices of different centring types may have the same point symmetry. Two space groups belong to the same Bravais class or Bravais type, if their lattices have the same holohedry and are of the same centring type. Again, equivalence of Bravais classes of space groups is implied by the existence of a matrix Q, such that the transformation eqn (3.27a) applied to the holohedry of one space group is identical to the holohedry of the other space group. Different centring types of lattice are distinguished by restricting Q to unimodular matrices.

TABLE 3.5. Space groups in dimensions 1–6 and (3+d)-dimensional superspace groups for $d = 1$, 2 and 3. Given are the numbers of Bravais lattices, the number of crystal classes and the number of space group types (space groups). The latter refer to space groups counting enantiomorphic pairs as a single space group type. Classifying enantiomorphic pairs as two different space group types results in 230 space groups in dimension 3 and 775 (3+1)-dimensional superspace groups.

Classification	Dimension of space or superspace								
	1	2	3	4	5	6	3+1	3+2	3+3
Bravais lattices	1	5	14	64	189	826	24	83	217
Crystal classes	2	10	32	227	955	7104	31	75	137
Space groups	2	17	219	4 783	222 018	28 927 922	756	3 355	11 764

3.5.2 *Definition of superspace groups*

With the definitions in Section 3.5.1, space groups in arbitrary dimensions can be analysed (Engel, 1986). While properties like symmetry elements, special positions ('orbits') and sub- and supergroups have been tabulated for all 230 space groups of dimension three, just enumerating space groups of dimensions higher than three is already an enormous task. Brown *et al.* (1978) have produced a table of all 4783 space groups of dimension four. Plesken and coworkers studied space groups in still higher dimensions (Opgenorth *et al.*, 1998; Plesken and Schulz, 2000), for example showing that there exist more than 28 million space groups of dimension six (Table 3.5). However, not all space groups of dimensions higher than three are of importance for aperiodic crystals. In Sections 3.2 and 3.4 it was argued that the point group of the n-dimensional space group of an aperiodic crystal must be isomorphous to a point group of three-dimensional space. Consequently, the number of physically relevant point groups remains limited. In contrast, the number of crystallographic point groups increases strongly with increasing dimension of space (the number of crystal classes in Table 3.5). Retaining only those space groups that are based on a three-dimensional point group results in a much smaller number of space groups. For example, only 371 physically relevant space groups exist in space of dimension four (Janssen *et al.*, 1995).

A second modification to the theory of higher-dimensional space groups pertains to the definition of equivalence [eqn (3.27)]. Since only a selection of point group operators is relevant to aperiodic crystals, it seems most natural to restrict the coordinate transformations Q to such operators too. The result is an increase in the number of space group types, from 371 towards 775 in dimension four (Table 3.5). The resulting special space groups of dimension n are called superspace groups.

Point-symmetry operators of superspace groups as well as coordinate transformations Q defining equivalence of superspace groups are matrices of the form of eqn (3.11). These operations are said to be $(3, d)$-reducible (with $n = 3 + d$),

TABLE 3.6. Bravais classes of (3+1)-dimensional superspace groups.

No.	Symbol	No.	Symbol
Triclinic		13	$Cmmm(0\,0\,\sigma_3)00\bar{1}$
1	$P\bar{1}(\sigma_1\,\sigma_2\,\sigma_3)\bar{1}$	14	$Cmmm(1\,0\,\sigma_3)00\bar{1}$
Monoclinic		15	$Ammm(0\,0\,\sigma_3)00\bar{1}$
2	$P2/m(\sigma_1\,\sigma_2\,0)\bar{1}0$	16	$Ammm(\frac{1}{2}\,0\,\sigma_3)00\bar{1}$
3	$P2/m(\sigma_1\,\sigma_2\,\frac{1}{2})\bar{1}0$	17	$Fmmm(0\,0\,\sigma_3)00\bar{1}$
4	$B2/m(\sigma_1\,\sigma_2\,0)\bar{1}0$	18	$Fmmm(1\,0\,\sigma_3)00\bar{1}$
5	$P2/m(0\,0\,\sigma_3)0\bar{1}$	Tetragonal	
6	$P2/m(\frac{1}{2}\,0\,\sigma_3)0\bar{1}$	19	$P4/mmm(0\,0\,\sigma_3)0\bar{1}00$
7	$B2/m(0\,0\,\sigma_3)0\bar{1}$	20	$P4/mmm(\frac{1}{2}\,\frac{1}{2}\,\sigma_3)0\bar{1}00$
8	$B2/m(\frac{1}{2}\,0\,\sigma_3)0\bar{1}$	21	$I4/mmm(0\,0\,\sigma_3)0\bar{1}00$
Orthorhombic		Trigonal	
9	$Pmmm(0\,0\,\sigma_3)00\bar{1}$	22	$R\bar{3}m(0\,0\,\sigma_3)\bar{1}0$
10	$Pmmm(0\,\frac{1}{2}\,\sigma_3)00\bar{1}$	23	$R\bar{3}m(\frac{1}{3}\,\frac{1}{3}\,\sigma_3)\bar{1}0$
11	$Pmmm(\frac{1}{2}\,\frac{1}{2}\,\sigma_3)00\bar{1}$	Hexagonal	
12	$Immm(0\,0\,\sigma_3)00\bar{1}$	24	$P6/mmm(0\,0\,\sigma_3)0\bar{1}00$

i.e. the additional coordinate axes $\mathbf{a}_{s(3+j)}$ $(j \geqslant 1)$ are not mixed into the three physical-space directions, and the physical-space directions are only mixed into $\mathbf{a}_{s(3+j)}$ as long as they correspond to rational components of modulation wave vectors. Conditions for superspace groups are even more stringent: a single basis of Σ must exist, on which all elements of the superspace group simultaneously are in $(3,d)$-reduced form. In order to distinguish superspace groups from general n-dimensional space groups, superspace groups in space of dimension n are denoted by $(3+d)$-dimensional superspace groups. Notice that (3+1)-dimensional superspace groups are different from (2+2)-dimensional superspace groups, although both are derived from four-dimensional space groups. The number of superspace groups is much smaller than the number of space groups of the same dimension, as is evident from Table 3.5. (3+1)-Dimensional superspace groups have been tabulated by de Wolff *et al.* (1981), and they are compiled in the *International Tables for Crystallography Vol. C* (Janssen *et al.*, 1995). A list of (3+2)-dimensional and (3+3)-dimensional superspace groups has been produced by Yamamoto (2001, 2005). Bravais classes of superspace groups up to dimension 3+3 have been given by Janner *et al.* (1983). A listing of the 24 Bravais classes of (3+1)-dimensional superspace groups is given in Table 3.6.

An example might illustrate the importance of the distinction between superspace groups and space groups of n-dimensional space. Consider two crystals with basic-structure space groups Pm and $P2$, respectively (monoclinic, $\mathbf{a}_3$ unique). Both crystals have a one-dimensional modulation with $\mathbf{q} = (0,\, 0,\, \sigma_3)$. Matrices R_s for the operators m_z and 2^z are [eqn (3.16)],

$$(m_z, \bar{1}) = \begin{pmatrix} 1 & 0 & 0 & 0 \\ 0 & 1 & 0 & 0 \\ 0 & 0 & \bar{1} & 0 \\ 0 & 0 & 0 & \bar{1} \end{pmatrix} ; \qquad (2^z, 1) = \begin{pmatrix} \bar{1} & 0 & 0 & 0 \\ 0 & \bar{1} & 0 & 0 \\ 0 & 0 & 1 & 0 \\ 0 & 0 & 0 & 1 \end{pmatrix} . \tag{3.28}$$

It is an easy exercise to show that a coordinate transformation by the unimodular matrix

$$Q = Q^{-1} = \begin{pmatrix} 0 & 0 & 1 & 0 \\ 0 & 0 & 0 & 1 \\ 1 & 0 & 0 & 0 \\ 0 & 1 & 0 & 0 \end{pmatrix} \tag{3.29}$$

transforms $(m_z, \bar{1})$ into $(2^z, 1)$ [eqn (3.27a)]. Space group theory then determines that $(m_z, \bar{1})$ and $(2^z, 1)$ are equivalent operators, and that the corresponding four-dimensional space groups are equivalent too (Section 3.5.1). However, mirrors and twofold rotations are clearly different kinds of symmetry operators, and modulated structures based on basic structures with Pm and $P2$ space groups should be assigned different designations for their symmetries. Superspace groups do this job. The matrix Q in eqn (3.29) is not in $(3, 1)$-reduced form, and it is not an admissible matrix for defining equivalence of superspace groups. Thus the (3+1)-dimensional superspace group based on $(m_z, \bar{1})$ is different from the (3+1)-dimensional superspace group based on $(2^z, 1)$, in accordance with a physically relevant classification of the symmetry of modulated structures.

A different classification of superspace groups is possible by selecting only the $(3, d)$-reducible space groups of $(3+d)$-dimensional space, while keeping the most general form of the equivalence relation. In (3+1)-dimensional space this gives the previously cited number of 371 $(3, 1)$-reducible, (3+1)-dimensional space groups instead of 775 (3+1)-dimensional superspace groups. This classification has been favoured by Mermin (1992) and Mermin and Lifshitz (1992), who argued that a lattice of main reflections cannot be singled out for all aperiodic crystals, and that the definition of equivalence thus should allow any matrix Q. Following Mermin (1992), the superspace groups based on $(m_z, \bar{1})$ and $(2^z, 1)$ would be equivalent, although a crystallographic analysis of the crystal structure of a modulated crystal does require a distinction to be made between these situations. Therefore, the finer classification of the symmetry of modulated crystals according to the superspace groups of de Wolff *et al.* (1981) is the most useful classification for structural analysis as well as for studying the symmetry of physical properties.

3.5.3 *Symbols for superspace groups*

Symmetry operators that may belong to superspace groups are based on three-dimensional point group operators R. Matrices ϵ and $\mathbf{n}^*$ are uniquely determined for each operator R, once the set M of $3+d$ basis vectors has been selected. These properties of superspace symmetry have been used in Section 3.4 to develop a notation for point symmetry operators in (3+1)-dimensional superspace, that consists of the usual symbol for the three-dimensional operator R, combined

TABLE 3.7. Prefix symbols indicating rational components of the modulation wave vectors for (3+1)-dimensional superspace groups in the two-line notation.

Symbol	$\mathbf{q}_r$	Symbol	$\mathbf{q}_r$	Symbol	$\mathbf{q}_r$
P	(0, 0, 0)			R	$(\frac{1}{3}, \frac{1}{3}, 0)$
A	$(\frac{1}{2}, 0, 0)$	L	(1, 0, 0)	U	$(0, \frac{1}{2}, \frac{1}{2})$
B	$(0, \frac{1}{2}, 0)$	M	(0, 1, 0)	V	$(\frac{1}{2}, 0, \frac{1}{2})$
C	$(0, 0, \frac{1}{2})$	N	(0, 0, 1)	W	$(\frac{1}{2}, \frac{1}{2}, 0)$

with the value of ϵ into the symbol (R, ϵ). Screw axes and glide planes in (3+1)-dimensional superspace are described by a similar symbol, where R is replaced by the corresponding symbol for three-dimensional screw axes and glide planes, and the symbol 1 for ϵ is replaced by a special symbol indicating the fourth component of the intrinsic translation vector (Table 3.4). The symbol $\bar{1}$ for ϵ is retained, because $\epsilon = -1$ implies that the fourth component of the intrinsic translation is zero.

Symbols for superspace groups can be constructed by combining the symbols (R, ϵ) of its elements. Only the generators need to be specified, but—in accordance with symbols for three-dimensional space groups—more than the minimum number of generators is often given. Each superspace operator (R, ϵ) contains the symbol R for a symmetry operator in three-dimensional space. Therefore, the symbol for a superspace group is based on the symbol of the space group of its basic structure. The meaning of these symbols follows the convention as it is defined in the *International Table for Crystallography Vol. A*. In the original publication by de Wolff *et al.* (1981), a two-line symbol has been introduced, that is obtained by writing below each symbol for the three-dimensional operator R, the symbol $\bar{1}$ in case of $\epsilon = -1$ or the appropriate symbol indicating the fourth component of the intrinsic translation (Table 3.4). Instead of 0 (used in this book) they employed 1 in cases where the intrinsic translational component along $\mathbf{a}_{s4}$ is zero. The rational components of $\mathbf{q}_r$ are indicated by a capital letter that is put in front of the two-line symbol (Table 3.7). For example, the symbol

$$P\,\begin{matrix} Cmcm \\ s\,1\,\bar{1} \end{matrix}$$

indicates an orthorhombic superspace group with space group $Cmcm$ for the basic structure. The modulation wave vector is $(0, 0, \sigma_3)$, because P indicates that $\mathbf{q}_r = \mathbf{0}$ (Table 3.7), and the symbol

$$\begin{matrix} m \\ \bar{1} \end{matrix} = (m, \bar{1})$$

indicates that $\epsilon = -1$ for m_z, from which σ_3 follows as the only possible irrational component of $\mathbf{q}$ (Section 3.2). Symbols (m, s) and $(c, 1)$ have a meaning that can be derived from Table 3.4 and the discussion in Section 3.4. A centring $(\frac{1}{2}, \frac{1}{2}, 0)$ of the lattice of the basic structure is indicated by the capital letter C on the top line of the two-line symbol. Two-line symbols for superspace groups imply that the fourth component of centring translations is zero, and C indicates a centring $(\frac{1}{2}, \frac{1}{2}, 0, 0)$ of the superspace lattice as well. Although a setting can always be chosen such that this condition is fulfilled, it is an unnecessary and undesirably restriction of the two-line symbol. Especially for composite crystals, settings of superspace groups can be preferred that involve centring translations with non-zero fourth components, *e.g.* a C'-centring $(\frac{1}{2}, \frac{1}{2}, 0, \frac{1}{2})$ instead of the C-centring (see Section 3.6.2 and Chapter 4).

A one-line symbol was introduced in the *International Tables for Crystallography Vol. C* (Janssen et al., 1995). This symbol is obtained by separating the R and ϵ parts of all operators (R, ϵ), and by placing the group of symbols for R and the group of symbols for ϵ on a single line. The two groups of symbols are separated by the components of the modulation wave vector, and they are preceded by a symbol for the centring of the superspace lattice. For example, the superspace group just described has a one-line symbol

$$Cmcm(0\,0\,\sigma_3)s1\bar{1}\,.$$

Janssen *et al.* (1995) introduced three other novelties. Firstly, the symbol 1 for ϵ is replaced by zero, thus properly representing the zero value of the fourth component of the intrinsic translation. This convention is also followed in this book. The superspace group symbol becomes

$$Cmcm(0\,0\,\sigma_3)s0\bar{1}\,. \tag{3.30}$$

Secondly, Janssen *et al.* (1995) noticed that the fourth component of an intrinsic translation is zero if $\epsilon = -1$. Accordingly, they replaced $\bar{1}$ by 0, resulting in the symbol

$$Cmcm(0\,0\,\sigma_3)s00$$

that can be found in the *International Tables for Crystallography Vol. C.* Finally, they removed any designation for ϵ from the symbol for the superspace group, if the fourth components of intrinsic translations are zero for all operators. For example, the following symbols indicate the same (3+1)-dimensional superspace group:

$$P\,{}^{Cmcm}_{\;\;11\bar{1}}\,; \qquad Cmcm(0\,0\,\sigma_3)00\bar{1}; \qquad Cmcm(0\,0\,\sigma_3)000; \qquad Cmcm(0\,0\,\sigma_3)\,.$$

The value of ϵ of each operator R can easily be derived by analysing the effect of R on the incommensurate part of the modulation wave vector. However, the value of ϵ is even more easily incorporated into the one-line symbol, specifying the fourth component of intrinsic translations according to Table 3.4 or giving $\bar{1}$ if $\epsilon = -1$. This extended one-line symbol of the type of eqn (3.30) provides more

TABLE 3.8. Selected (3+1)-dimensional superspace groups with a two-line symbol according to De Wolff *et al.* (1981), an one-line symbol according to the *International Tables for Crystallography Vol. C* (ITVolC), and the extended one-line symbol employed in this book. Numbers refer to the number of the superspace group in the *International Tables for Crystallography Vol. C.*

No.	Two-line symbol	ITVolC	Extended symbol
4.2	$P\,{}^{P2_1}_{1}$	$P2_1(0\,0\,\sigma_3)$	$P2_1(0\,0\,\sigma_3)0$
13.2	$C\,{}^{P2/b}_{\bar{1}\;1}$	$P2/b(\sigma_1\,\sigma_2\,\frac{1}{2})$	$P2/b(\sigma_1\,\sigma_2\,\frac{1}{2})\bar{1}0$
62.2	$P\,{}^{Pnma}_{1\,s\,\bar{1}}$	$Pnma(0\,0\,\sigma_3)0s0$	$Pnma(0\,0\,\sigma_3)0s\bar{1}$
65.6	$L\,{}^{Cmmm}_{s\;s\;\bar{1}}$	$Cmmm(1\,0\,\sigma_3)ss0$	$Cmmm(1\,0\,\sigma_3)ss\bar{1}$
124.3	$W\,{}^{P4/mcc}_{1\;\bar{1}11}$	$P4/mcc(\frac{1}{2}\,\frac{1}{2}\,\sigma_3)$	$P4/mcc(\frac{1}{2}\,\frac{1}{2}\,\sigma_3)0\bar{1}00$
151.1	$R\,{}^{P3_112}_{1\,1\bar{1}}$	$P3_112(\frac{1}{3}\,\frac{1}{3}\,\sigma_3)$	$P3_112(\frac{1}{3}\,\frac{1}{3}\,\sigma_3)00\bar{1}$
155.2	$P\,{}^{R32}_{t\bar{1}}$	$R32(0\,0\,\sigma_3)t0$	$R32(0\,0\,\sigma_3)t\bar{1}$
191.4	$P\,{}^{P6/mmm}_{s\;\bar{1}\;1\;s}$	$P6/mmm(0\,0\,\sigma_3)s00s$	$P6/mmm(0\,0\,\sigma_3)s\bar{1}0s$

information about the superspace group than the conventional symbol does, but it is equally simple. Therefore the extended symbol is employed in this book. A comparison of symbols for selected superspace groups is compiled in Table 3.8.

The capital letter of the one-line symbol indicates the centring of the superspace lattice. The centring of the three-dimensional lattice of the basic structure is obtained by restricting each centring translation to its first three components. Standard symbols P, A, B, C, I, F, R and H have a meaning similar to that for three-dimensional space groups, with the fourth component of each centring translation equal to zero. Symbols for centrings with non-zero fourth components are defined by adding a prime to the capital letter, if the physical-space centring is replaced by the corresponding centring with $v_{s4} = \frac{1}{2}$. A subscript a is added to the capital letter, if the set of centring translations is extended by the translation $(\frac{1}{2},\ 0,\ 0,\ \frac{1}{2})$, with similar definitions for subscripts b and c. This notation is redundant, because different symbols may indicate identical centrings, like $I'_a = A_a$, $B_a = B_c$ and $I_b = B'_b$. Alternatively, the symbol X has been used to indicate non-standard centring translations, that then must be explicitly specified. Possible non-standard centrings are presented in Table 3.9.

A rudimentary notation of symmetry operators and space group symbols in

TABLE 3.9. Standard and non-standard centrings of lattices of (3+1)-dimensional superspace groups. R refers to the obverse setting, and R_r refers to the reverse setting of the rhombohedral lattice.

Symbol	Centring translations	Symbol	Centring translations
P	none	X	*ad hoc* specification
A	$(0, \frac{1}{2}, \frac{1}{2}, 0)$	A'	$(0, \frac{1}{2}, \frac{1}{2}, \frac{1}{2})$
B	$(\frac{1}{2}, 0, \frac{1}{2}, 0)$	B'	$(\frac{1}{2}, 0, \frac{1}{2}, \frac{1}{2})$
C	$(\frac{1}{2}, \frac{1}{2}, 0, 0)$	C'	$(\frac{1}{2}, \frac{1}{2}, 0, \frac{1}{2})$
I	$(\frac{1}{2}, \frac{1}{2}, \frac{1}{2}, 0)$	I'	$(\frac{1}{2}, \frac{1}{2}, \frac{1}{2}, \frac{1}{2})$
P_a	$(\frac{1}{2}, 0, 0, \frac{1}{2})$	A_a	$(0, \frac{1}{2}, \frac{1}{2}, 0)$, $(\frac{1}{2}, 0, 0, \frac{1}{2})$, $(\frac{1}{2}, \frac{1}{2}, \frac{1}{2}, \frac{1}{2})$
B_a	$(\frac{1}{2}, 0, \frac{1}{2}, 0)$, $(\frac{1}{2}, 0, 0, \frac{1}{2})$, $(0, 0, \frac{1}{2}, \frac{1}{2})$	A'_a	$(0, \frac{1}{2}, \frac{1}{2}, \frac{1}{2})$, $(\frac{1}{2}, 0, 0, \frac{1}{2})$, $(\frac{1}{2}, \frac{1}{2}, \frac{1}{2}, 0)$
P_b	$(0, \frac{1}{2}, 0, \frac{1}{2})$	B_b	$(\frac{1}{2}, 0, \frac{1}{2}, 0)$, $(0, \frac{1}{2}, 0, \frac{1}{2})$, $(\frac{1}{2}, \frac{1}{2}, \frac{1}{2}, \frac{1}{2})$
C_b	$(\frac{1}{2}, \frac{1}{2}, 0, 0)$, $(0, \frac{1}{2}, 0, \frac{1}{2})$, $(\frac{1}{2}, 0, 0, \frac{1}{2})$	B'_b	$(\frac{1}{2}, 0, \frac{1}{2}, \frac{1}{2})$, $(0, \frac{1}{2}, 0, \frac{1}{2})$, $(\frac{1}{2}, \frac{1}{2}, \frac{1}{2}, 0)$
P_c	$(0, 0, \frac{1}{2}, \frac{1}{2})$	C_c	$(\frac{1}{2}, \frac{1}{2}, 0, 0)$, $(0, 0, \frac{1}{2}, \frac{1}{2})$, $(\frac{1}{2}, \frac{1}{2}, \frac{1}{2}, \frac{1}{2})$
A_c	$(0, \frac{1}{2}, \frac{1}{2}, 0)$, $(0, 0, \frac{1}{2}, \frac{1}{2})$ $(0, \frac{1}{2}, 0, \frac{1}{2})$	C'_c	$(\frac{1}{2}, \frac{1}{2}, 0, \frac{1}{2})$, $(0, 0, \frac{1}{2}, \frac{1}{2})$, $(\frac{1}{2}, \frac{1}{2}, \frac{1}{2}, 0)$
F	$(0, \frac{1}{2}, \frac{1}{2}, 0)$, $(\frac{1}{2}, 0, \frac{1}{2}, 0)$, $(\frac{1}{2}, \frac{1}{2}, 0, 0)$	F'	$(0, \frac{1}{2}, \frac{1}{2}, \frac{1}{2})$, $(\frac{1}{2}, 0, \frac{1}{2}, 0)$, $(\frac{1}{2}, \frac{1}{2}, 0, \frac{1}{2})$
F''	$(0, \frac{1}{2}, \frac{1}{2}, 0)$, $(\frac{1}{2}, 0, \frac{1}{2}, \frac{1}{2})$, $(\frac{1}{2}, \frac{1}{2}, 0, \frac{1}{2})$	F'''	$(0, \frac{1}{2}, \frac{1}{2}, \frac{1}{2})$, $(\frac{1}{2}, 0, \frac{1}{2}, \frac{1}{2})$, $(\frac{1}{2}, \frac{1}{2}, 0, 0)$
H	$(\frac{1}{3}, \frac{2}{3}, 0, 0)$, $(\frac{2}{3}, \frac{1}{3}, 0, 0)$	H'	$(\frac{1}{3}, \frac{2}{3}, 0, \frac{2}{3})$, $(\frac{2}{3}, \frac{1}{3}, 0, \frac{1}{3})$
		H''	$(\frac{1}{3}, \frac{2}{3}, 0, \frac{1}{3})$, $(\frac{2}{3}, \frac{1}{3}, 0, \frac{2}{3})$
R	$(\frac{1}{3}, \frac{2}{3}, \frac{2}{3}, 0)$, $(\frac{2}{3}, \frac{1}{3}, \frac{1}{3}, 0)$	R'	$(\frac{1}{3}, \frac{2}{3}, \frac{2}{3}, \frac{1}{3})$, $(\frac{2}{3}, \frac{1}{3}, \frac{1}{3}, \frac{2}{3})$
		R''	$(\frac{1}{3}, \frac{2}{3}, \frac{2}{3}, \frac{2}{3})$, $(\frac{2}{3}, \frac{1}{3}, \frac{1}{3}, \frac{1}{3})$
R_r	$(\frac{1}{3}, \frac{2}{3}, \frac{1}{3}, 0)$, $(\frac{2}{3}, \frac{1}{3}, \frac{2}{3}, 0)$	R'_r	$(\frac{1}{3}, \frac{2}{3}, \frac{1}{3}, \frac{1}{3})$, $(\frac{2}{3}, \frac{1}{3}, \frac{2}{3}, \frac{2}{3})$
		R''_r	$(\frac{1}{3}, \frac{2}{3}, \frac{1}{3}, \frac{2}{3})$, $(\frac{2}{3}, \frac{1}{3}, \frac{2}{3}, \frac{1}{3})$

(3+d)-dimensional superspace ($d = 2, 3$) has been proposed by Janner *et al.* (1983). Symbols for point group operators and space group operators in four-dimensional and higher-dimensional spaces have been proposed by Weigel and coworkers (Weigel *et al.*, 1987; Phan *et al.*, 1989; Grebille *et al.*, 1990). However, they considered n-dimensional space groups in general, and they did not allow for the special equivalence relation defining superspace groups. I will not pursue the problem of notation for (3+d)-dimensional space ($d \geqslant 2$), because numerous symbols need to be defined and memorized, while misinterpretations easily will occur. For symmetries in higher-dimensional superspace it is advisable to use an *ad hoc* notation, *e.g.* the symbol for the three-dimensional space group of the basic structure, together with an explicit definition of the superspace operators $\{R_s|\mathbf{v}_s\}$, and the modulation wave vectors through the $d \times 3$ matrix σ of their components with respect to Λ^*.

3.5.4 *Symbols for Bravais classes*

Bravais classes contain different superspace groups, that have a common holohedry and centring of their lattices. The symmetry of the lattice defining the Bravais class is itself a superspace group. Therefore, a unique designation of Bravais classes of superspace groups is obtained by the superspace group defining the symmetry of the lattice. Alternatively, the *International Tables for Crystallography Vol. C* have introduced a symbol for Bravais classes of (3+1)-dimensional superspace groups, that is a reordering of the symbol of the corresponding superspace group. The latter symbol does not contain information about intrinsic translations, that are zero by definition, nor about the value of ϵ. Again, I prefer an extended symbol to designate the Bravais class, that includes a designation for the values of ϵ, and the symbol of the superspace group that gives the lattice symmetry is used to denote the Bravais class (Table 3.6).

3.6 Symmetry of modulation functions

3.6.1 *Transformation of atomic coordinates*

Superspace groups provide symmetry of the generalized electron density. This property is illustrated in Fig. 3.3 for a simple structure with two atoms in the unit cell, that are related by the single non-trivial symmetry operator $(m_z, \bar{1})$ of the (3+1)-dimensional superspace group $Pm(0\,0\,\sigma_3)\bar{1}$. The intrinsic translation of $(m_z, \bar{1})$ is zero, because $\epsilon = -1$. The origin of the coordinate system is chosen to lie on the symmetry element $(m_z, \bar{1})$, implying that the origin-dependent translation is zero too. The symmetry operator in superspace, including intrinsic and origin-dependent translations, is given by the Seitz symbol $\{m_z\bar{1}|0,0,0,0\}$. Coordinates $\mathbf{x}_s(1)$ in superspace are transformed towards coordinates $\mathbf{x}_s(2)$ by $\{m_z\bar{1}|0,0,0,0\}$ according to [eqn (3.18)],

$$\begin{pmatrix} x_{s1}(2) \\ x_{s2}(2) \\ x_{s3}(2) \\ x_{s4}(2) \end{pmatrix} = \begin{pmatrix} 1\,0\,0\,0 \\ 0\,1\,0\,0 \\ 0\,0\,\bar{1}\,0 \\ 0\,0\,0\,\bar{1} \end{pmatrix} \begin{pmatrix} x_{s1}(1) \\ x_{s2}(1) \\ x_{s3}(1) \\ x_{s4}(1) \end{pmatrix} = \begin{pmatrix} x_{s1}(1) \\ x_{s2}(1) \\ -x_{s3}(1) \\ -x_{s4}(1) \end{pmatrix}. \tag{3.31}$$

Figure 3.3 gives a structure for which this symmetry is fulfilled. Point A1, defined by $\mathbf{x}(1) = (x_{s1}(1), x_{s2}(1), x_{s3}(1), x_{s4}(1))$ is mapped onto point A2 with coordinates $\mathbf{x}(2) = (x_{s1}(1), x_{s2}(1), -x_{s3}(1), -x_{s4}(1))$, while both points occur in section $t = 0$ of superspace; point B1 in a section $t \neq 0$ is mapped onto B2 in section $-t$; and point C1 with a negative value of x_{s4} is mapped onto C2 with positive x_{s4}.

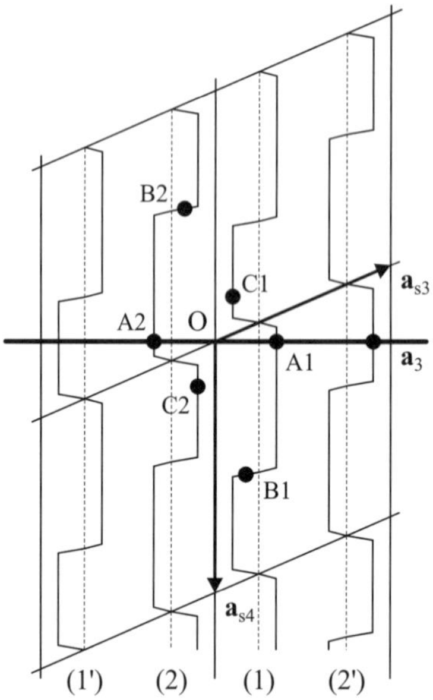

FIG. 3.3. Superspace model for a modulated structure with two atoms in the unit cell and superspace group $Pm(0,0,\sigma_3)\bar{1}$. Dashed lines indicate basic-structure positions of the atoms.

Symmetry of $\rho_s(\mathbf{x}_s)$ follows from application of the symmetry operators $\{R_s|\mathbf{v}_s\}$ to the superspace coordinates, such that

$$\rho_s(\{R_s|\mathbf{v}_s\}\mathbf{x}_s) = \rho_s(\mathbf{x}_s)$$

for all points $\mathbf{x}_s$ of superspace. However, the generalized electron density of modulated crystals is not defined by a structure model that directly specifies the electron density as a function of $\mathbf{x}_s$. Instead, modulated structures are defined by N atoms in the unit cell, with for each atom μ a set of basic-structure coordinates $\bar{\mathbf{x}}(\mu) = [x_1^0(\mu), x_2^0(\mu), x_3^0(\mu)]$ and a set of modulation functions $\mathbf{u}^\mu(\bar{x}_{s4})$, that are functions of the basic structure coordinates through their arguments $\bar{x}_{s4} = t + \mathbf{q}\cdot\bar{\mathbf{x}}$. The problem then arises of how the structural parameters will be transformed by the symmetry operators.

First it is noticed that any symmetry $\{R_s|\mathbf{v}_s\} = \{R, \epsilon|\mathbf{v}_s\}$ of the generalized electron density defines symmetry of the basic structure. A superspace model for the latter is obtained by setting modulation functions equal to zero, resulting in a collection of straight lines parallel to $\mathbf{a}_{s4}$ replacing the strings of the modulated structure (compare the dashed lines in Fig. 3.3). The coordinates $\bar{\mathbf{x}}_s(1)$ of atom 1 in the basic structure transform towards coordinates $\bar{\mathbf{x}}_s(2)$ of atom 2, according to [eqn (2.20) and eqn (3.18)],

$$\begin{pmatrix} \bar{x}_{s1}(2) \\ \bar{x}_{s2}(2) \\ \bar{x}_{s3}(2) \\ \bar{x}_{s4}(2) \end{pmatrix} = \begin{pmatrix} \bar{x}_1(2) \\ \bar{x}_2(2) \\ \bar{x}_3(2) \\ \bar{x}_4(2) \end{pmatrix} = R_s \begin{pmatrix} \bar{x}_{s1}(1) \\ \bar{x}_{s2}(1) \\ \bar{x}_{s3}(1) \\ \bar{x}_{s4}(1) \end{pmatrix} + \begin{pmatrix} v_{s1} \\ v_{s2} \\ v_{s3} \\ v_{s4} \end{pmatrix} . \tag{3.32}$$

Because of the special form of R_s [eqn (3.11)], the three-dimensional part, $\{R|\mathbf{v}\}$, of $\{R_s|\mathbf{v}_s\}$ provides symmetry of the basic structure in physical space,

$$\begin{pmatrix} \bar{x}_1(2) \\ \bar{x}_2(2) \\ \bar{x}_3(2) \end{pmatrix} = R \begin{pmatrix} \bar{x}_1(1) \\ \bar{x}_2(1) \\ \bar{x}_3(1) \end{pmatrix} + \begin{pmatrix} v_1 \\ v_2 \\ v_3 \end{pmatrix} . \tag{3.33}$$

The modulated structure is obtained by combining basic structure positions of the atoms with modulation functions [eqn (2.21)]. Modulation functions are functions of the fourth superspace coordinate of the basic structure, $\bar{x}_{s4}$. In general, they are functions of the basic structure coordinates $\bar{\mathbf{x}}_s$, that happen to be independent of the coordinates $\bar{x}_{s1}$, $\bar{x}_{s2}$ and $\bar{x}_{s3}$. A transformation of the basic structure coordinates by $\{R_s|\mathbf{v}_s\}$ is known to imply a transformation of a function of these coordinates as (Maradudin and Vosko, 1968),

$$\begin{aligned} p^2(\bar{x}_{s4}) &= p^1(\{R_s|\mathbf{v}_s\}^{-1}\bar{x}_{s4}) \\ \mathbf{u}^2(\bar{x}_{s4}) &= R\mathbf{u}^1(\{R_s|\mathbf{v}_s\}^{-1}\bar{x}_{s4}) \\ U^2(\bar{x}_{s4}) &= RU^1(\{R_s|\mathbf{v}_s\}^{-1}\bar{x}_{s4})R^t , \end{aligned} \tag{3.34}$$

where R^t is the transpose of R. $p^\mu(\bar{x}_{s4})$ is a scalar function describing occupational modulation; $\mathbf{u}^\mu(\bar{x}_{s4})$ is a vector describing displacement modulations; and $U^\mu(\bar{x}_{s4})$ is a tensor (3×3 matrix) describing a modulation of the anisotropic temperature factor.

The inverse of the superspace group operator $\{R_s|\mathbf{v}_s\}$ is

$$\{R_s|\mathbf{v}_s\}^{-1} = \{R_s^{-1}| - R_s^{-1}\mathbf{v}_s\} .$$

Employing the special form of superspace operators R_s, gives for the application of $\{R_s|\mathbf{v}_s\}^{-1}$ to the basic-structure coordinates $\bar{\mathbf{x}}(2)$ in superspace [compare eqn (3.33)],

$$\begin{pmatrix} \bar{x}_{s1}(1) \\ \bar{x}_{s2}(1) \\ \bar{x}_{s3}(1) \end{pmatrix} = R^{-1} \begin{pmatrix} \bar{x}_{s1}(2) \\ \bar{x}_{s2}(2) \\ \bar{x}_{s3}(2) \end{pmatrix} - R^{-1} \begin{pmatrix} v_{s1} \\ v_{s2} \\ v_{s3} \end{pmatrix} \tag{3.35}$$

and

$$\bar{x}_{s4}(1) = \mathbf{m}^* \cdot (\bar{\mathbf{x}}(2) - \mathbf{v}) + \epsilon^{-1}(\bar{x}_{s4}(2) - v_{s4}) , \tag{3.36}$$

where $\mathbf{m}^*$ follows as $\mathbf{m}^* = \left(\sigma R^{-1} - \epsilon^{-1}\sigma\right)$ [eqn (3.3)] or by direct computation of R_s^{-1} [eqn (3.12)].

The transformed fourth superspace coordinate of the basic structure appears to depend on the first three coordinates, if the modulation wave vector has non-zero rational components. Structural parameters of modulated structures are the basic-structure coordinates of the N atoms in the unit cell of the basic structure and their modulation functions [eqns (1.9), (2.20) and (2.18)]. The transformation of these parameters by the superspace symmetry operator $\{R_s|\mathbf{v}_s\}$ follows as,

$$\begin{pmatrix} \bar{x}_1(2) \\ \bar{x}_2(2) \\ \bar{x}_3(2) \end{pmatrix} = R \begin{pmatrix} \bar{x}_1(1) \\ \bar{x}_2(1) \\ \bar{x}_3(1) \end{pmatrix} + \begin{pmatrix} v_1 \\ v_2 \\ v_3 \end{pmatrix} \tag{3.37a}$$

$$\mathbf{u}^2[\bar{x}_{s4}(2)] = R\,\mathbf{u}^1\left[\mathbf{m}^* \cdot (\bar{\mathbf{x}}(2) - \mathbf{v}) + \epsilon^{-1}(\bar{x}_{s4}(2) - v_{s4})\right], \tag{3.37b}$$

with $\bar{x}_{s4}(2) = t + \mathbf{q}\cdot\bar{\mathbf{x}}(2)$. Equation (3.37) is the transformation used in the crystallographic analysis of modulated structures. The transformation of the modulation function can be simplified in the important case of $\mathbf{q}_r = \mathbf{0}$, resulting in

$$\mathbf{u}^2[\bar{x}_{s4}(2)] = R\,\mathbf{u}^1\left[\epsilon^{-1}(\bar{x}_{s4}(2) - v_{s4})\right]. \tag{3.38}$$

For the example of Fig. 3.3, with $\mathbf{v}_s = \mathbf{0}$ for $(m_z, \bar{1})$, the modulation function transforms as

$$\begin{pmatrix} u_1^2(\bar{x}_{s4}) \\ u_2^2(\bar{x}_{s4}) \\ u_3^2(\bar{x}_{s4}) \end{pmatrix} = \begin{pmatrix} u_1^1(-\bar{x}_{s4}) \\ u_2^1(-\bar{x}_{s4}) \\ -u_3^1(-\bar{x}_{s4}) \end{pmatrix}, \tag{3.39}$$

where it is understood that the arguments of both functions $\mathbf{u}^1$ and $\mathbf{u}^2$ are $\bar{x}_{s4}$ of atom 2. With $\bar{x}_{s4}(1) = t + \sigma_3\,\bar{x}_3(1)$ and $-\bar{x}_{s4}(2) = -t - \sigma_3\,\bar{x}_3(2) = -t + \sigma_3\,\bar{x}_3(1)$, the displacement along $\mathbf{a}_3$ of atom 2 in section $-t$ of superspace is found to be opposite to the displacement along $\mathbf{a}_3$ of atom 1 in section t, in accordance with the displacements of pairs of points A, B and C in Fig. 3.3. The displacements of atoms 1 and 2 in section t of superspace are related by a relative phase shift of their modulation waves of magnitude $2t$.

3.6.2 *Special positions*

Application of all symmetry operators of a space group to a point results in g points in the unit cell, where g is the order of the point group of the crystal class to which the space group belongs. By definition, these g points are equivalent by symmetry.

It may happen that a symmetry operator $\{R_s|\mathbf{v}_s\}$ maps a point onto itself. If the order of R_s is n, then only g/n different points in the unit cell are obtained by application of all symmetry operators. Points with this property are called special points of the space group. The special points of three-dimensional space groups (also called Wyckoff positions) are tabulated in the *International Tables for Crystallography Vol. A*. They include points on mirror planes and on axes of rotation as well as the points defined by inversion centres or rotoinversion operators. Furthermore, Wyckoff positions may correspond to points that are simultaneously on two or more symmetry elements. For example, a point on a twofold axis, that is the line of intersection of two mirror planes, simultaneously lies on the twofold axis and on both mirror planes. The site symmetry of this point is designated by $mm2$.

A point with arbitrary values of the coordinates $(\bar{x}_1, \bar{x}_2, \bar{x}_3)$ defines the general position in a structure with three-dimensional translational symmetry. Special positions have coordinates that obey certain restrictions. The latter follow from the requirement that the transformed coordinates are equal to the original coordinates $(\bar{x}_1, \bar{x}_2, \bar{x}_3)$ of this point [eqn (3.37a)], and they are obtained by solving the equation

$$\begin{pmatrix} \bar{x}_1 \\ \bar{x}_2 \\ \bar{x}_3 \end{pmatrix} = R \begin{pmatrix} \bar{x}_1 \\ \bar{x}_2 \\ \bar{x}_3 \end{pmatrix} + \begin{pmatrix} v_1 \\ v_2 \\ v_3 \end{pmatrix} . \tag{3.40}$$

Equation (3.40) can never be fulfilled for screw axes and glide planes. In case of a mirror m_z, eqn (3.40) leads to the condition

$$\begin{pmatrix} \bar{x}_1 \\ \bar{x}_2 \\ \bar{x}_3 \end{pmatrix} = \begin{pmatrix} \bar{x}_1 \\ \bar{x}_2 \\ -\bar{x}_3 \end{pmatrix} . \tag{3.41}$$

This condition is fulfilled for $\bar{x}_3 = 0$, and special positions on the mirror m_z are points $(\bar{x}_1, \bar{x}_2, 0)$.

Each symmetry operator $\{R_s|\mathbf{v}_s\}$ of a modulated structure implies symmetry $\{R|\mathbf{v}\}$ of the basic structure [eqn (3.33)]. Therefore, a special position in superspace must also be a special position of the basic structure in physical space. Conversely, special positions of the three-dimensional space group of the basic structure define special positions of the superspace group of the modulated structure. However, superspace atoms are strings instead of points, and the requirement on special points in three-dimensional space need to be replaced by the requirement in superspace, that a string is mapped onto itself by $\{R_s|\mathbf{v}_s\}$. That is, one point of the string can be mapped onto a different point belonging the same string. These requirements lead to the condition of eqn (3.40) on the basic-structure coordinates of the atom, and the condition on the modulation functions,

$$\mathbf{u}(\bar{x}_{s4}) = R\,\mathbf{u}\big[\mathbf{m}^* \cdot (\bar{\mathbf{x}} - \mathbf{v}) + \epsilon^{-1}(\bar{x}_{s4} - v_{s4})\big] . \tag{3.42}$$

This condition cannot be fulfilled for modulation functions of general shapes. Two strings with the same basic-structure position define atomic positions that are too close to each other in physical space and thus cannot be part of the same crystal structure.[1] Instead, eqn (3.42) provides restrictions on the shapes of the modulation functions.

[1]Modulation functions can be defined, such that two symmetry-related atoms possess the same set of basic structure coordinates, while these atoms never come too close to each other. This especially occurs for occupational modulation functions that have value zero for different ranges of $\bar{x}_{s4}$ (Section 7.6).

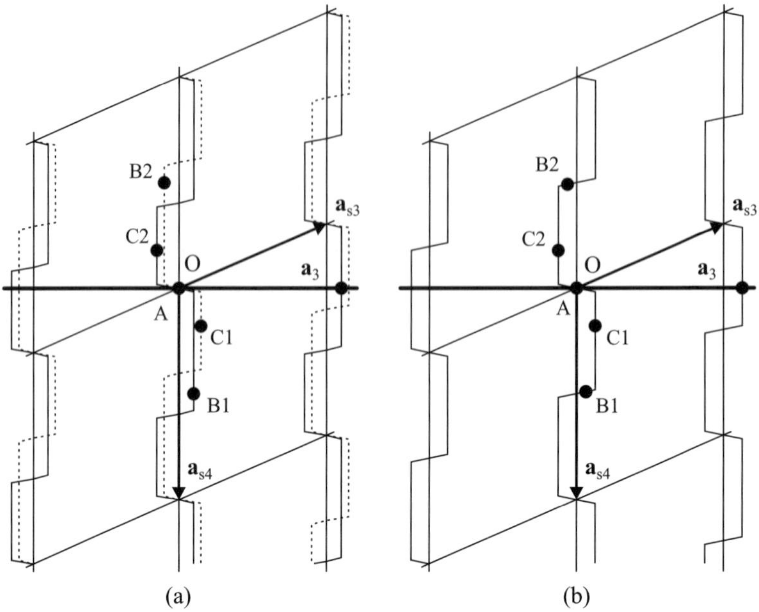

FIG. 3.4. Superspace model for a modulated structure with one atom in the unit cell on the special position $(\bar{x}_1,\ \bar{x}_2,\ 0)$ of the superspace group $Pm(0,0,\sigma_3)\bar{1}$. (a) General shape of the modulation function (full line) and its mirror image (dashed line). (b) Symmetry adapted modulation function.

For the example of Fig. 3.3 with $\{R_s|\mathbf{v}_s\} = \{m_z, \bar{1}|0,0,0,0\}$, eqn (3.40) determines that the basic-structure coordinates of an atom on this superspace mirror are restricted to $(\bar{x}_1, \bar{x}_2, 0)$. Modulation functions of Fig. 3.3 shifted towards $(\bar{x}_1, \bar{x}_2, 0)$ would obey this symmetry [Fig. 3.4(a)], but the intersecting strings clearly violate the accepted properties of matter, that a minimum separation between atoms must exist. The condition that a string be mapped onto itself leads to a condition on the modulation functions of,

$$\begin{pmatrix} u_1(\bar{x}_{s4}) \\ u_2(\bar{x}_{s4}) \\ u_3(\bar{x}_{s4}) \end{pmatrix} = \begin{pmatrix} u_1(-\bar{x}_{s4}) \\ u_2(-\bar{x}_{s4}) \\ -u_3(-\bar{x}_{s4}) \end{pmatrix}. \tag{3.43}$$

The components of $\mathbf{u}(\bar{x}_{s4})$ along $\mathbf{a}_1$ and $\mathbf{a}_2$ are found to be even functions, while $u_3(\bar{x}_{s4})$ is an odd function. The symmetry $(m_z, \bar{1})$ of an odd function centred on $(\bar{x}_1, \bar{x}_2, 0)$ is easily checked by comparing points B1 with B2 and C1 with C2 in Fig. 3.4(b).

A second example is a modulated structure with $\mathbf{q} = (0,\ 0,\ \sigma_3)$ and symmetry $(2,0) = \{2^z, 1|0,0,0,0\}$. Special positions on the twofold axis have basic-structure coordinates $(0,0,\bar{x}_3)$ and conditions on the modulation functions,

$$\begin{pmatrix} u_1(\bar{x}_{s4}) \\ u_2(\bar{x}_{s4}) \\ u_3(\bar{x}_{s4}) \end{pmatrix} = \begin{pmatrix} -u_1(\bar{x}_{s4}) \\ -u_2(\bar{x}_{s4}) \\ u_3(\bar{x}_{s4}) \end{pmatrix} . \tag{3.44}$$

Components $u_1(\bar{x}_{s4})$ and $u_2(\bar{x}_{s4})$ must be zero, while $u_3(\bar{x}_{s4})$ is not restricted.

As a final example the symmetry operator $(2, s) = \{2^z, 1|0, 0, 0, \frac{1}{2}\}$ is considered. $v_{s4} = \frac{1}{2}$ is the only non-zero component of the intrinsic translation. The coordinates of the basic structure position on $(2, s)$ are again $(0, 0, \bar{x}_3)$. Restrictions on the modulation functions now are,

$$\begin{pmatrix} u_1(\bar{x}_{s4}) \\ u_2(\bar{x}_{s4}) \\ u_3(\bar{x}_{s4}) \end{pmatrix} = \begin{pmatrix} -u_1(\bar{x}_{s4} - \frac{1}{2}) \\ -u_2(\bar{x}_{s4} - \frac{1}{2}) \\ u_3(\bar{x}_{s4} - \frac{1}{2}) \end{pmatrix} . \tag{3.45}$$

Different restrictions are obtained for odd and even harmonics of the modulation functions [compare eqn (1.10)], because

$$\begin{aligned} \sin\left[2\pi n(\bar{x}_{s4} + \tfrac{1}{2})\right] &= (-1)^n \sin[2\pi n(\bar{x}_{s4})] \\ \cos\left[2\pi n(\bar{x}_{s4} + \tfrac{1}{2})\right] &= (-1)^n \cos[2\pi n(\bar{x}_{s4})] . \end{aligned} \tag{3.46}$$

Evaluation of the restrictions for the odd and even harmonics then gives,

$$\begin{aligned} &A_3^n = B_3^n = 0 && \text{for} \quad n = \text{odd} \\ &A_1^n = B_1^n = A_2^n = B_2^n = 0 && \text{for} \quad n = \text{even} . \end{aligned} \tag{3.47}$$

A few important properties of symmetry-adapted modulation functions are demonstrated by these examples. Atoms in mirror planes of the basic structure can have non-zero displacements in the modulated structure into directions perpendicular to the mirror planes. In this case, restrictions apply to the shapes of the modulation functions, *i.e.* the displacement perpendicular to the mirror is described by a modulation that is an odd function of $\bar{x}_{s4}$. Alternatively, symmetry restrictions on modulation functions may imply that displacements in certain directions are zero. Finally, it is found that symmetry restrictions may still be obtained for symmetry elements with non-zero intrinsic translations along $\mathbf{a}_{s4}$. This point will be further discussed in Section 3.9, in relation to different settings of superspace groups.

3.7 Symmetry of the structure factor

A modulated crystal with N atoms in the unit cell of the basic structure has an atomic structure that is described by $3N$ basic structure coordinates, $6N$ atomic displacement parameters and $3N$ modulation functions for displacive modulation. Rotational symmetry, including glide and screw operators, reduces the number of independent parameters, either by restrictions on the parameters of a single atom (Section 3.6.2) or by defining a relation between the structural parameters of two atoms (Section 3.6.1).

Consider a crystal with a superspace group for which the point group of its crystal class has order g. The N atoms in the unit cell split into a group of N_a crystallographically independent atoms and $N - N_a$ atoms, which are dependent on the first N_a atoms. Obviously, $N_a \geqslant N/g$, where the equal sign applies if all atoms are on general positions. The multiplicity, m_μ, of an atomic site is defined as the number of different positions in the unit cell of the basic structure, that are generated by application of all symmetry operators to the original atom; $m_\mu = g$ for a general position and $m_\mu < g$ for special positions.

The structure factor is a function of basic-structure positions and modulation functions of the N atoms in the unit cell of the basic structure. Employing the symmetry relations between these parameters, the expression for the structure factor can be rewritten to depend on the parameters of the N_a unique atoms. Consider an atom μ' that follows from atom μ by application of the symmetry operator $\{R_s|\mathbf{v}_s\}$ [eqn (3.37)]. The modulation scattering factor, $g_{\mu'}(\mathbf{S}_s)$, of atom μ' [eqn (2.34b)] can then be written as a function of the parameters of atom μ,

$$g_{\mu'}(\mathbf{S}_s) = \int_0^1 \exp\big[2\pi i \left(\mathbf{S}\cdot R\,\mathbf{u}^\mu\big[\mathbf{m}^*\cdot R\bar{\mathbf{x}}(\mu) + \epsilon^{-1}(\tau - v_{s4})\big] + S_4\,\tau\right)\big]\,\mathrm{d}\tau\,.$$

Introduction of a new integration variable τ' with $\mathrm{d}\tau' = \mathrm{d}\tau$ and [eqn (3.13)]

$$\begin{aligned}\tau' &= \epsilon^{-1}\left(\epsilon\mathbf{m}^*\cdot R\bar{\mathbf{x}}(\mu) + \tau - v_{s4}\right)\\ &= \epsilon^{-1}\left(\tau - \mathbf{n}^*\cdot\bar{\mathbf{x}}(\mu) - v_{s4}\right)\end{aligned}$$

gives

$$\begin{aligned}g_{\mu'}(\mathbf{S}_s) = &\int_0^1 \exp[2\pi i\left(\mathbf{S}\cdot R\,\mathbf{u}^\mu(\tau') + S_4\,\epsilon\,\tau'\right)]\,\mathrm{d}\tau'\\ &\times \exp[2\pi i S_4\left(\mathbf{n}^*\cdot\bar{\mathbf{x}}(\mu) + v_{s4}\right)]\,,\end{aligned}$$

where the periodicity of modulation functions has been used. The integral in this expression is recognized as $g_\mu(\mathbf{S}_s R_s)$, because the projection of $\mathbf{S}_s R_s$ onto physical space is equal to the transformation by R of the projection $\mathbf{S}$ of $\mathbf{S}_s$ onto physical space. The structure factor follows as [eqn (2.34a)],

$$\begin{aligned}F(\mathbf{S}_s) = &\sum_{\mu=1}^{N_a}\sum_{\{R_s|\mathbf{v}_s\}}^{g} \frac{m_\mu}{g}\, f_\mu(\mathbf{S}R)\, g_\mu(\mathbf{S_s}R_s)\\ &\times \exp\big[2\pi i(\mathbf{S} - S_4\sigma)\cdot\left(R\mathbf{x}^0(\mu) + \mathbf{v}\right) + S_4\left(\mathbf{n}^*\cdot\bar{\mathbf{x}}(\mu) + v_{s4}\right)\big]\,,\end{aligned} \qquad (3.48)$$

where the sum over $\{R_s|\mathbf{v}_s\}$ extends over g superspace group operators for which the corresponding operators R_s form the point group of the crystal class. With the relation $\mathbf{n}^* - \sigma R = -\epsilon\sigma$ [eqn (3.14)], the structure factor becomes,

$$F(\mathbf{S}_s) = \sum_{\mu=1}^{N_a} \sum_{\{R_s|\mathbf{v}_s\}}^{g} \frac{m_\mu}{g} f_\mu(\mathbf{S}R)\, g_\mu(\mathbf{S_s}R_s) \times \exp\big[2\pi i \left(\mathbf{S}R - S_4\epsilon\sigma\right)\cdot\mathbf{x}^0(\mu) + \mathbf{S}_s\cdot\mathbf{v}_s)\big]\,. \tag{3.49}$$

The effect of superspace symmetry on the structure factor of modulated crystals appears to be a direct generalization of the effect of space group symmetry on the structure factor of periodic crystals. The contribution of atom μ' to the structure factor is obtained as the transformation of the contribution of atom μ by $\{R_s|\mathbf{v}_s\}$,

$$F^{\mu'}(\mathbf{S}_s) = F^{\mu}(\mathbf{S}_s R_s)\exp[2\pi i\mathbf{S}_s\cdot\mathbf{v}_s]\,. \tag{3.50}$$

Employing this relation, it is a straightforward exercise to show that structure factors obey the following symmetry relation,

$$F(\mathbf{S}_s R_s) = F(\mathbf{S}_s)\exp[-2\pi i\mathbf{S}_s\cdot\mathbf{v}_s]\,, \tag{3.51}$$

for any superspace operator $\{R_s|\mathbf{v}_s\}$ representing symmetry of the modulated structure.

3.8 Symmetry of the diffraction pattern

Diffracted intensities are proportional to the square of the absolute value of the structure factor. Symmetry of the structure factor according to eqn (3.51) then implies symmetry of the diffracted intensities given by

$$I(\mathbf{S}_s R_s) = I(\mathbf{S}_s)\,, \tag{3.52}$$

for any superspace operator $\{R_s|\mathbf{v}_s\}$ representing symmetry of the modulated structure. Employing the property that the projection of $\mathbf{S}_s R_s$ onto physical space is equal to the transformation by R of the projection of $\mathbf{S}_s$ onto physical space, equation (3.52) implies that the diffraction pattern has pure rotational symmetry according to the point group of the crystal class of the superspace group.

Inspection of the expressions for the structure factor and the modulation scattering factor reveals that

$$F(-\mathbf{S}_s) = F^*(\mathbf{S}_s)\,, \tag{3.53}$$

where * indicates complex conjugate. Accordingly, the scattered intensities $I(\mathbf{S}_s) \propto F^*(\mathbf{S}_s)F(\mathbf{S}_s)$ possess inversion symmetry,

$$I(-\mathbf{S}_s) = I(\mathbf{S}_s)\,, \tag{3.54}$$

even if the crystal structure is acentric. In complete analogy to periodic crystals, Friedel's law is found to be valid for modulated crystals. And—if anomalous scattering is taken into account—Friedel's law is violated.

While the point symmetry of the diffraction pattern is independent of the translational parts of the symmetry operators, non-zero intrinsic translations lead to systematic extinctions of Bragg reflections. Consider a symmetry operator $\{R_s|\mathbf{v}_s\}$ of a modulated crystal. The structure factor of any reflection with $\mathbf{S}_s R_s = \mathbf{S}_s$ must obey the condition [eqn (3.51)],

$$F(\mathbf{S}_s) = F(\mathbf{S}_s)\exp[-2\pi i \mathbf{S}_s \cdot \mathbf{v}_s] .$$

Either $F(\mathbf{S}_s) = 0$ (an extinction) or the phase factor must be equal to one. The latter condition provides a condition for a possible non-zero values of structure factors, given by,

$$\mathbf{S}_s \cdot \mathbf{v}_s = n , \tag{3.55}$$

with n = integer. It is stressed that eqn (3.55) need only be fulfilled for those reflections that have scattering vectors invariant under the symmetry operator R_s, *i.e.*

$$\mathbf{S}_s R_s = \mathbf{S}_s . \tag{3.56}$$

One property of symmetry operators is that $\mathbf{S}_s \cdot \mathbf{v}_s = 0$ if the intrinsic components of $\mathbf{v}_s$ are zero and $\mathbf{S}_s$ is an invariant vector. Equation (3.55) thus involves the intrinsic translation of $\{R_s|\mathbf{v}_s\}$ only.

Invariant points of a reciprocal lattice in three-dimensional space are a reciprocal lattice line coinciding with a rotation axis or a reciprocal lattice plane coinciding with a mirror plane. Invariant reciprocal lattice points of R_s in superspace can be derived from eqn (3.56) in a way similar to the analysis of periodic crystals. However, the relation between superspace and physical space determines that invariant points $\mathbf{S}_s$ of R_s in superspace are invariant points $\mathbf{S}$ of R in physical space, where $\mathbf{S}$ is the projection of $\mathbf{S}_s$ onto physical space. Thus, for $R_s = (R, \epsilon)$ it is sufficient to determine the reciprocal points $\mathbf{S}$ that lie on the symmetry element R of physical space. These are reciprocal points on the line defined by an axis of rotation or reciprocal points on the plane defined by a mirror plane.

As an example consider an orthorhombic crystal with a one-dimensional modulation given by $\sigma = (0, 0, \sigma_3)$. Invariant reciprocal points of $(2^z, 1)$ are $(0 0 h_3 h_4)$, as is easily checked by inspection of Fig. 3.1(a). The mirror $(m_x, 1)$ has invariant points $(0\ h_2\ h_3\ h_4)$, while invariant points of $(m_z, \bar{1})$ are $(h_1\ h_2\ 0\ 0)$. In general, for crystals with a one-dimensional modulation, $\epsilon = -1$ implies that $h_4 = 0$ for invariant reciprocal points of the operator $(R, \bar{1})$, *i.e.* satellite reflections are never invariant points of these operators. Invariant reciprocal points of operators $(R, 1)$ include satellite reflections, if $\mathbf{q}_r = 0$. Special attention is required for modulations with $\mathbf{q}_r \neq 0$ (see Section 3.9).

Reflection conditions are conditions on the reflection indices of invariant reciprocal points, that need to be fulfilled for a reflection to have non-zero intensity. They are obtained by substitution of the indices of invariant points $\mathbf{S}_s$ and of

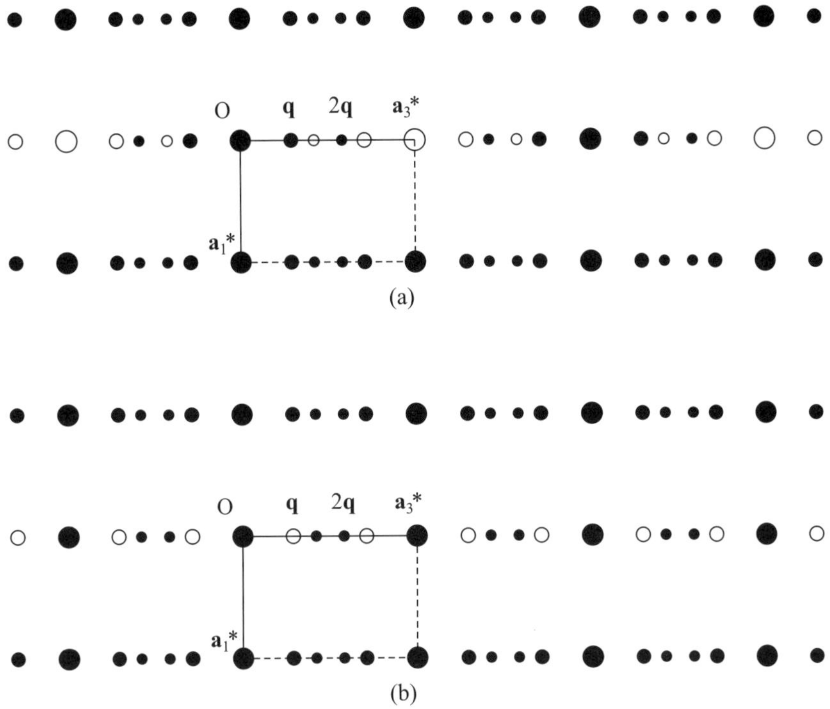

FIG. 3.5. Sections $h_2 = 0$ of reciprocal lattice planes in physical space of orthorhombic crystals with $\mathbf{q} = (0,\, 0,\, \sigma_3)$. (a) Symmetry $(2_1, 0)$. (b) Symmetry $(2, s)$. Filled circles represent scattering vectors of main reflections and satellite reflections up to second order. Open circles represent scattering vectors of extinct reflections.

the components of the intrinsic translation into eqn (3.55). For the example of Fig. 3.1(a) and the symmetry operator $(2_1^z, 0) = \{2^z, 1|0, 0, \frac{1}{2}, 0\}$ this gives,

$$(0\,0\,h_3\,h_4)\ :\ h_3 = 2n\,,$$

while $(2^z, s) = \{2^z, 1|0, 0, 0, \frac{1}{2}\}$ results in the reflection condition,

$$(0\,0\,h_3\,h_4)\ :\ h_4 = 2n\,.$$

The two situations are compared in Fig. 3.5.

In analogy to the analysis of space group symmetry of periodic crystals, intrinsic translation vectors, $\mathbf{v}_s^{\mathrm{I}}$, can be obtained from reflection conditions

$$p_1 h_1 + p_2 h_2 + p_3 h_3 + p_4 h_4 = p_0 n \tag{3.57}$$

(p_i, h_i, n integers) according to,

TABLE 3.10. Reflection conditions on $(h_1\, h_2\, h_3\, h_4)$ for selected lattices with centred unit cells.

Symbol	Centring translations	Reflection conditions
C	$(\frac{1}{2}, \frac{1}{2}, 0, 0)$	$h_1 + h_2 = 2n$
C'	$(\frac{1}{2}, \frac{1}{2}, 0, \frac{1}{2})$	$h_1 + h_2 + h_4 = 2n$
P_a	$(\frac{1}{2}, 0, 0, \frac{1}{2})$	$h_1 + h_4 = 2n$
F'	$(0, \frac{1}{2}, \frac{1}{2}, \frac{1}{2})$, $(\frac{1}{2}, 0, \frac{1}{2}, 0)$,	$h_2 + h_3 + h_4 = 2n$, $h_1 + h_3 = 2n$
	$(\frac{1}{2}, \frac{1}{2}, 0, \frac{1}{2})$	$h_1 + h_2 + h_4 = 2n$
H'	$(\frac{1}{3}, \frac{2}{3}, 0, \frac{2}{3})$, $(\frac{2}{3}, \frac{1}{3}, 0, \frac{1}{3})$	$h_1 - h_2 - h_4 = 3n$
R	$(\frac{1}{3}, \frac{2}{3}, \frac{2}{3}, 0)$, $(\frac{2}{3}, \frac{1}{3}, \frac{1}{3}, 0)$	$h_1 - h_2 - h_3 = 3n$
R'	$(\frac{1}{3}, \frac{2}{3}, \frac{2}{3}, \frac{1}{3})$, $(\frac{2}{3}, \frac{1}{3}, \frac{1}{3}, \frac{2}{3})$	$h_1 - h_2 - h_3 + h_4 = 3n$
R_r	$(\frac{1}{3}, \frac{2}{3}, \frac{1}{3}, 0)$, $(\frac{2}{3}, \frac{1}{3}, \frac{2}{3}, 0)$	$h_1 - h_2 + h_3 = 3n$

$$\mathbf{v}_s^{\mathrm{I}} = \left(\frac{p_1}{p_0}, \frac{p_2}{p_0}, \frac{p_3}{p_0}, \frac{p_4}{p_0}\right), \tag{3.58}$$

where $v_{si}^{\mathrm{I}} = p_i = 0$ if the corresponding reflection index h_i equals zero for the invariant reciprocal point.

Extinctions due to glide and screw operators are fundamentally different from those due to lattice centrings, because the latter can always be removed by choosing a primitive unit cell for the centred lattice. Nevertheless, reflection conditions resulting from the choice of a centred unit cell can be analysed in exactly the same manner as those for glide planes and screw axes. Centring translations are based on the unit operator $R_s = (E, 1)$, and invariant reciprocal points comprise of all reflections $(h_1\, h_2\, h_3\, h_4)$. Reflection conditions on indices h_i must be valid for the complete diffraction pattern. A few examples of reflection conditions corresponding to centring translations of Table 3.9 are given in Table 3.10, while the others can be obtained by application of eqn (3.55) or eqn (3.57) to the centring translations.

3.9 Equivalent settings of superspace groups

3.9.1 *Modulation wave vectors with rational components*

Modulations with $\mathbf{q}_r \neq 0$ define centred lattices in superspace. Indexing Bragg reflections on the basis of a modulation wave vector with a non-zero rational part (Table 3.2) corresponds to the choice of a primitive unit cell for the lattice in superspace. Crystallographic analyses are simplified, if an alternative indexing is chosen, for which the embedding in superspace results in the appropriate centred unit cell. The required transformation is a transformation towards a supercell of the basic structure, such that the transformed $\mathbf{q}_r$ has integer components. Subsequently, $\mathbf{q}_i$ is chosen as new modulation wave vector and reflection indices h_1, h_2 and h_3 are modified to account for $m\mathbf{q}_r$.

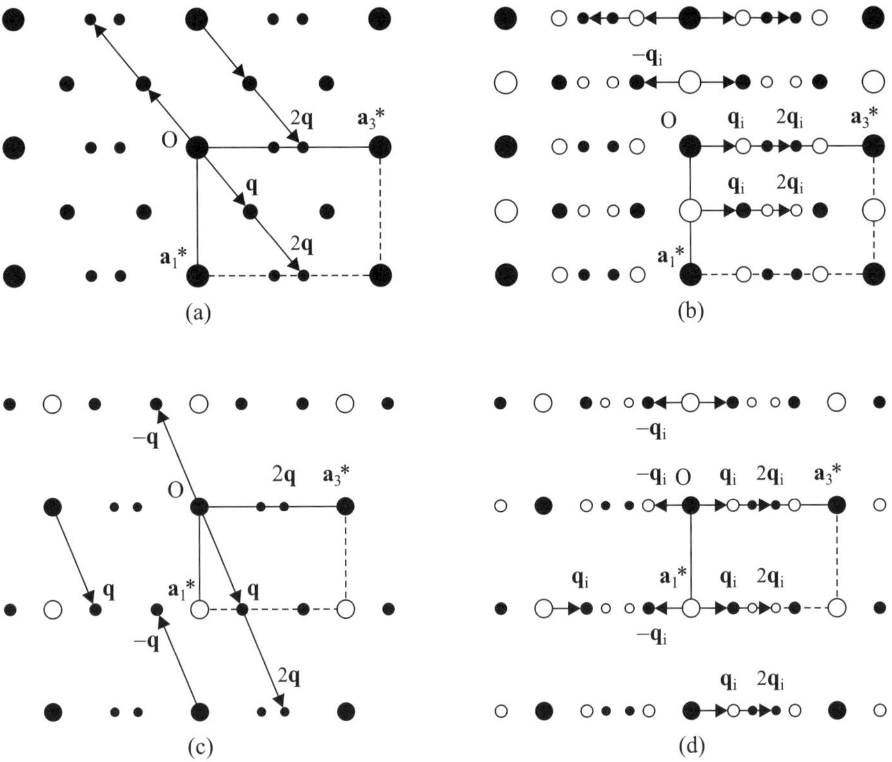

FIG. 3.6. Sections $h_2 = H_2 = 0$ of reciprocal lattices in physical space of orthorhombic crystals with one-dimensional modulations. (a) Bravais class $Pmmm(\frac{1}{2}\,0\,\sigma_3)00\bar{1}$ with indexing $(h_1\;h_2\;h_3\;h_4)$ on $\mathbf{q}$. (b) Alternative setting $P_a mmm(00\sigma_3)00\bar{1}$ with indexing $(H_1\,H_2\,H_3\,H_4)$ on $\mathbf{q}_i = (0,\,0,\,\sigma_3)$. Open circles are reciprocal points that violate the reflection condition $H_1 + H_4 = 2n$. (c) $Cmmm(10\sigma_3)00\bar{1}$ with indexing $(h_1\,h_2\,h_3\,h_4)$ on $\mathbf{q}$. (d) $C'mmm(00\sigma_3)00\bar{1}$ with indexing $(H_1\;H_2\;H_3\;H_4)$ on $\mathbf{q}_i = (0,\,0,\,\sigma_3)$. Open circles are reciprocal points that violate the reflection condition $H_1 + H_2 + H_4 = 2n$.

As an example consider the Bravais class $Pmmm(\frac{1}{2}\,0\,\sigma_3)00\bar{1}$ [Fig. 3.6(a)]. The required supercell is

$$\mathbf{A}_1 = 2\mathbf{a}_1 \qquad \mathbf{A}_2 = \mathbf{a}_2 \qquad \mathbf{A}_3 = \mathbf{a}_3\,.$$

Reflection indices $(H_1\,H_2\,H_3\,H_4)$ with respect to $\mathbf{A}_1^*$, $\mathbf{A}_2^*$, $\mathbf{A}_3^*$ and $\mathbf{q}_i = (0,\,0,\,\sigma_3)$ are obtained from the indexing $(h_1\;h_2\;h_3\;h_4)$ on $\mathbf{a}_1^*$, $\mathbf{a}_2^*$, $\mathbf{a}_3^*$ and $\mathbf{q}$ by the transformation,

$$H_1 = 2h_1 + h_4 \qquad H_2 = h_2 \qquad H_3 = h_3 \qquad H_4 = h_4\,.$$

Capital letters are used to indicate basis vectors and reflection indices of the transformed unit cell. The transformed indices obey the reflection condition [Fig. 3.6(b)],

$$(H_1\, H_2\, H_3\, H_4)\ :\ H_1 + H_4 = 2n\,.$$

With eqn (3.57) this condition is found to correspond to the P_a centring of the superspace lattice (Table 3.10), while the basic-structure unit cell $\{\mathbf{A}_1,\, \mathbf{A}_2,\, \mathbf{A}_3\}$ is centred by the translation $(\frac{1}{2},\, 0,\, 0)$. The supercell only depends on $\mathbf{q}_r$ and the lattice type of the basic structure. Therefore, the same transformation applies to all superspace groups belonging to a single Bravais class. They are listed in Table 3.11.

The basic-structure unit cell is not changed if $\mathbf{q}_r$ has integer components. This situation is only found for symmetries with a centred lattice of the basic structure (compare Table 3.6). Then $\mathbf{A}_1 = \mathbf{a}_1$, $\mathbf{A}_2 = \mathbf{a}_2$ and $\mathbf{A}_3 = \mathbf{a}_3$, but the indexing $(H_1\ H_2\ H_3\ H_4)$ is different from $(h_1\ h_2\ h_3\ h_4)$, because the indexing on the transformed unit cell employs the modulation wave vector $\mathbf{q}_i$ instead of $\mathbf{q}$ as is used for $(h_1\ h_2\ h_3\ h_4)$. For example, $\mathbf{q}_i = (0, 0, \sigma_3)$ for Bravais class $Cmmm(1\,0\,\sigma_3)00\bar{1}$, and the reflection indices transform as

$$H_1 = h_1 + h_4 \qquad H_2 = h_2 \qquad H_3 = h_3 \qquad H_4 = h_4\,.$$

The reflection condition $h_1 + h_2 = 2n$ for the C-centring leads to a condition on the transformed indices of

$$H_1 + H_2 + H_4 = 2n\,.$$

This condition implies a C' centring $(\frac{1}{2}, \frac{1}{2}, 0, \frac{1}{2})$ of the superspace lattice [Fig. 3.6(c,d)]. Similarly, $Fmmm(10\sigma_3)00\bar{1}$ can be transformed towards the alternative setting $F''mmm(0\,0\,\sigma_3)00\bar{1}$, with centring translations of the superspace lattice given in Table 3.9. $P\bar{3}1m(\frac{1}{3}\,\frac{1}{3}\,\sigma_3)\bar{1}00$ transforms to settings $H'\bar{3}m1(0\,0\,\sigma_3)\bar{1}00$ or $H''\bar{3}m1(0\,0\,\sigma_3)\bar{1}00$, depending on the choice of the supercell (Table 3.11). In Chapter 4 It will be argued that a non-standard centring of the type C' or F' can be the preferred setting for composite crystals.

Symbols for lattice centrings have been introduced in Table 3.9. However, an accepted convention is not available for symbols of non-standard settings, and the many possibilities for the choice of the basic-structure unit cell and the modulation wave vector easily lead to errors. Therefore it is recommended to use these symbols only in conjunction with an explicit definition of the centring translations, or to use the symbol X to denote a centred lattice, again combined with an explicit definition of the centring translations.

Centred settings of superspace groups are particularly useful for analysing reflection conditions for modulated structures with non-zero $\mathbf{q}_r$. For example, superspace group $P222_1(\frac{1}{2}0\sigma_3)\bar{1}\bar{1}0$ belongs to the Bravais class $Pmmm(\frac{1}{2}0\sigma_3)\bar{1}\bar{1}0$. The centred setting is $P_a222_1(0\,0\,\sigma_3)\bar{1}\bar{1}0$ with a supercell defined by $\mathbf{A}_1 = 2\mathbf{a}_1$, and a centring translation $(\frac{1}{2}, 0, 0, \frac{1}{2})$. Invariant reciprocal points of 2^z are points

TABLE 3.11. Centred unit cells of superspace lattices for one-dimensionally modulated structures with wave vectors with non-zero rational components. See text for the definition of $\mathbf{A}_1$, $\mathbf{A}_2$, $\mathbf{A}_3$, H_1, H_2 and H_3; $H_4 = h_4$.

No.	Bravais class	$\mathbf{A}_1$	$\mathbf{A}_2$	$\mathbf{A}_3$	H_1	H_2	H_3	Centrings	Refl. conditions
3	$P2/m(\sigma_1\,\sigma_2\,\frac{1}{2})\bar{1}0$	$\mathbf{a}_1$	$\mathbf{a}_2$	$2\mathbf{a}_3$	h_1	h_2	$2h_3+h_4$	$(0,0,\frac{1}{2},\frac{1}{2})$	$H_3+H_4=2n$
6	$P2/m(\frac{1}{2}0\sigma_3)0\bar{1}$	$2\mathbf{a}_1$	$\mathbf{a}_2$	$\mathbf{a}_3$	$2h_1+h_4$	h_2	h_3	$(\frac{1}{2},0,0,\frac{1}{2})$	$H_1+H_4=2n$
8	$B2/m(\frac{1}{2}0\sigma_3)0\bar{1}$	$\mathbf{a}_1$	$2\mathbf{a}_2$	$\mathbf{a}_3$	h_1	$2h_2+h_4$	h_3	$(0,\frac{1}{2},0,\frac{1}{2})$	$H_2+H_4=2n$
								$(\frac{1}{2},0,\frac{1}{2},0)$	$H_1+H_3=2n$
								$(\frac{1}{2},\frac{1}{2},\frac{1}{2},\frac{1}{2})$	$H_1+H_2+H_3$ $+H_4=2n$
10	$Pmmm(0\frac{1}{2}\sigma_3)00\bar{1}$	$\mathbf{a}_1$	$2\mathbf{a}_2$	$\mathbf{a}_3$	h_1	$2h_2+h_4$	h_3	$(0,\frac{1}{2},0,\frac{1}{2})$	$H_2+H_4=2n$
11	$Pmmm(\frac{1}{2}\frac{1}{2}\sigma_3)00\bar{1}$	$2\mathbf{a}_1$	$2\mathbf{a}_2$	$\mathbf{a}_3$	$2h_1+h_4$	$2h_2+h_4$	h_3	$(\frac{1}{2},\frac{1}{2},0,0)$	$H_1+H_2=2n$
								$(\frac{1}{2},0,0,\frac{1}{2})$	$H_1+H_4=2n$
								$(0,\frac{1}{2},0,\frac{1}{2})$	$H_2+H_4=2n$
14	$Cmmm(10\sigma_3)00\bar{1}$	$\mathbf{a}_1$	$\mathbf{a}_2$	$\mathbf{a}_3$	h_1+h_4	h_2	h_3	$(\frac{1}{2},\frac{1}{2},0,\frac{1}{2})$	$H_1+H_2+H_4=2n$
16	$Ammm(\frac{1}{2}0\sigma_3)00\bar{1}$	$2\mathbf{a}_1$	$\mathbf{a}_2$	$\mathbf{a}_3$	$2h_1+h_4$	h_2	h_3	$(\frac{1}{2},0,0,\frac{1}{2})$	$H_1+H_4=2n$
								$(0,\frac{1}{2},\frac{1}{2},0)$	$H_2+H_3=2n$
								$(\frac{1}{2},\frac{1}{2},\frac{1}{2},\frac{1}{2})$	$H_1+H_2+H_3$ $+H_4=2n$
18	$Fmmm(10\sigma_3)00\bar{1}$	$\mathbf{a}_1$	$\mathbf{a}_2$	$\mathbf{a}_3$	h_1+h_4	h_2	h_3	$(\frac{1}{2},\frac{1}{2},0,\frac{1}{2})$	$H_1+H_2+H_4=2n$
								$(\frac{1}{2},0,\frac{1}{2},\frac{1}{2})$	$H_1+H_3+H_4=2n$
								$(0,\frac{1}{2},\frac{1}{2},0)$	$H_2+H_3=2n$
20	$P4/mmm(\frac{1}{2}\frac{1}{2}\sigma_3)0\bar{1}00$	$\mathbf{a}_1-\mathbf{a}_2$	$\mathbf{a}_1+\mathbf{a}_2$	$\mathbf{a}_3$	h_1-h_2	$h_1+h_2+h_4$	h_3	$(\frac{1}{2},\frac{1}{2},0,\frac{1}{2})$	$H_1+H_2+H_4=2n$
23	$R\bar{3}m(\frac{1}{3}\frac{1}{3}\sigma_3)\bar{1}0$	$\mathbf{a}_1-\mathbf{a}_2$	$\mathbf{a}_1+2\mathbf{a}_2$	$\mathbf{a}_3$	h_1-h_2	$h_1+2h_2+h_4$	h_3	$(\frac{1}{3},\frac{2}{3},0,\frac{1}{3})$	$H_1-H_2+H_4=3n$
23	$R\bar{3}m(\frac{1}{3}\frac{1}{3}\sigma_3)\bar{1}0$	$2\mathbf{a}_1+\mathbf{a}_2$	$-\mathbf{a}_1+\mathbf{a}_2$	$\mathbf{a}_3$	$2h_1+h_2$ $+h_4$	$-h_1+h_2$	h_3	$(\frac{1}{3},\frac{2}{3},0,\frac{2}{3})$	$H_1-H_2-H_4=3n$

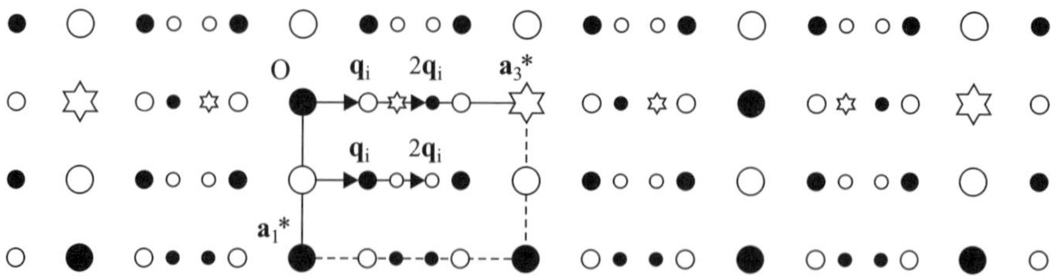

FIG. 3.7. Section $h_2 = H_2 = 0$ of the reciprocal lattice in physical space of an orthorhombic crystal with superspace group $P222_1(\frac{1}{2}\;0\;\sigma_3)\bar{1}\bar{1}0$. Indexing $(H_1 H_2 H_3 H_4)$ on $\mathbf{q}_i = (0,\,0,\,\sigma_3)$. Open circles are reciprocal points that violate the reflection condition $H_1 + H_4 = 2n$. Star symbols indicate reflections that are extinct due to the $(2_1, 0)$ screw axis.

in reciprocal physical space on the line defined by $\mathbf{a}_3^*$. Inspection of Fig. 3.6(d) shows these points to be,

$$(0\,0\,H_3\,H_4)\ :\ H_4 = 2n\,,$$

where the reflection condition is the result of the condition $(H_1 H_2 H_3 H_4)$: $H_1 + H_4 = 2n$, corresponding to the P_a centring. The screw axis of $P222_1(\frac{1}{2}\;0\;\sigma_3)\bar{1}\bar{1}0$ is $(2_1, 0) = \{2^z, 1|0, 0, \frac{1}{2}, 0\}$. The corresponding reflection condition is,

$$(0\,0\,H_3\,H_4)\ :\ H_3 = 2n\,.$$

As implied by $h_4 = 2n$, the reflection condition for $(2_1, 0)$ involves main reflections and satellites of even orders only. Extinct reflections are represented by stars in Fig. 3.7. The two conditions $H_3 = 2n$ and $H_4 = 2n$ imply a third condition,

$$(0\,0\,H_3\,H_4)\ :\ H_3 + H_4 = 2n\,.$$

This corresponds to an intrinsic translation $(0, 0, \frac{1}{2}, \frac{1}{2})$ for 2^z [eqn (3.58)], that is the operator $(2_1, s)$. Indeed, both $(2_1, s)$ and $(2_1, 0)$ belong to the superspace group $P222_1(\frac{1}{2}0\sigma_3)\bar{1}\bar{1}0$. The operator $(2_1, 0) = \{2^z, 1|0, 0, \frac{1}{2}, 0\}$ can be combined with the P_a centring translation to result in,

$$(2_1, s) = \{2^z, 1|\tfrac{1}{2}, 0, \tfrac{1}{2}, \tfrac{1}{2}\}\,.$$

Alternatively, $(2_1, s) = \{2^z, 1|0, 0, \frac{1}{2}, \frac{1}{2}\}$ combined with $(\frac{1}{2}, 0, 0, \frac{1}{2})$ gives,

$$(2_1, 0) = \{2^z, 1|\tfrac{1}{2}, 0, \tfrac{1}{2}, 0\}\,.$$

The origin of the superspace lattice can be chosen to lie on either $(2_1, 0)$ or $(2_1, s)$, but both elements are part of the superspace group. Thus, superspace groups $P222_1(\frac{1}{2}0\sigma_3)\bar{1}\bar{1}0$ and $P222_1(\frac{1}{2}0\sigma_3)\bar{1}\bar{1}s$ are equivalent superspace groups, that differ by a shift of the origin of the coordinate system. On the other hand, superspace groups $P222_1(0\,0\,\sigma_3)\bar{1}\bar{1}0$ and $P222_1(0\,0\,\sigma_3)\bar{1}\bar{1}s$ are different, because a superspace centring translation is not available here, that could have made $(2_1, 0)$ and $(2_1, s)$ equivalent.

This kind of equivalence is a general property of superspace symmetry. It is one reason that fewer superspace group types exist with $\mathbf{q}_r \neq \mathbf{0}$ than exist with $\mathbf{q}_r = \mathbf{0}$. However, the equivalence of $P222_1(\frac{1}{2}\;0\;\sigma_3)\bar{1}\bar{1}0$ and $P222_1(\frac{1}{2}\;0\;\sigma_3)\bar{1}\bar{1}s$ does not imply that the latter superspace group would be invalid. Depending on an—initially arbitrary—choice of the origin of the coordinate system, either $P222_1(\frac{1}{2}\,0\,\sigma_3)\bar{1}\bar{1}0$ or $P222_1(\frac{1}{2}\,0\,\sigma_3)\bar{1}\bar{1}s$ can be found as the superspace group that correctly describes the crystal with the chosen indexing of its Bragg reflections.

The superior performance of the centred setting for analysing reflection conditions is illustrated by consideration of the standard setting of $P222_1(\frac{1}{2}0\sigma_3)\bar{1}\bar{1}0$. Figure 3.6(a) shows that invariant points of $(2^z, 1)$ are points $(-h_1\,0\,h_3\,2h_1)$. The reflection condition corresponding to $(2_1, 0)$ is $(-h_1 0 h_3 2h_1) : h_3 = 2n$. Although $h_4 = 2n$ must be valid for invariant points of 2^z, and thus provides a second reflection condition, it is not a straightforward analysis to conclude towards the presence of $(2_1, s)$ (as is correct) in favour of $(2, s)$ (as is not correct).

3.9.2 *Choice of the modulation wave vector*

Satellite order is a genuine property of reflections. Main reflections are main reflections and satellite reflections of order m are satellite reflections of order m in any imaginable indexing of the Bragg reflections. However, an infinite number of different values exist for the indices h_1, h_2 and h_3 of a particular reflection: firstly, because an infinite number of different unit cells can be chosen for the basic structure—of which all but a few will be discarded as unreasonable; secondly, because of the arbitrariness of the modulation wave vector. A modulation given by $\mathbf{q}$ can alternatively be described by $\mathbf{q}' = \mathbf{G} \pm \mathbf{q}$, where $\mathbf{G}$ is a reciprocal lattice vector of the basic structure. Again, most possibilities will appear unreasonable, like $\mathbf{q}' = (220, -5, 0.371)$ or $\mathbf{q}' = (0, 0, 541.371)$. Restricting choices for $\mathbf{q}$ to the few reasonable possibilities provides alternative settings of superspace groups, that might not be immediately clear to be different settings of a single superspace group type.

As an example consider a modulated structure with $\mathbf{q} = (0, 0, \sigma_3)$. An alternatively indexing of the Bragg reflections is possible with

$$\mathbf{q}' = \mathbf{c}^* - \mathbf{q} = (0, 0, 1 - \sigma_3)\,.$$

Reflection indices transform as,

$$h_1' = h_1 \qquad h_2' = h_2 \qquad h_3' = h_3 + h_4 \qquad h_4' = h_4\,.$$

A symmetry operator $(2_1^z, 0)$ has reflection conditions [Fig. 3.5(a)],

$$(0\,0\,h_3\,h_4)\;:\;h_3 = 2n\,.$$

For the alternative indexing $(h_1'\,h_2'\,h_3'\,h_4')$, the reflection condition becomes (Fig. 3.8),

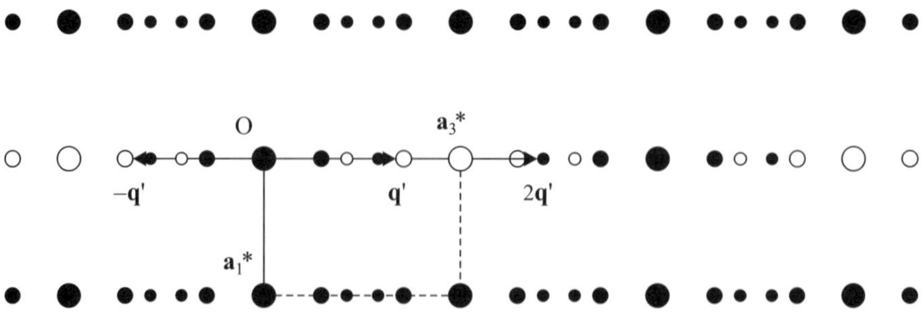

FIG. 3.8. Section $h_2 = 0$ of the reciprocal lattice in physical space of the same orthorhombic crystal as in Fig. 3.5(a). Indexing $(h'_1\, h'_2\, h'_3\, h'_4)$ on $\mathbf{q}' = \mathbf{c}^* - \mathbf{q} = (0,\, 0,\, 1 - \sigma_3)$. Open circles are reciprocal points that violate the reflection condition $h'_1 + h'_4 = 2n$.

$$(0\,0\,h'_3\,h'_4)\ :\ h'_3 + h'_4 = 2n\,.$$

This condition corresponds to an intrinsic translation $(0, 0, \frac{1}{2}, \frac{1}{2})$ for the twofold axis along $\mathbf{a}_3$, which defines the symmetry element $(2_1, s)$. It appears that $(2_1, 0)$ and $(2_1, s)$ are equivalent symmetry operators. The presence of one or the other depends on the choice of the modulation wave vector. Similar relations can be derived for screw axes on the basis of fourfold, threefold and sixfold rotations (Table 3.12).

So-called intrinsic translations of superspace symmetry elements of aperiodic crystals are not truly intrinsic, because the fourth components of these translations may depend on the modulation wave vector used for indexing of the Bragg reflections. Inspection of Table 3.12 allows some general properties of equivalence relations to be derived. The fourth component of an intrinsic translation can only assume different values in different settings if the symmetry element is a screw axis or glide plane in physical space and $\epsilon = 1$, *i.e.* the reflection condition must apply to satellite reflections as well as main reflections. Consequently, $(2, 0)$ and $(2, s)$ are not equivalent, and $(4, 0)$, $(4, q)$ and $(4, s)$, $(3, 0)$ and $(3, t)$, and $(6, 0)$, $(6, h)$, $(6, t)$ and $(6, s)$ are not equivalent as well. Different values of the fourth components of intrinsic translations may still imply inequivalent operators for 4_2, 6_2 and 6_3 screw axes (Table 3.12). By the same type of reasoning, mirror planes $(m_x, 0)$ and (m_x, s) are inequivalent, while the fourth component of intrinsic translations of a, b, c, n and d glide planes with $\epsilon = 1$ will depend on the choice of the modulation wave vector.

Screw axes 3_1 and 3_2 lead to the identical reflection conditions $h_3 = 3n$ for the reciprocal points $(0\,0\,h_3)$ in physical space. Space groups $P3_1$ and $P3_2$ form an enantiomorphous pair, because they are equivalent by a transformation Q with $\det(Q) = -1$ [eqn. (3.27a)]. Crystal structures in $P3_1$ and $P3_2$ are not identical, but they are each others mirror image instead. Determination of the correct space group amounts to a determination of the absolute configuration of the crystal structure, *e.g.* of optically active substances. It appears that reflection conditions

TABLE 3.12. Transformations of superspace screw axes for various transformations of the modulation wave vector. For each screw axis (R) along $\mathbf{a}_3$ of physical space, the corresponding screw axis in superspace (top line) and the reflection condition on $(0\,0\,h_3\,h_4)$ are indicated.

R	Indexing on modulation wave vector			
	$\mathbf{q}$	$\mathbf{c}^* - \mathbf{q}$	$2\mathbf{c}^* - \mathbf{q}$	$3\mathbf{c}^* - \mathbf{q}$
2_1	$(2_1, 0)$	$(2_1, s)$		
	$h_3 = 2n$	$h_2 + h_4 = 2n$		
4_1	$(4_1, 0)$	$(4_1, q)$	$(4_1, s)$	$(4_1, \bar{q})$
	$h_3 = 4n$	$h_3 + h_4 = 4n$	$h_3 + 2h_4 = 4n$	$h_3 + 3h_4 = 4n$
4_2	$(4_2, 0)$	$(4_2, s)$		
	$h_3 = 2n$	$h_3 + h_4 = 2n$		
	$(4_2, q)$	$(4_2, q)$	$(4_2, \bar{q})$	
	$2h_3 + h_4 = 4n$	$2h_3 + h_4 = 4n$	$2h_3 + 3h_4 = 4n$	
4_3	$(4_3, 0)$	$(4_3, \bar{q})$	$(4_3, s)$	$(4_3, q)$
	$h_3 = 4n$	$h_3 + h_4 = 4n$	$h_3 + 2h_4 = 4n$	$h_3 + 3h_4 = 4n$
3_1	$(3_1, 0)$	$(3_1, t)$	$(3_1, \bar{t})$	
	$h_3 = 3n$	$h_3 + h_4 = 3n$	$h_3 + 2h_4 = 6n$	
3_2	$(3_1, 0)$	$(3_2, \bar{t})$	$(3_2, t)$	
	$h_3 = 3n$	$h_3 + h_4 = 3n$	$h_3 + 2h_4 = 6n$	
6_1	$(6_1, 0)$	$(6_1, h)$	$(6_1, t)$	$(6_1, s)$
	$h_3 = 6n$	$h_3 + h_4 = 6n$	$h_3 + 2h_4 = 6n$	$h_3 + 3h_4 = 6n$
6_2	$(6_2, 0)$	$(6_2, t)$	$(6_2, \bar{t})$	
	$h_3 = 3n$	$h_3 + h_4 = 3n$	$h_3 + 2h_4 = 3n$	
	$(6_2, h)$	$(6_2, h)$	$(6_2, s)$	$(6_2, \bar{h})$
	$2h_3 + h_4 = 6n$	$2h_3 + h_4 = 6n$	$2h_3 + 3h_4 = 6n$	$2h_3 + 5h_4 = 6n$
6_3	$(6_3, 0)$	$(6_3, s)$		
	$h_3 = 2n$	$h_3 + h_4 = 2n$		
	$(6_3, h)$	$(6_3, t)$	$(6_2, \bar{h})$	
	$3h_3 + h_4 = 6n$	$3h_3 + 2h_4 = 6n$	$3h_3 + 5h_4 = 6n$	
6_4	$(6_4, 0)$	$(6_4, \bar{t})$	$(6_4, t)$	
	$h_3 = 3n$	$h_3 + h_4 = 3n$	$h_3 + 2h_4 = 3n$	
	$(6_4, \bar{h})$	$(6_4, \bar{h})$	$(6_4, s)$	$(6_4, h)$
	$2h_3 + h_4 = 6n$	$2h_3 + h_4 = 6n$	$2h_3 + 3h_4 = 6n$	$2h_3 + 5h_4 = 6n$
6_5	$(6_5, 0)$	$(6_5, \bar{h})$	$(6_5, \bar{t})$	$(6_5, s)$
	$h_3 = 6n$	$h_3 + h_4 = 6n$	$h_3 + 2h_4 = 6n$	$h_3 + 3h_4 = 6n$

cannot be used for this purpose. In a similar way, $P3(0 0 \sigma_3)t$ and $P3(0 0 \sigma_3)\bar{t}$ cannot be distinguished on the basis of the reflection condition $(0\,0\,h_3\,h_4) : h_4 = 3n$. The 'absolute configuration' of the modulation is the handedness of the spiral defined by **q**, and—as shown here—it cannot be determined from reflection conditions. Despite their equivalence, $(3_1, t)$ and $(3_1, \bar{t})$ lead to different reflection conditions, $(0\,0\,h_3\,h_4) : h_3 \pm h_4 = 3n$, respectively (Table 3.12), Accordingly, superspace groups $P3_1(0\,0\,\sigma_3)t$ and $P3_1(0\,0\,\sigma_3)\bar{t}$ can be distinguished on the basis of observed extinctions of Bragg reflections. On the other hand, superspace groups $P3_1(0\,0\,\sigma_3)t$ and $P3_2(0\,0\,\sigma_3)\bar{t}$ cannot be distinguished on the basis of reflection conditions (Table 3.12). Thus, the absolute configuration of modulated structures cannot be determined from reflection conditions, in accordance with properties of periodic crystals. However, the relative absolute configuration of basic structure and modulation is uniquely determined by the symmetry of the diffraction pattern. These properties have been applied to the analysis of composite crystals formed by urea/alkane inclusion compounds, where the relative absolute configuration of the two subsystems could be described (van Smaalen and Harris, 1996).

This example serves as an illustration of the (obvious) fact that the arbitrariness of the choice of the modulation wave vector can be used only once. Although $(2_1, s)$ is equivalent to $(2_1, 0)$, and $(c, 0)$ is equivalent to (c, s), superspace groups $Pca2_1(0\,0\,\sigma_3)000$ and $Pca2_1(0\,0\,\sigma_3)0ss$ are not equivalent. The transformation by $\mathbf{q}' = \mathbf{c}^* - \mathbf{q}$ simultaneously transforms $(2, 0)$ towards $(2, s)$ and $(c, 0)$ towards (c, s), leading to superspace groups $Pca2_1(0\,0\,\sigma_3)s0s$ and $Pca2_1(0\,0\,\sigma_3)ss0$, respectively.

Finally, an important property of equivalence relations is that only those reciprocal lattice vectors of the basic structure may be used, that are not extinct. Thus **q** and $\mathbf{c}^* - \mathbf{q}$ are not equivalent in the I-centred lattice, and $I2cb(0 0 \sigma_3)s0s$ and $I2cb(0\,0\,\sigma_3)0s0$ are inequivalent superspace groups.

4

INCOMMENSURATE COMPOSITE CRYSTALS

4.1 Introduction

Incommensurate composite crystals are compounds that can be characterized as the intergrowth of several incommensurately modulated structures (Section 1.1.3). Each incommensurately modulated structure is called a subsystem, and it contains a finite fraction of the atoms of the intergrowth compound. The number of subsystems (N_{sub}) is not restricted, but presently known compounds have $N_{\text{sub}} = 2$ or $N_{\text{sub}} = 3$. Subsystems are enumerated by the parameter ν, that can assume values $1, \cdots, N_{\text{sub}}$.

Atomic structures of composite crystals can be described through specification of the structural parameters of subsystem ν with respect to the lattice Λ_ν [eqn (1.11)],

$$\Lambda_\nu = \{\mathbf{a}_{\nu 1}, \mathbf{a}_{\nu 2}, \mathbf{a}_{\nu 3}\} . \tag{4.1}$$

Λ_ν defines the translational symmetry of the basic structure of subsystem ν. The position of atom μ with respect to the unit cell of subsystem ν is [eqn (1.1)]

$$\mathbf{x}_\nu^0(\mu) = x_{\nu 1}^0(\mu)\,\mathbf{a}_{\nu 1} + x_{\nu 2}^0(\mu)\,\mathbf{a}_{\nu 2} + x_{\nu 3}^0(\mu)\,\mathbf{a}_{\nu 3} . \tag{4.2}$$

Lattice vectors of Λ_ν are denoted by [eqn (1.2)]

$$\mathbf{L}_\nu = l_{\nu 1}\,\mathbf{a}_{\nu 1} + l_{\nu 2}\,\mathbf{a}_{\nu 2} + l_{\nu 3}\,\mathbf{a}_{\nu 3} , \tag{4.3}$$

where $l_{\nu i}$ are integers. Modulation functions for the displacive modulation of atom μ are [eqn (1.8)]

$$\mathbf{u}_\nu^\mu(\bar{x}_{\nu s4}) = u_{\nu 1}^\mu(\bar{x}_{\nu s4})\,\mathbf{a}_{\nu 1} + u_{\nu 2}^\mu(\bar{x}_{\nu s4})\,\mathbf{a}_{\nu 2} + u_{\nu 3}^\mu(\bar{x}_{\nu s4})\,\mathbf{a}_{\nu 3} . \tag{4.4}$$

As will be discussed in detail below, $\bar{x}_{\nu s4}$ is defined in analogy with $\bar{x}_{s4} = \bar{x}_4$ of modulated structures [eqns (1.5), (1.13) and (2.20)]. Other structural parameters obtain ν as additional subscript or superscript too. For example, $p_\nu^\mu(\bar{x}_{\nu s4})$ is the occupational modulation function of atom μ in subsystem ν [eqn (2.37)]; $A_{\nu i}^n(\mu)$ and $B_{\nu i}^n(\mu)$ are Fourier amplitudes of displacive modulation functions of atom μ in subsystem ν [eqn (1.10)]; and $U_{\nu ij}(\mu)$ are anisotropic displacement parameters.

Subsystems are not independent structures. Foremost, atoms cannot be closer to each other than some minimum distance of approach. This property must be valid for any pair of atoms, including the case where the atoms are taken from different subsystems. Implications are building principles of composite crystals, as they have been illustrated by the examples in Section 1.1.3: coplanar layers of

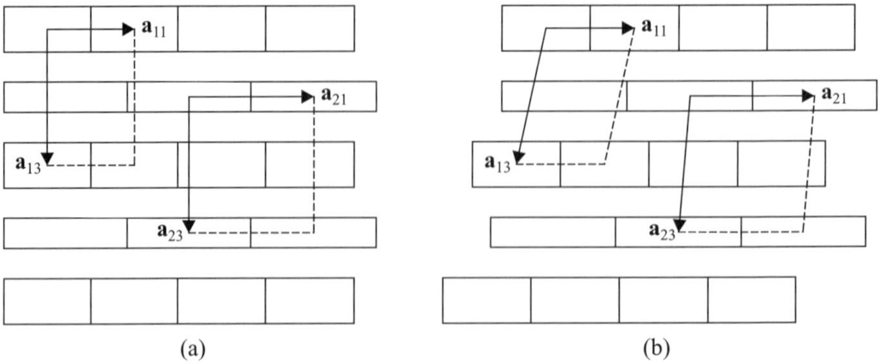

FIG. 4.1. Schematic representation of the $(\mathbf{a}_{\nu1}, \mathbf{a}_{\nu3})$ lattice plane of composite crystals. (a) Orthorhombic subsystem lattices with $\mathbf{a}_{23} = \mathbf{a}_{13}$. (b) Monoclinic subsystem lattices with $\mathbf{a}_{23} \neq \mathbf{a}_{13}$.

two subsystems are stacked alternatingly, collinear columns or chains of atoms of different subsystems are arranged periodically on some two-dimensional grid, or columns of atoms are arranged in channels of framework structures, with collinear channel and column axes.[2]

Coplanarity and collinearity can be expressed in the form of relations between reciprocal lattice vectors of the subsystems. Two periodic systems have a specific interaction in a certain direction, if reciprocal periodicities match in this direction. For example, each subsystem lattice of a layered compound possesses two basis vectors parallel to the layers, while the third basis vector of each lattice connects one layer with the next layer of the same subsystem (Fig. 4.1). Coplanarity implies that the plane defined by $\{\mathbf{a}_{11}, \mathbf{a}_{12}\}$ is equal to the plane defined by $\{\mathbf{a}_{21}, \mathbf{a}_{22}\}$, but specific relations do not exist between the basis vectors of Λ_1 and those of Λ_2. Instead, the distance between consecutive layers of subsystem 1 is equal to the corresponding distance of subsystem 2. Together with coplanarity, this implies $\mathbf{a}_{13}^* = \mathbf{a}_{23}^*$. The building principle does not require further relations between direct or reciprocal lattices of the subsystems. A single common basis vector of the reciprocal lattices of the basic structures of the subsystems is sufficient for composite crystals to exist. However, thermodynamic stability of incommensurate composite crystals is achieved through bonding interactions between atoms, that may be optimized by specifically favourable interactions in more than one direction. Consequently, one additional common basis vector is found for the reciprocal lattices of the basic structures of the subsystems. (Three common reciprocal basis vectors would imply a periodic crystal.) All known incommensurate composite crystals have a common reciprocal lattice plane of the subsystems, *e.g.* $\mathbf{a}_2^* = \mathbf{a}_{12}^* = \mathbf{a}_{22}^*$ and $\mathbf{a}_3^* = \mathbf{a}_{13}^* = \mathbf{a}_{23}^*$. The direction perpendicular

[2]With the exception of layered compounds, structures of ever increasing complexity can be imagined, in which collinear columns of atoms exist in two or even more different directions.

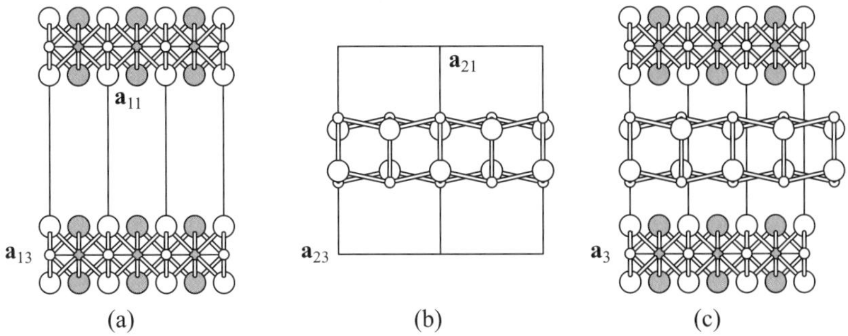

FIG. 4.2. Layered incommensurate composite crystal. (a) Atoms of subsystem 1 with three unit cells indicated. (b) Atoms of subsystem 2 with two unit cells indicated. (c) All atoms with three unit cells of subsystem 1 indicated. $\mathbf{a}_{11}$ and $\mathbf{a}_{21}$ are mutually incommensurate [eqn (4.5)]. $\mathbf{a}_3 = \mathbf{a}_{13} = \mathbf{a}_{23}$ (orthorhombic lattice).

to this reciprocal lattice plane is the direction of mutual incommensurability of the subsystems, with $\mathbf{a}_{11}$ parallel to $\mathbf{a}_{21}$, and

$$\mathbf{a}_{11} = \sigma_1 \, \mathbf{a}_{21} \, . \tag{4.5}$$

In a similar way, channel and columnar composite crystals possess common reciprocal lattice planes perpendicular to the column axis. The direction parallel to the channels and columns is the direction of mutual incommensurability of the subsystems [eqn (4.5)].

Unit cells of different subsystems provide alternative coverings of space. Atoms of the periodic basic structure of subsystem 1 do not fill the unit cell of Λ_1. Space is left empty, in which the atoms of subsystem 2 are accommodated (Fig. 4.2). This empty space is repeated in every unit cell of Λ_1, again leading to the requirement of at least one common reciprocal basis vector between the two subsystems.

Any point in space can be described by coordinates with respect to Λ_1 or alternatively by coordinates with respect to Λ_2. It is stressed that the subscript ν in eqns (4.2)–(4.4) indicates that the coordinates are with respect to the basis Λ_ν. Whether they are coordinates of an atom of subsystem 1 or subsystem 2 is usually encoded by a unique number μ for each atom of the basic structure, *e.g.* $\mu = 1, \cdots, N_1$ for atoms of subsystem 1, and $\mu = N_1 + 1, \cdots, N_1 + N_2$ for atoms of subsystem 2.

4.2 Diffraction by composite crystals

Composite crystals based on the intergrowth of periodic subsystems give rise to Bragg reflections in their diffraction at the nodes of the reciprocal lattices

$$\Lambda_\nu^* = \{\mathbf{a}_{\nu 1}^*, \, \mathbf{a}_{\nu 2}^*, \, \mathbf{a}_{\nu 3}^*\} \tag{4.6}$$

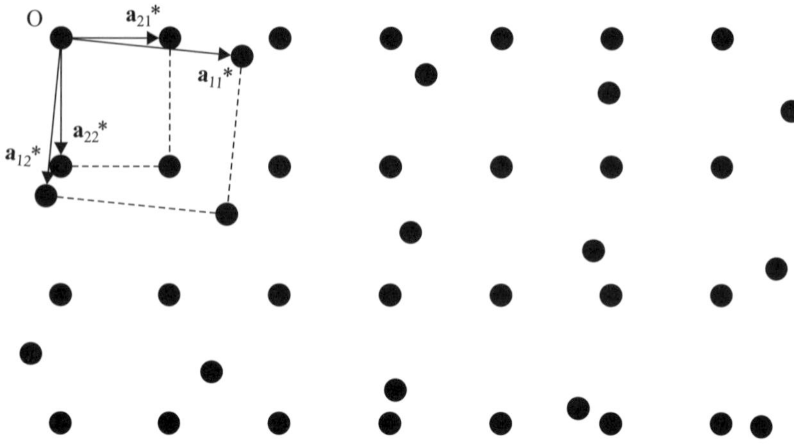

FIG. 4.3. Interpenetrating reciprocal lattices of two periodic subsystems.

of the subsystems [compare eqn (2.2)]. A maximum of $3N_{\rm sub}$ reciprocal basis vectors is sufficient for an integer indexing of the Bragg reflections,

$$\mathbf{H} = \sum_{\nu=1}^{N_{\rm sub}} \sum_{i=1}^{3} h_{\nu i}\, \mathbf{a}_{\nu i}^{*} \,. \tag{4.7}$$

For each reflection, only one triplet of indices $(h_{\nu 1}\, h_{\nu 2}\, h_{\nu 3})$ contains non-zero values, and a composite crystal with periodic subsystems cannot be distinguished from a twinned crystal on the basis of the geometry of the diffraction pattern (Fig. 4.3).

Real composite crystals are composed of incommensurately modulated subsystems, where the modulation is the result of interactions between mutually incommensurate, periodic structures. According to the principle that specific interactions between periodic systems require reciprocal periodicities to match, the modulations of subsystem ν are determined by reciprocal lattice vectors of subsystems ν' with $\nu' \neq \nu$. Thus, reciprocal basis vectors of subsystem 2 act as modulation wave vectors of subsystem 1. Bragg reflections of composite crystals with modulated subsystems can be indexed on the same $3N_{\rm sub}$ reciprocal basis vectors as were used for periodic subsystems, but the modulations imply that every scattering vector of eqn (4.7) may represent a Bragg reflection with non-zero intensity. Bragg reflections that exist by virtue of modulations are satellite reflections, as opposed to main reflections $(h_{\nu 1}\, h_{\nu 1}\, h_{\nu 3})_\nu$ of subsystem ν, that are already present for periodic subsystems.

Two subsystems share a reciprocal lattice plane (Section 4.1). Consequently, an integer indexing of Bragg reflections is achieved by less than $3N_{\rm sub}$ reciprocal basis vectors. For $N_{\rm sub} = 2$ only four and for $N_{\rm sub} = 3$ four or five reciprocal basis vectors are required. In analogy to the analysis of modulated crystals (Section 2.2), the set

$$M = \{\mathbf{a}_1^*,\, \mathbf{a}_2^*,\, \mathbf{a}_3^*,\, \mathbf{a}_{3+1}^*, \cdots, \mathbf{a}_{3+d}^*\} \tag{4.8}$$

is defined to contain the minimum number of reciprocal basis vectors, that is required for an integer indexing of the Bragg reflections. The convention is employed that the first three vectors in M are linearly independent, thus defining a reciprocal lattice in physical space,

$$\Lambda^* = \{\mathbf{a}_1^*,\, \mathbf{a}_2^*,\, \mathbf{a}_3^*\}\,. \tag{4.9}$$

Notice that Λ^* is not necessarily a reciprocal lattice of a basic structure of any of the subsystems. The number of additional reciprocal vectors is d. Specific cases of practical importance have $d = 1$ and $d = 2$, while $d = 3$ has been found for very few compounds only. The additional vectors, $\mathbf{a}_{3+j}^*$ $(j = 1, \cdots, d)$, need not be modulation wave vectors of any subsystem. They can be defined through their components with respect to Λ^* [compare eqn (2.3)],

$$\mathbf{a}_{3+j}^* = \sum_{i=1}^{3} \sigma_{ji}\, \mathbf{a}_i^*\,. \tag{4.10}$$

The components of the additional reciprocal vectors form a $d \times 3$ matrix. Incommensurability requires that each row of σ (each vector $\mathbf{a}_{3+j}^*$) contains at least one irrational number. Bragg reflections are indexed by $3+d$ integers [eqn (2.1)],

$$\mathbf{H} = \sum_{k=1}^{3+d} h_k\, \mathbf{a}_k^*\,. \tag{4.11}$$

By way of construction, the basis vectors of the reciprocal lattice of the basic structure of subsystem ν follow as integer linear combinations of basis vectors in M [eqn (4.8)]. Modulation wave vectors of subsystem ν ($\mathbf{q}^{\nu j} = \mathbf{a}_{\nu,3+j}^*$) are integer linear combinations of vectors $\mathbf{a}_k^*$ $(k = 1, \cdots, 3+d)$ too, such that

$$M_\nu = \{\mathbf{a}_{\nu 1}^*,\, \mathbf{a}_{\nu 2}^*,\, \mathbf{a}_{\nu 3}^*,\, \mathbf{a}_{\nu,3+1}^*, \cdots, \mathbf{a}_{\nu,3+d}^*\} \tag{4.12}$$

is a set of $3 + d$ reciprocal basis vectors that is again suitable for an integer indexing of all Bragg reflections of the composite crystal. The transformation between M and M_ν is given by a non-singular, $(3+d) \times (3+d)$ integer matrix W^ν, defined by,

$$\mathbf{a}_{\nu i}^* = \sum_{k=1}^{3+d} W_{ik}^\nu\, \mathbf{a}_k^* \qquad (i = 1, 2, 3) \tag{4.13a}$$

$$\mathbf{a}_{\nu,3+j}^* = \sum_{k=1}^{3+d} W_{jk}^\nu\, \mathbf{a}_k^* \qquad (j = 1, \cdots, d)\,. \tag{4.13b}$$

The inverse of W^ν exists, and reflection indices transform as,

$$(h_{\nu 1} \;\cdots\; h_{\nu,3+d})_\nu = (h_1 \;\cdots\; h_{3+d})(W^\nu)^{-1}\,. \tag{4.14}$$

Of course, one could start with an indexing $M = M_1$, resulting in W^1 being the $(3+d) \times (3+d)$ unit matrix. However, the matrices W^ν $(\nu \neq 1)$ then are always different from the unit matrix, because modulation wave vectors of subsystem 1 ($\mathbf{q}^{1j}$) correspond to reciprocal periodicities of the basic structures of the other subsystems.

Calculations are simplified by consideration of a partitioning of W^ν into 3×3 and $d \times d$ submatrices, together with off-diagonal $d \times 3$ and $3 \times d$ submatrices,

$$W^\nu = \begin{pmatrix} W^{\nu(33)} & W^{\nu(3d)} \\ W^{\nu(d3)} & W^{\nu(dd)} \end{pmatrix} . \tag{4.15}$$

The components, σ^ν, of the modulation wave vectors $\mathbf{q}^{\nu j}$ with respect to the reciprocal lattice Λ_ν^* of subsystem ν then follow as

$$\sigma^\nu = \left(W^{\nu(d3)} + W^{\nu(dd)}\sigma\right)\left(W^{\nu(33)} + W^{\nu(3d)}\sigma\right)^{-1} , \tag{4.16}$$

where σ^ν, like σ, is a $d \times 3$ matrix [eqn (4.10)].

As an example a composite crystal is considered with two subsystems and primitive orthorhombic lattices [Fig. 4.1(a)]. Bragg reflections are indexed on the basis of M with $d = 1$ [eqn (4.8)]. The common reciprocal lattice plane is defined by

$$\mathbf{a}_2^* = \mathbf{a}_{12}^* = \mathbf{a}_{22}^* \qquad \text{and} \qquad \mathbf{a}_3^* = \mathbf{a}_{13}^* = \mathbf{a}_{23}^* .$$

Incommensurability is expressed by eqn (4.5), which for orthorhombic lattices results in

$$\mathbf{a}_4^* = \mathbf{a}_{21}^* = \sigma_1\, \mathbf{a}_{11}^* = \sigma_1\, \mathbf{a}_1^* .$$

With these definitions one finds

$$W^1 = \begin{pmatrix} 1\,0\,0\,0 \\ 0\,1\,0\,0 \\ 0\,0\,1\,0 \\ 0\,0\,0\,1 \end{pmatrix} \qquad \text{and} \qquad W^2 = \begin{pmatrix} 0\,0\,0\,1 \\ 0\,1\,0\,0 \\ 0\,0\,1\,0 \\ 1\,0\,0\,0 \end{pmatrix} . \tag{4.17}$$

Main reflections of subsystem ν have subsystem indices $(h_{\nu1}\; h_{\nu2}\; h_{\nu3}\; 0)_\nu$ with respect to M_ν. Application of eqn (4.14) to both matrices W^ν shows that the corresponding indices with respect to M are:

$$\begin{array}{lll} (h_1\, h_2\, h_3\, 0) & \text{for} & \text{subsystem 1} \\ (0\, h_2\, h_3\, h_4) & \text{for} & \text{subsystem 2}, \end{array}$$

with $h_4 = h_{21}$. The common reciprocal lattice plane thus is found to contain the reciprocal points with indices [Fig. 4.4(a)]:

$$(0\, h_2\, h_3\, 0) \qquad \text{common main reflections}.$$

The orthorhombic geometry of both lattices implies that the reciprocal lattice plane $(h_1\; h_2\; 0\; 0)$ of subsystem 1 coincides with the reciprocal lattice plane

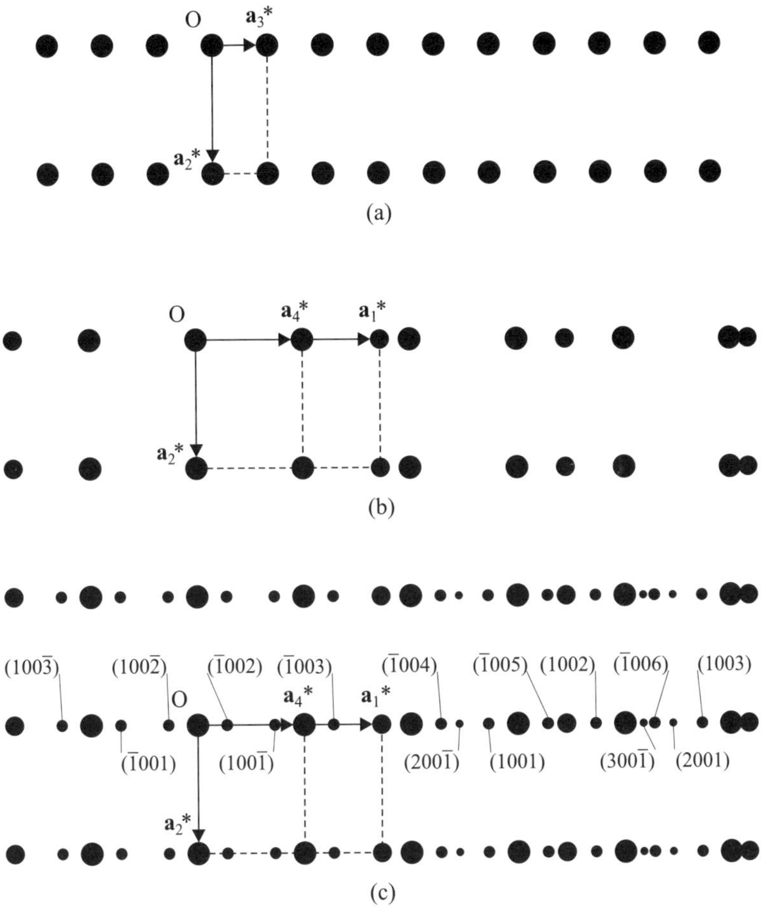

FIG. 4.4. Reciprocal lattice of an orthorhombic composite crystal with two subsystems and $\sigma = (\sigma_1,\, 0,\, 0)$. (a) Common reciprocal lattice plane $(0\, h_2\, h_3\, 0)$. (b) Main reflections $(h_1\, h_2\, 0\, 0)$ of subsystem 1 (smaller circles) and $(0\, h_2\, 0\, h_4)$ of subsystem 2 (larger circles). (c) Main reflections and first-order satellite reflections $(h_1\, h_2\, 0\, h_4)$.

$(0\, h_2\, 0\, h_4)$ of subsystem 2 [Fig. 4.4(b)]. Similar to the direct lattices, the reciprocal lattices offer alternative tilings of reciprocal space, that share a common origin. Incommensurability then determines that main reflections of one subsystem will be arbitrary close to main reflections of the other subsystem for certain combinations of reflection indices. Usually this happens for reflections with long scattering vectors. However, in the example of Fig. 4.4(b), scattering vectors $(3\, h_2\, 0\, 0)$ are close to scattering vectors $(0\, h_2\, 0\, 5)$, and pairs of main reflections $(3\, h_2\, h_3\, 0)$ and $(0\, h_2\, h_3\, 5)$ might be too close to each other, in order to be resolved in a diffraction experiment.

Satellite reflections have indices $(h_1\, h_2\, h_3\, h_4)$ with both $h_1 \neq 0$ and $h_4 \neq 0$. Accordingly, they do not appear in the section of reciprocal space defined by the common reciprocal lattice plane $(0\, h_2\, h_3\, 0)$ [Fig. 4.4(a)]. But they are present in the $(\mathbf{a}_1^*, \mathbf{a}_2^*)$-section of reciprocal space, that contains all reflections $(h_1\, h_2\, 0\, h_4)$. Figure 4.4(c) shows all main reflections, $(h_1\, h_2\, 0\, 0)$ and $(0\, h_2\, 0\, h_4)$, together with satellites of order 1. A satellite of subsystem 1 of order m has $|h_4| = m$, while satellite reflections of subsystem 2 of order m have $|h_1| = m$. Satellite reflections of order m can then be defined as reflections with

$$\text{Minimum}\,(|h_1|, |h_4|) = m\,. \tag{4.18}$$

Satellite reflections $(\pm 1\, h_2\, h_3\, \pm 1)$ are first-order satellite reflections common to the subsystems. Reflections $(h_1 h_2 h_3 \pm 1)$ are first-order satellite reflections of subsystem 1 and simultaneously $|h_1|^{th}$-order satellites of the second subsystem. In general, each reflection has contributions to its intensity from both subsystems, because any reciprocal point is a satellite or main reflection of either subsystem. In particular, main reflections of one subsystem are satellite reflections of the other subsystem at the same time.

Diffraction patterns of incommensurate composite crystals are obtained as the superposition of the diffraction patterns of N_{sub} incommensurately modulated structures [compare Fig. 4.4(c) with Fig. 2.1(a)]. Complications arise because main reflections of one subsystem are main reflections or satellite reflections of the other subsystem at the same time. A more serious problem for practical purposes is the partial overlap of reflections, that occurs due to the mutual incommensurability of the subsystems.

4.3 Reciprocal superspace

With an integer indexing of Bragg reflections of composite crystals according to eqn (4.11), the reciprocal superspace lattice can be defined in complete analogy to the case of incommensurately modulated crystals [eqns (2.6) and (2.7)]. Differences arise, because the sets of reflections with appreciable non-zero intensities are different for composite crystals and modulated crystals. A section $(h_1\, 0\, 0\, h_4)$ of the reciprocal superspace lattice corresponding to the example of Fig. 4.4 is depicted in Fig. 4.5. Main reflections of subsystem 1 are on the reciprocal lattice line defined by $\mathbf{a}_{s1}^*$ and those of subsystem 2 occur along $\mathbf{a}_{s4}^*$ in Fig. 4.5. In general, main reflections of subsystem 1 are reciprocal lattice points $(h_1\, h_2\, h_3\, 0)$ in superspace. These reciprocal lattice points of Σ^* have already been encountered as main reflections of a simple modulated structure. In addition, main reflections—now of subsystem 2—are represented by the reciprocal lattice points $(0\, h_2\, h_3\, h_4)$. Similarly, satellite reflections of order 1 are not restricted to reciprocal lattice points $(h_1\, h_2\, h_3\, \pm 1)$, but they also included $(\pm 1\, h_2\, h_3\, h_4)$. Figure 4.5 shows this increased number of reflections with non-zero intensities, by large circles for main reflections, circles of intermediate size for first-order satellites and small

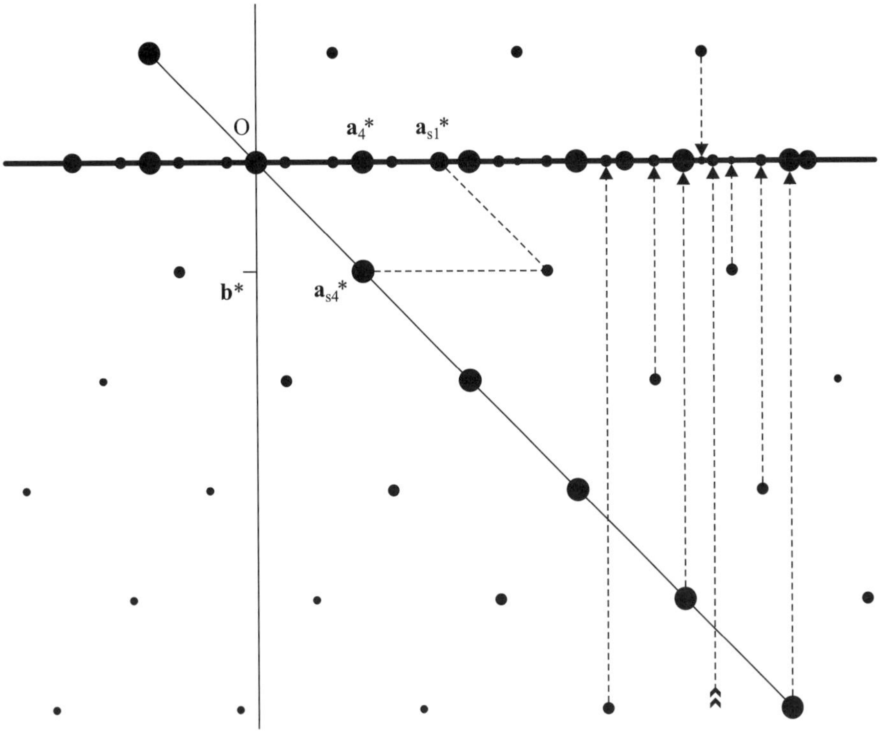

FIG. 4.5. $(h_1\,0\,0\,h_4)$-Section of the reciprocal superspace lattice of a composite crystal according to Fig. 4.4. Main reflections, first-order and higher-order satellite reflections are indicated by circles of decreasing sizes. Compare with Fig. 2.1.

dots for higher-order satellite reflections (compare to reciprocal superspace for modulated crystals in Fig. 2.1).

The matrices W^ν ($\nu = 1, \cdots, N_{\text{sub}}$) extract the reciprocal basis vectors and modulation wave vectors of subsystem ν from the $3 + d$ basis vectors of M [eqn (4.13)]. Alternatively, W^ν can be considered as a matrix defining the transformation between the sets M and M_ν, both of which provide valid integer indexings of all Bragg reflections [eqn (4.12)]. Accordingly, W^ν defines a coordinate transformation in superspace, between the superspace setting based on M and a superspace setting based on M_ν. The latter is the standard superspace for subsystem ν, that is defined in analogy to Σ^* as [eqns (2.7) and (4.12)],

$$\Sigma^*_\nu : \begin{cases} \mathbf{a}^*_{\nu si} = (\mathbf{a}^*_{\nu i}, 0) & i = 1, 2, 3 \\ \mathbf{a}^*_{\nu s4} = (\mathbf{a}^*_{\nu 4}, \mathbf{b}^*_\nu)\,, \end{cases} \tag{4.19}$$

for $d = 1$. Main reflections of subsystem ν have indices $(h_{\nu 1}\, h_{\nu 2}\, h_{\nu 3}\, 0)_\nu$ in this setting. The direct lattice of standard superspace of subsystem ν is [eqn (2.9)],

$$\Sigma_\nu : \begin{cases} \mathbf{a}_{\nu s i} = (\mathbf{a}_{\nu i}, -\sigma_i^\nu \mathbf{b}_\nu) & i = 1, 2, 3 \\ \mathbf{a}_{\nu s 4} = (0, \mathbf{b}_\nu) . \end{cases} \tag{4.20}$$

Standard superspace of subsystem ν is obtained by application of W^ν as coordinate transformation to Σ and Σ^*. These N_{sub} settings of superspace are precisely the settings needed for the description of atomic structures of subsystems by coordinates with respect to Λ_ν and $\mathbf{q}^{\nu j}$.

Composite crystals do not posses periodicities solely representing modulations.[3] Any of the $3 + d$ reciprocal basis vectors will act as a basic periodicity of some subsystem. Therefore, the length of $\mathbf{b}^*$ is not arbitrary, but should be chosen equal to the length of $\mathbf{a}_4^*$ or $\mathbf{a}_{\nu 4}^*$ in physical space, in accordance with the use of $\mathbf{b}_\nu^*$ and $\mathbf{b}_\nu$.

4.4 Structure in superspace

Reciprocal superspace of composite crystals is formally equivalent to reciprocal superspace of modulated crystals. Structure factors $F(h_1 \cdots h_{3+d})$ of Bragg reflections are assigned to reciprocal lattice points $(h_1 \cdots h_{3+d})$ in superspace. The generalized electron density is defined as the inverse Fourier transform of the structure factors in superspace, employing the same formula as for modulated crystals [eqn (2.12)]. Concomitantly, the generalized electron density of composite crystals is a periodic function according to the superspace lattice Σ [eqn (2.9)].

The generalized electron densities for composite crystals and modulated crystals differ, because of the different distributions of scattered intensities over the nodes of the reciprocal superspace lattice Σ^*. Composite crystals consist of several subsystems, each one being an incommensurately modulated structure. Accordingly, the generalized electron density contains strings of high density that define superspace atoms in a way similar to the superspace atoms of modulated crystals. However, the strings are not necessary parallel to $\mathbf{a}_{s4}$, as it would be the case for modulated crystals. Instead, the string representing an atom of subsystem ν is parallel to the direct lattice line corresponding to the reciprocal superspace direction $\mathbf{a}_{\nu 4}^* = \mathbf{q}^\nu$. The latter direction can be any vector of the set M, or even some linear combination of vectors of M, as it is defined by the matrices W^ν in eqn (4.13b). For the example of Fig. 4.4 and eqn (4.17), the generalized electron density contains strings parallel to $\mathbf{a}_{s4}$ representing atoms of subsystem 1, because W^1 is the unit matrix and $\mathbf{q}^1 = \mathbf{a}_4^*$. The superspace unit cell contains one string for each atom in the basic-structure unit cell of subsystem 1. In addition, the superspace unit cell contains one string for each atom in the basic-structure unit cell of subsystem 2, but these strings are parallel to $\mathbf{a}_{s1}$, because $\mathbf{q}^2 = \mathbf{a}_1^*$ (Fig. 4.6).

[3] It is always possible to include additional modulations that do not correspond to any of the basic periodicities of the subsystems. This case is obtained as a straightforward extension of the theory presented in this chapter.

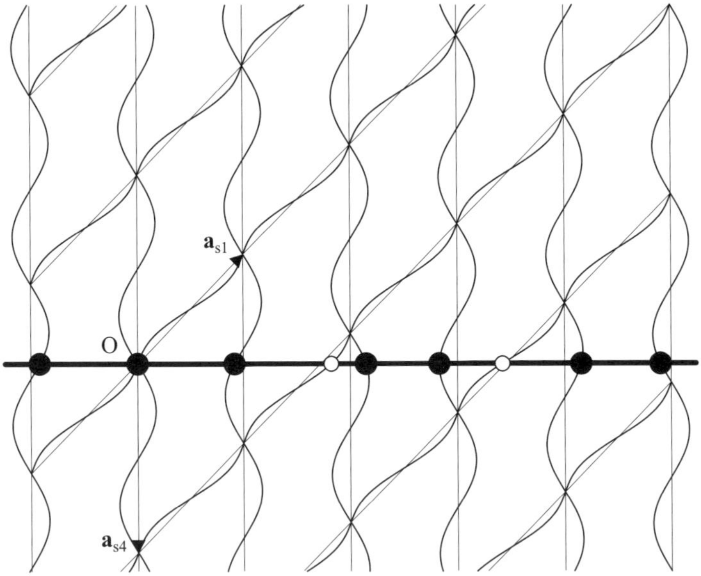

FIG. 4.6. Generalized electron density of a composite crystal projected along the reciprocal directions common to the subsystems. Wavy lines correspond to strings defining the superspace atoms. Atoms in physical space are indicated by filled circles ($\nu = 1$) or open circles ($\nu = 2$). The superspace lattice is highlighted by grid lines.

The superspace unit cell is found to contain N strings defining superspace atoms corresponding to the N atoms in the basic structure of the composite crystal. The N_1 strings of subsystem 1 are parallel to one direction ($\mathbf{a}_{s4}$ in the example), while the N_2 strings of subsystem 2 are parallel to another direction ($\mathbf{a}_{s1}$ in the example). Strings cannot intersect, because different atoms cannot occupy the same position in space. Atoms of different subsystems must have different coordinates along at least one direction corresponding to the common reciprocal lattice plane of the subsystems (coordinates $x_{\nu 2}$ and $x_{\nu 3}$ in the example). This property is apparent from a projection of the superspace structure onto a plane defined by the common reciprocal lattice plane of Σ^*. While atoms of a single subsystem may project onto positions at arbitrarily small distances from each other, atoms of different subsystems are always well separated in this projection [compare Figs. 1.4(b) and 1.4(d)].

Atomic structures of subsystems are defined by coordinates with respect to Λ_ν and by modulations based on $\mathbf{q}^\nu$ (Section 4.1). These lattices and modulation wave vectors provide independent superspace descriptions of the subsystems, with superspace coordinates of subsystem ν with respect to the subsystem superspace lattice Σ_ν given by [eqns (2.20) and (2.21)],

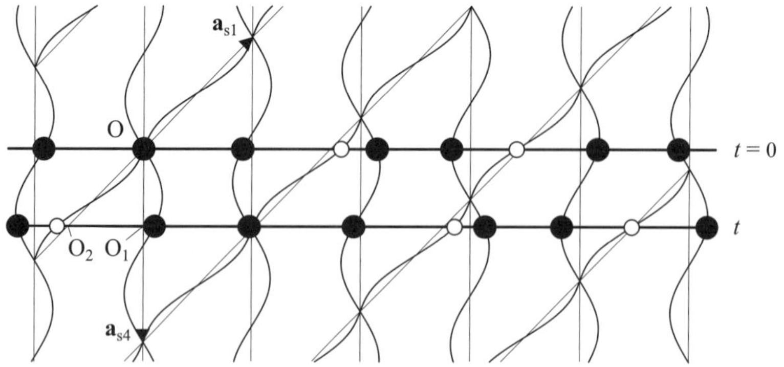

FIG. 4.7. Superspace of a composite crystal with one atom in each of two subsystems, at the origin of the respective basic-structure unit cells. O is the common origin of the subsystem lattices in physical-space section $t = 0$. O_1 and O_2 are the origins in section t.

$$\begin{aligned}
\bar{x}_{\nu si}(j) &= \bar{x}_{\nu i}(j) = l_{\nu i} + x^0_{\nu i}(\mu) && i = 1,2,3 \\
\bar{x}_{\nu s4}(j) &= \bar{x}_{\nu 4}(j) = t^\nu + \mathbf{q}^\nu \cdot \bar{\mathbf{x}}_\nu(j) && \\
x_{\nu si}(j) &= x_{\nu i}(j) = \bar{x}_{\nu si}(j) + u^\mu_{\nu i}(\bar{x}_{\nu s4}) && i = 1,2,3 \\
x_{\nu s4}(j) &\qquad\quad\;\; = \bar{x}_{\nu s4}(j) + \mathbf{q}^\nu \cdot \mathbf{u}^\mu_\nu(\bar{x}_{\nu s4}),
\end{aligned} \tag{4.21}$$

for $d = 1$. Atom j is defined as atom μ in unit cell $\mathbf{L}_\nu$ of subsystem ν.

The phase of the modulation wave of an incommensurately modulated crystal is characterized by the parameter t. The structure in physical space is obtained by choosing a specific value for t, but different values of t give physical-space structures related by translations—*i.e.* they are equivalent representations of a single structure. The superspace description interprets t as the parameter characterizing sections of superspace perpendicular to $\mathbf{b}$. Different sections give equivalent descriptions of physical space. Subsystems of composite crystals fit into this scheme, if t^ν is taken as the phase parameter for subsystem ν [eqn (4.21)]. Superspace based on M provides one more phase parameter, which will be denoted by t. This parameter corresponds to a relative shift of the subsystems along the direction of mutual incommensurability, as is apparent from a comparison of the sections $t = 0$ and t in Fig. 4.7. An immediate consequence of the incommensurability is, that structures of composite crystals are independent of this relative shift of subsystems. Different but equivalent representations of physical space are obtained for different sections t.

The independent subsystem approach thus gives $N_{\text{sub}} + 1$ independent phase parameters, t^ν and t. However, subsystems are not independent. The relation between subsystems is obvious for the directions defined by the common reciprocal lattice plane: coordinates of different subsystems assume complementary values, such that atoms of different subsystems occupy different regions of space (compare Fig. 4.2, which shows $\bar{x}_{13} \approx 0$ for atoms of subsystem 1 and $\bar{x}_{23} \approx 0.5$ for atoms of subsystem 2). A relation also exists between phases of modulation waves and basic-structure coordinates along the mutually incommensurate direction. This can be understood from the fact that the modulation in one subsystem is defined by the interactions with atoms of the other subsystems. A relative shift of the basic structures of the subsystems along the incommensurate direction (change of the value of t) necessarily brings about a shift of phases of the modulation waves (change of the values of t^ν). The result is that just one independent phase parameter exists for a composite crystal, from which the remaining $N_{\rm sub}$ parameters are derived in a unique way.

The existence of a relation between t and t^ν follows from the relation between superspace lattices Σ and Σ_ν, as it is provided by the coordinate transformations W^ν. Employing the partitioning of W^ν [eqn (4.15)], the relation between t^ν and t becomes,

$$\mathbf{t}^\nu = \left(W^{\nu(dd)} - \sigma^\nu W^{\nu(3d)}\right) \mathbf{t}\,. \tag{4.22}$$

The distance between the origins of coordinate systems Λ_ν and Λ is

$$W^{\nu(3d)}\,\mathbf{t}\,,$$

for coordinates specified in eqn (4.21). Coordinates of subsystems can be referred to lattices Λ_ν with a common origin (O_1 in Fig. 4.7), if basic-structure coordinates

$$\begin{aligned} \bar{x}'_{\nu si}(j) &= \qquad l_{\nu i} + x^0_{\nu i}(\mu) - \left(W^{\nu(3d)}\,t\right)_i \qquad \text{for} \qquad i = 1,2,3 \\ \bar{x}'_{\nu s4}(j) &= \bar{x}_{\nu s4}(j) = W^{\nu(dd)}t + \mathbf{q}^\nu \cdot \bar{\mathbf{x}}'_\nu(j) \end{aligned} \tag{4.23}$$

are used instead of eqn (4.21). The relations between t^ν and t as well as between subsystem lattices Λ_ν and Λ are important for the application of symmetry to the structure (Section 4.5), for the computation of the structure factor (Section 4.6) and for the computation of interatomic distances and other structural properties (Section 4.8).

4.5 Symmetry of composite crystals

(3+d)-Dimensional superspace groups have been defined as a variation on n-dimensional space groups ($n = 3 + d$). An element of a space group in n-dimensional space consists of a rotation followed by a translation as summarized in the symbol $\{R_s|\mathbf{v}_s\}$ (Section 3.4). The superspace concept restricts admissible space groups to those groups for which the point group is isomorphous to a point group in physical space. Consequently, a set of n basis vectors in superspace can be chosen, such that R_s is in $(3,d)$-reduced form for all elements of the superspace group. For modulated crystals, the basis vectors of the superspace lattice Σ are an appropriate basis for the $(3,d)$-reduced form, and the symmetry of a modulated crystal is given by a superspace group. For composite crystals a unique lattice does not exist for the basic structure in physical space. Valid choices for M do not necessarily lead to a $(3,d)$-reduced form for the operators R_s. Nevertheless, the fact that the point group of a n-dimensional space group is isomorphous to a crystallographic point group in three-dimensional space, suggests that a reciprocal lattice basis M can always be chosen such that the n-dimensional space group is in $(3,d)$-reduced form, *i.e.* it is one of the superspace groups. A proof of this property has not been given, but it is valid for all known composite crystals.

The second foundation of the theory of superspace groups is the definition of equivalence (Section 3.5.2). Two groups are equivalent, if their elements $\{R_s^{(1)}|\mathbf{v}_s^{(1)}\}$ and $\{R_s^{(2)}|\mathbf{v}_s^{(2)}\}$ are pairwise equal after the application of a suitable coordinate transformation to the elements $\{R_s^{(2)}|\mathbf{v}_s^{(2)}\}$ [eqn (3.27)]. Superspace groups are obtained if coordinate transformations are restricted to unimodular, $(3,d)$-reduced matrices Q, while space group equivalence in general allows any unimodular matrix Q. Superspace groups have been tabulated in the *International Tables for Crystallography Vol. C*, and they uniquely characterize the symmetry of a modulated crystal.

Superspace groups do not provide a classification of symmetries of composite crystals. As described above, the symmetry of a composite crystal is completely determined by a superspace group, but different, equally valid choices for M may lead to superspace groups that are not equivalent. That is, they are equivalent as n-dimensional space groups, but not as $(3+d)$-dimensional superspace groups. In particular, the series of standard subsystem superspaces—as defined by M_ν, Σ_ν^* and Σ_ν (Section 4.3)—provide different settings of superspace that are not related by the restricted equivalence of superspace groups. An essential property of the coordinate transformations W^ν is that they are not in $(3,d)$-reduced form, because they interchange basic periodicities and modulation periodicities between the subsystems.

A complete description of the symmetry of modulated crystals is given by the superspace group. The superspace group implicitly contains information on the symmetry of the lattice of the basic structure and on the modulation wave vectors. For composite crystals this is not true anymore, because the rôle of the reciprocal basis vectors in M depends on the subsystem. Even the number of subsystems cannot be derived from the dimension of superspace. Therefore, the specification of the set of matrices W^ν is essential for a complete characterization of the symmetry of composite crystals.

These considerations have led to the concept of subsystem superspace groups and subsystem space groups (van Smaalen, 1991). In complete analogy to the theory in Chapter 3, the superspace group G_s of a composite crystal has elements $\{R_s|\mathbf{v}_s\}$, if M is chosen as reciprocal basis [eqn (4.8)]. The matrix representations of symmetry operators with respect to the standard subsystem lattices Σ_ν then are obtained by application of the coordinate transformations W^ν to $\{R_s|\mathbf{v}_s\}$,

$$\begin{aligned} R_s^{\nu} &= W^{\nu} R_s \left(W^{\nu}\right)^{-1} \\ \mathbf{v}_{\nu s} &= W^{\nu} \mathbf{v}_s \,. \end{aligned} \tag{4.24}$$

Operators $\{R_s^{\nu}|\mathbf{v}_{\nu s}\} = \{R^{\nu}, \epsilon^{\nu}|\mathbf{v}_{\nu s}\}$ define a superspace group G_s^{ν}, that need not be equivalent to G_s. In particular, G_s^{ν} provides symmetry of subsystem ν with respect to its standard superspace description as modulated structure. Therefore, G_s^{ν} is called the subsystem superspace group. Symmetry of the basic structure of modulated crystals has been defined by extraction of operators $\{R|\mathbf{v}\}$ transforming three-dimensional space from operators $\{R_s|\mathbf{v}_s\}$ (Section 3.4). The subsystem space group, G_{ν}, of subsystem ν is defined as the collection of operators $\{R^{\nu}|\mathbf{v}_{\nu}\}$ that have been obtained by restriction of the operators $\{R_s^{\nu}|\mathbf{v}_{\nu s}\}$ to three-dimensional space. An example illustrating these concepts is provided in Section 4.7.

Coordinates of atoms of subsystem ν have been defined with respect to Σ_{ν} [eqn (4.21)]. The subsystem superspace group thus provides the appropriate setting for application of the symmetry to these coordinates [eqn (3.27)]:

$$\begin{pmatrix} \bar{x}_{\nu 1}(2) \\ \bar{x}_{\nu 2}(2) \\ \bar{x}_{\nu 3}(2) \end{pmatrix} = R^{\nu} \begin{pmatrix} \bar{x}_{\nu 1}(1) \\ \bar{x}_{\nu 2}(1) \\ \bar{x}_{\nu 3}(1) \end{pmatrix} + \begin{pmatrix} v_{\nu 1} \\ v_{\nu 2} \\ v_{\nu 3} \end{pmatrix} \tag{4.25a}$$

and

$$\mathbf{u}_{\nu}^2(\bar{x}_{\nu s4}(2)) = R^{\nu}\, \mathbf{u}_{\nu}^1\left[\mathbf{m}_{\nu}^* \cdot (\bar{\mathbf{x}}_{\nu}(2) - \mathbf{v}_{\nu}) \,+\, (\epsilon^{\nu})^{-1}\, (\bar{x}_{\nu s4}(2) - v_{\nu s4})\right], \tag{4.25b}$$

with $\mathbf{m}_{\nu}^*$ defined by [eqn (3.3)],

$$\mathbf{m}_{\nu}^* = \sigma^{\nu}\, (R^{\nu})^{-1} - (\epsilon^{\nu})^{-1}\, \sigma^{\nu} \,. \tag{4.26}$$

The shift towards a common origin of the subsystem lattices Λ_{ν} needs to be made after symmetry has been applied. Accordingly, the primed coordinates of atom 2 are [eqn (4.23)]

$$\bar{\mathbf{x}}'_{\nu s}(j) = \bar{\mathbf{x}}_{\nu s}(j) - W^{\nu(3d)}\, t \,. \tag{4.27}$$

A different treatment is required for symmetry elements that map one subsystem onto another subsystem. This type of composite crystal structure has been found for the compound $[Hg]_{3-\delta}[AsF_6]$ (Janner and Janssen, 1980*b*; van Smaalen, 1991). The first subsystem has composition AsF_6 and it has orthorhombic symmetry. The second and third subsystems are monoclinic. They consist of chains of Hg atoms, that are aligned in different directions for the two subsystems (Fig. 4.8). The symmetry elements of the orthorhombic superspace group of the composite crystal, that do not belong to the monoclinic superspace groups of the mercury subsystems, provide symmetry relating the second and third subsystems. Conversely, subsystem superspace groups are defined to contain only those operators $\{R^{\nu}|\mathbf{v}_{\nu}\}$ that map subsystem ν onto itself.

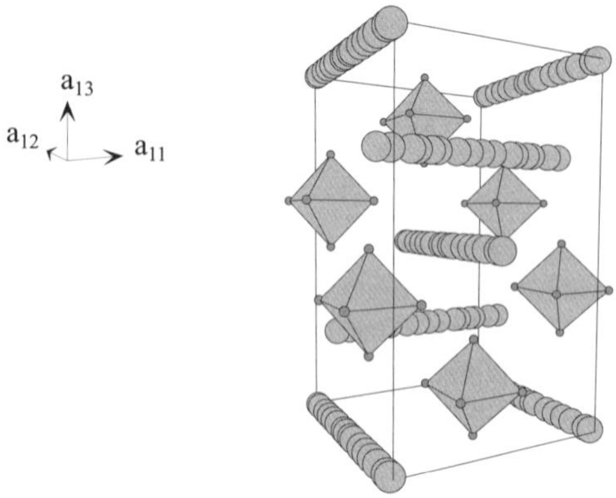

FIG. 4.8. Perspective view of the crystal structure of the composite crystal $[Hg]_{3-\delta}[AsF_6]$. One unit cell of the AsF_6 subsystem is shown. Chains of Hg atoms in two different directions represent subsystems 2 and 3, respectively. Coordinates from Tun and Brown (1986).

The existence of at least one symmetry operator that maps subsystem ν onto subsystem ν' implies that all atoms of subsystem ν' are related by symmetry to atoms of subsystem ν. The coordinates of atoms in subsystem ν' with respect to the lattice $\Sigma_{\nu'}$ can be obtained from the coordinates of atoms in subsystem ν with respect to Σ_ν by application of a transformation similar to eqn (4.25), but employing the matrices,

$$\begin{aligned} R_s^{\nu\nu'} &= W^{\nu'} R_s \, (W^{\nu})^{-1} \\ \mathbf{v}_{\nu' s} &= W^{\nu'} \mathbf{v}_s \,. \end{aligned} \tag{4.28}$$

Notice that $R_s^{\nu\nu'}$ is not a proper matrix representation of R_s, but only serves the purpose of computational convenience.

4.6 Structure factor

The structure factor provides a quantitative description of the intensities of Bragg reflections. An expression for the structure factor of incommensurately modulated structures has been derived from the atomic positions in physical space (Section 1.3), and alternatively by Fourier transform of the generalized electron density in one unit cell of superspace (Section 2.5). Superspace symmetry has been incorporated into the expression of the structure factor in Section 3.7. Bragg reflections of composite crystals and modulated crystals are given by similar sets of points, based on $3+d$ reciprocal basis vectors M [eqns (2.4) and (4.8)]. Subsystems of composite crystals are incommensurately modulated structures. Therefore, their contributions, $F_\nu(\mathbf{H}_{\nu s})$, to the structure factor $F(\mathbf{H}_s)$ of Bragg reflection $\mathbf{H}_s$ are given by structure factors of modulated structures [eqn (3.49)],

$$F_\nu(\mathbf{H}_{\nu s}) = \sum_{\mu=1}^{N_{a\nu}} \sum_{\{R_s|\mathbf{v}_s\}}^{g} f_\mu(\mathbf{H}_\nu R^\nu)\, g_\mu(\mathbf{H}_{\nu s} R_s^\nu)\, m_\mu^\nu$$
$$\times \exp\big[2\pi i \left([\mathbf{H}_\nu R^\nu - h_{\nu 4}\epsilon^\nu \sigma^\nu]\cdot \mathbf{x}^0(\mu) + \mathbf{H}_{\nu s}\cdot \mathbf{v}_{\nu s}\right)\big], \quad (4.29)$$

for the case $d = 1$, and with straightforward extensions towards composite crystals with $d > 1$. Summations are over all $N_{a\nu}$ unique atoms in the basic-structure unit cell of subsystem ν, and over the g symmetry operators of the superspace group G_s. m_μ^ν is the multiplicity of site μ with respect to the subsystem space group G_ν of the basic structure of subsystem ν. $\mathbf{H}_\nu$ is the projection of $\mathbf{H}_{\nu s}$ onto physical space. The subsystem indexing and subsystem symmetry operators are defined by $\mathbf{H}_{\nu s} = \mathbf{H}_s\,(W^\nu)^{-1}$, $R_s^\nu = W^\nu R_s\,(W^\nu)^{-1}$ and $\mathbf{v}_{\nu s} = W^\nu \mathbf{v}_s$ [eqns (4.14) and (4.24)]. Employing these relations, one can show that

$$\mathbf{H}_{\nu s}\cdot \mathbf{v}_{\nu s} = \mathbf{H}_s \cdot \mathbf{v}_s\,. \quad (4.30)$$

Each reflection $\mathbf{H}_s$ has contributions from all subsystems. Because an intergrowth compound is a single thermodynamic phase, amplitudes of scattered waves need to be combined rather than intensities as it would be correct for twinned crystals. The structure factor defines the scattered wave normalized to the scattering power of one unit cell. The latter is the volume of the unit cell of the basic structure of subsystem ν for a structure factor given by eqn (4.29). Different subsystems have unit cells with mutually incommensurate volumes, and $F_\nu[\mathbf{H}_{\nu s}]$ pertains to different scattering volumes (V_ν) for the different subsystems. The combined scattering of all subsystems is obtained by addition of the contributions $F_\nu[\mathbf{H}_{\nu s}]$ after renormalization to the same scattering volume. For example, each contribution $F_\nu[\mathbf{H}_{\nu s}]$ can be divided by the corresponding volume V_ν to arrive at the scattering per unit of volume. Alternatively, the scattering can be normalized to the scattering of one unit cell of the superspace lattice Σ. Employing the coordinate transformations W^ν, the weight of subsystem ν is obtained as the Jacobian of the transformation of t_ν towards t [eqn (4.22)],

$$J_\nu = \left|\det\left(W^{\nu(dd)} - \sigma^\nu W^{\nu(3d)}\right)\right|^{-1}. \quad (4.31)$$

The structure factor of a composite crystal with N_{sub} subsystems then is

$$F(\mathbf{H}_s) = \sum_{\nu=1}^{N_{\mathrm{sub}}} J_\nu\, F_\nu\left[\mathbf{H}_s\,(W^\nu)^{-1}\right]. \quad (4.32)$$

The coherence of the scattered waves between subsystems implies that relative positions of subsystems can be derived from the diffracted intensities. The measured intensities of the common reciprocal lattice plane of main reflections completely determine the corresponding coordinates of all atoms. Any structure model that provides a good fit to the diffraction data will obey the building

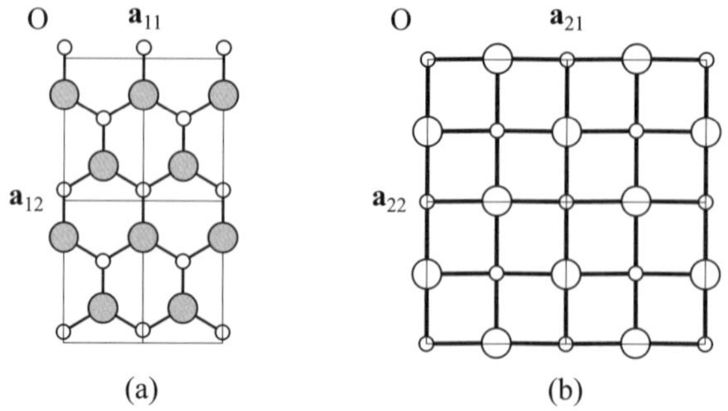

FIG. 4.9. Projection of the structure of $[LaS]_{1.14}[NbS_2]$ along $\mathbf{a}_3^*$, the stacking direction of the layers. (a) Atoms of one NbS_2 layer. (b) Atoms of one LaS layer. Large circles denote sulfur atoms and small circles depict metal atoms. White and grey circles denote atoms with different values of the projected coordinate (compare Fig. 1.4). Coordinates from Jobst and van Smaalen (2002).

principles of composite crystals (Fig. 4.2). A relative shift of the subsystems is allowed along the incommensurate direction. Contributions of all subsystems to all reflections then determine the phases of the modulations in dependence on the relative positions of the subsystems.

4.7 Common reciprocal lattice plane

A common reciprocal lattice plane of the subsystem reciprocal lattices Λ_ν^* is used as leading principle for constructing incommensurate composite crystals. Two subsystems with primitive lattices and a common reciprocal lattice plane $(h_{\nu 1}\, h_{\nu 2}\, 0)$ provide a straightforward application of the theory, with $\mathbf{a}_{23}^*$ acting as modulation wave vector for subsystem 1 and $\mathbf{q}^2 = \mathbf{a}_{13}^*$. However, a common reciprocal lattice plane of Bragg reflections may also exist if subsystem lattices have different centrings. The intricacies that are the result of combining centred lattices are illustrated in this section by a detailed discussion of the symmetries of inorganic misfit layer compounds, and in particular of $[LaS]_{1.14}[NbS_2]$ (van Smaalen, 1992*b*; Wiegers, 1996).

$[LaS]_{1.14}[NbS_2]$ is a layered compound with $N_{\text{sub}} = 2$ and orthorhombic symmetry (Fig. 1.4). One layer LaS has a structure corresponding to a two-atom thick slice parallel to the (0 0 1) plane of a rocksalt-type structure. The two-dimensional lattice parallel to the layer is pseudo-tetragonal and centred. This implies a C-centring of the three-dimensional lattice of the basic structure of the LaS subsystem (Λ_2). One layer NbS_2 is three atoms thick and it has the same structure as one layer of the compound $2H$-NbS_2. The two-dimensional lattice parallel to the layers is pseudo-trigonal, but the interactions between the subsystems make it C-centred orthorhombic, with the orthohexagonal unit cell of the trigonal lattice (Fig. 4.9). Lattices of the subsystems are adapted to each

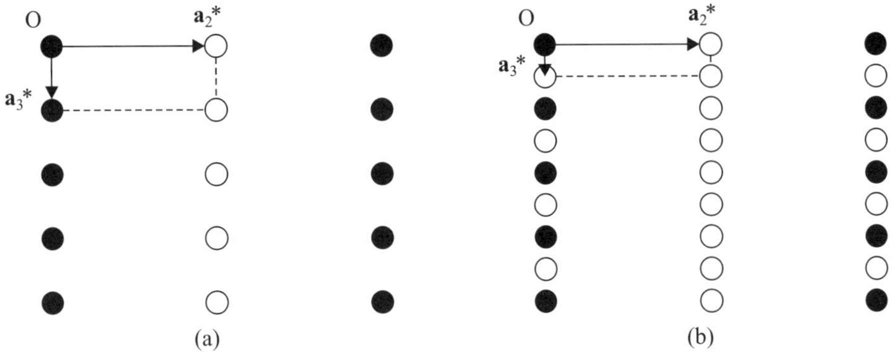

FIG. 4.10. Common reciprocal lattice plane of incommensurate misfit layer compounds. (a) C-centred lattice. (b) F-centred lattice. A perfect coincidence is observed of non-vanishing Bragg reflections (filled circles), while positions of extinct reflections (open circles) are different.

other such that $\mathbf{a}_{12}^* = \mathbf{a}_{22}^*$, which leads to $\mathbf{a}_{12} = \mathbf{a}_{22}$ because of the orthorhombic symmetry. The incommensurate direction is provided by $\mathbf{a}_{11} = \sigma_1 \mathbf{a}_{21}$.

The atomic structures of single layers imply centrings $(\frac{1}{2}, \frac{1}{2}, 0)$ for both subsystem lattices Λ_ν. Layers of a single subsystem may be stacked in different ways, thus resulting in C or F centred lattices for the subsystems. All four combinations CC, CF, FC and FF are compatible with the principle of a common reciprocal lattice plane. The C and F centrings differ only in the number of forbidden reflections within the common reciprocal lattice plane, while there is a perfect match of Bragg reflections with non-vanishing intensities (Fig. 4.10).

$[LaS]_{1.14}[NbS_2]$ crystallizes with a F-centred lattice for NbS_2 and a C-centred lattice for LaS. An appropriate choice for M is,

$$\mathbf{a}_1^* = \mathbf{a}_{11}^* \qquad \mathbf{a}_2^* = \mathbf{a}_{12}^* = \mathbf{a}_{22}^* \qquad \mathbf{a}_3^* = \mathbf{a}_{13}^* = \tfrac{1}{2}\mathbf{a}_{23}^* \qquad \mathbf{a}_4^* = \mathbf{a}_{21}^* \,. \tag{4.33}$$

Centrings of the subsystems uniquely determine the centring of the superspace lattice. Together with corresponding reflection conditions they are,

$$\begin{aligned} H_1 + H_2 + H_4 &= 2n \qquad & (\tfrac{1}{2}, \tfrac{1}{2}, 0, \tfrac{1}{2}) \\ H_1 + H_3 &= 2n \qquad & (\tfrac{1}{2}, 0, \tfrac{1}{2}, 0) \\ H_2 + H_3 + H_4 &= 2n \qquad & (0, \tfrac{1}{2}, \tfrac{1}{2}, \tfrac{1}{2}) \,. \end{aligned} \tag{4.34}$$

This centring is recognized as the F' centring in Table 3.9. The superspace group of $[LaS]_{1.14}[NbS_2]$ has been determined as $F'm2m(\sigma_1\, 0\, 0)\bar{1}\bar{1}s$ with $\sigma_1 = a_{11}/a_{21} = 0.57$. Obviously, this is a non-standard setting of superspace groups. Subsystem reciprocal lattices are recovered by a definition of W^ν matrices based on eqn (4.33). Modulation wave vectors of the subsystems can be chosen from the scattering vectors of 'true' (*i.e.* non-extinct) Bragg reflections, implying that

$\mathbf{a}_{11}^*$ and $\mathbf{a}_{21}^*$ are not suitable choices for modulation wave vectors, because they point towards extinct reflections. With correctly chosen modulation wave vectors, standard settings of the subsystem structures are obtained for

$$W^1 = \begin{pmatrix} 1\,0\,0\,0 \\ 0\,1\,0\,0 \\ 0\,0\,1\,0 \\ 0\,1\,0\,1 \end{pmatrix} \quad \text{and} \quad W^2 = \begin{pmatrix} 0\,0\,0\,1 \\ 0\,1\,0\,0 \\ 0\,0\,2\,0 \\ 1\,0\,1\,1 \end{pmatrix}. \tag{4.35}$$

Centrings of the subsystem lattices are obtained by application of W^ν to the F'-centring [eqn (4.24)], resulting in an F-centred lattice for subsystem 1 and a C-centred lattice for subsystem 2. Modulation wave vectors of the subsystems follow as [eqn (4.16)],

$$\sigma^1 = (\sigma_1,\, 1,\, 0) \qquad \sigma^2 = \left(1 + \frac{1}{\sigma_1},\, 0,\, \tfrac{1}{2}\right). \tag{4.36}$$

Symmetry operators $\{R_s|\mathbf{v}_s\}$ transform according to eqn (4.24), resulting in subsystem superspace groups and subsystem space groups,

$$\begin{aligned} G_s^1 &= Fm2m(\sigma_1,\, 1,\, 0)\bar{1}\bar{1}s & G_1 &= Fm2m \\ G_s^2 &= Cm2a(1 + \tfrac{1}{\sigma_1},\, 0,\, \tfrac{1}{2})\bar{1}\bar{1}s & G_2 &= Cm2a\,. \end{aligned} \tag{4.37}$$

This example shows that the standard settings of subsystem superspace groups require a careful selection of rather complicated W^ν matrices [eqn (4.35)]. Both subsystems possess modulation wave vectors with rational components. For many applications the non-standard setting with $\mathbf{q}_r = 0$ is preferred. These settings can be obtained directly from Σ by much simpler W^ν matrices,

$$W^1 = \begin{pmatrix} 1\,0\,0\,0 \\ 0\,1\,0\,0 \\ 0\,0\,1\,0 \\ 0\,0\,0\,1 \end{pmatrix} \quad \text{and} \quad W^2 = \begin{pmatrix} 0\,0\,0\,1 \\ 0\,1\,0\,0 \\ 0\,0\,1\,0 \\ 1\,0\,0\,0 \end{pmatrix}. \tag{4.38}$$

Modulation wave vectors are now restricted to their incommensurate components,

$$\sigma^1 = (\sigma_1,\, 0,\, 0) \qquad \sigma^2 = \left(\frac{1}{\sigma_1},\, 0,\, 0\right), \tag{4.39}$$

and subsystem symmetries are given by (Table 3.9),

$$\begin{aligned} G_s^1 &= F'm2m(\sigma_1,\, 1,\, 0)\bar{1}\bar{1}s \\ G_s^2 &= C_c'm2a(\tfrac{1}{\sigma_1},\, 0,\, 0)\bar{1}\bar{1}0\,. \end{aligned} \tag{4.40}$$

A second example is provided by $[\mathrm{Sr}]_{1.132}[\mathrm{TiS_3}]$ (Onoda *et al.*, 1993). Collinear chains of Sr and $\mathrm{TiS_3}$ are parallel to the unique axis of a hexagonal lattice, that provides the incommensurate direction (Fig. 1.4). Subsystem 1 (TiS_3) is rhombohedral with subsystem superspace group $R3m(00\sigma_3)0s$ (R-centring in Table 3.9). Subsystem 2 (Sr) is trigonal with subsystem superspace group $H'3c1(0\,0\,\sigma_3)000$ (Table 3.9) or in standard setting with subsystem superspace group $P31c(\frac{1}{3}\frac{1}{3}\sigma_3)000$. Despite these different symmetries, a perfect coincidence of non-vanishing Bragg reflections is observed in the common reciprocal plane $(h_1 h_2 00)$. This property follows directly from the reflection conditions in the R and H' settings. The R-centring corresponds to $-h_1 + h_2 + h_3 = 3n$ for $(h_1\, h_2\, h_3\, h_4)$, while the H'-centring corresponds to $-h_1 + h_2 + h_4 = 3n$. Restricted to the common reciprocal lattice plane $(h_1\; h_2\; 0\; 0)$ both centring conditions reduce to $-h_1 + h_2 = 3n$, *i.e.* non-vanishing Bragg reflections coincide.

4.8 Interatomic distances

t-Plots of interatomic distances of modulated crystals have been introduced in Section 2.4. Distances between atoms within a single subsystem of an incommensurate composite crystal can be analysed in exactly the same manner, because subsystems have incommensurately modulated structures. Distances and other structural properties are obtained as a function of t^ν from the atomic coordinates with respect to the subsystem lattices Λ_ν [eqn (4.21)]. Figure 4.11(a) shows distances between a pair of atoms of subsystem 1 for four physical-space sections characterized by different values of t [compare Fig. 2.8(a)]. Distances from La towards the five nearest-neighbour sulfur atoms of the LaS subsystem are given as a t-plot in Fig. 4.11(b). An important property of t-plots is the correlation between different curves. For example, Fig. 4.11(b) shows that at values of t where one La–S distance is shorter than average, other La–S distances are larger than average. In accordance with the discussion in Section 2.4.1, variations of distances compensate each other, such that the central atom has one particular valence throughout the modulated structure.

Sections of superspace representing physical space are characterized by the value of t. Employing the relation between t^ν and t [eqn (4.22)], modulation functions are found to possess mutually incommensurate periodicities in t for the different subsystems. Consequently, t-plots of different subsystems do not provide information about correlations between modulations in different subsystems. Distances between atoms of different subsystems are functions that are not periodic in t. Figure 4.11(c) illustrates this property for the distance between an atom of subsystem 1 with $\mathbf{x}_1^0 = (0, 0.26, 0.7)$ and an atom of subsystem 2 with $\mathbf{x}_2^0 = (0, 0, 0.33)$. In section $t = 0$ these atoms differ only by their $x_{\nu 2}$ and $x_{\nu 3}$ coordinates, and they are at a distance of minimum approach [point A_1 in Figs. 4.11(c), (d)]. The atoms become more separated for increasing values of $|t|$, such that the distance between this pair of atoms increases indefinitely for $|t| \to \infty$ [compare the curve in Fig. 4.11(d)]. Periodicity is restored if distances are considered between a single atom of one subsystem and all atoms of the other subsystem that are related to the first atom by translational symmetry of the basic structure [variation of $l_{\nu i}$ in eqn (4.21)]. Atom 1 of subsystem 1 with $\mathbf{L}_1 =$

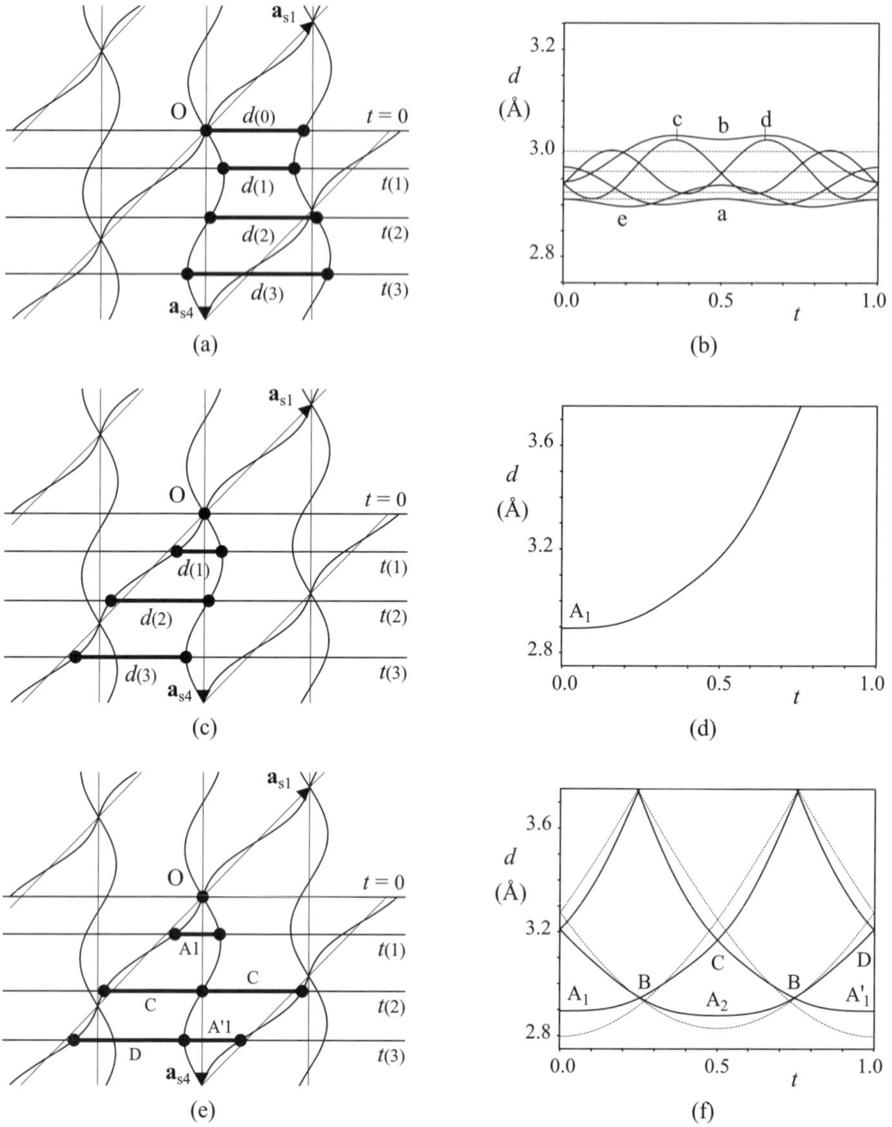

FIG. 4.11. Interatomic distances of the composite crystal $[LaS]_{1.14}[NbS_2]$. (a) Distances within the first subsystem for several sections t of superspace. (b) t-Plot of the five La–S2 distances within the LaS subsystem. (c) Distances between one atom each of the first and second subsystems. (d) t-Plot of the distance between La at $\mathbf{x}_1^0(\mathrm{La}) = (0, 0, 0.33)$ and S1 of the NbS_2 subsystem at $\mathbf{x}_2^0(\mathrm{S1}) = (0, 0.26, 0.07)$. (e) Distances between one atom of the first subsystem and several atoms of the second subsystem. (f) t-Plot of La–S1 distances for La as central atom. Dashed lines indicate distances in the basic structure. (b), (d) and (f) Reprinted from Jobst (2003) by courtesy of A. Jobst (Nürnberg).

(0, 0, 0) has been chosen as central atom in Fig. 4.11(e). For t close to zero, the shortest distance towards atoms of subsystem 2 is obtained for the atom with $\mathbf{L}_2 = (0, 0, 0)$, illustrated by the distance A_1 in Figs. 4.11(e), (f). For increasing t a point will be reached where the distances towards the atoms with $\mathbf{L}_2 = (0, 0, 0)$ and $\mathbf{L}_2 = (1, 0, 0)$ are equal to each other (point C at $t = 0.5$). For $t > 0.5$, the atom at (1, 0, 0) is nearest to the central atom (point A'_1). A periodic plot results, if distances are included between a central atom of one subsystem (fixed $\mathbf{x}_1^0$ and $\mathbf{L}_1$) and all atoms of the other subsystem that are closer than a certain cut-off distance (fixed $\mathbf{x}_2^0$ and all possible $\mathbf{L}_2$). The period of this plot is the period of the central atom. Thus, for each atom of interest, t-plots can be obtained for distances towards all atoms of all subsystems, such that all of them have the periodicity in t equal to the period of the modulation of the central atom. In this way, the complete environment of the central atom can be studied in dependence on t, including the atoms of other subsystems.

The computation of intersubsystem distances requires the use of a single coordinate system for coordinates in physical space. A suitable choice is the lattice Λ, that is the direct lattice corresponding to Λ^* in eqn (4.9). Coordinates of atoms in subsystem ν are defined relative to Λ_ν with the subsystem origin O_ν [eqn (4.21) and Fig. 4.7]. First, a shift towards a common origin is required, resulting in basic-structure coordinates $\mathbf{x}'_{\nu s}$ that depend on t [eqn (4.23)]. Subsequently, the coordinates with respect to Λ_ν must be transformed towards coordinates with respect to Λ. From the coordinate transformation W^ν in superspace, one can derive that this is achieved by

$$\mathbf{x} = \left(W^{\nu(33)} + W^{\nu(3d)}\sigma\right)^{-1} \mathbf{x}'_\nu \,. \qquad (4.41)$$

A linear dependence on t is obtained for the basic-structure coordinates [eqn (4.23)]. A consequence is that the distance between a single pair of atoms from different subsystems increases linearly with increasing $|t|$ for $|t| \to \infty$.

In Section 2.4.1 it was shown that the average distance in a modulated structure is not equal to the distance between average positions of the atoms, but the two values usually are close to each other. Modulation functions of atoms of different subsystems have different periodicities in t. Furthermore, the difference of their basic-structure positions depends on t. Consequently, the average distance does not need to be close to the distance in the basic structure. It might involve a shift into one direction, if—for example—the nearest neighbour of the central atom is considered. Compare solid and dotted curves in Fig. 4.11(f), that show that the modulation acts as to resolve the strain of too short intersubsystem contacts.

t-Plots can be made for other structural properties too. Figure 4.12 gives the variation of the modulated temperature factor of the La atom of $[LaS]_{1.14}[NbS_2]$ in dependence on t. Comparison with t-plots of the distances with La as central atom indicate that minima for U_{ii} occur at the same values of t at which the intersubsystem La–S bond is shortest [Fig. 4.11(f)]. The shortest intersubsystem

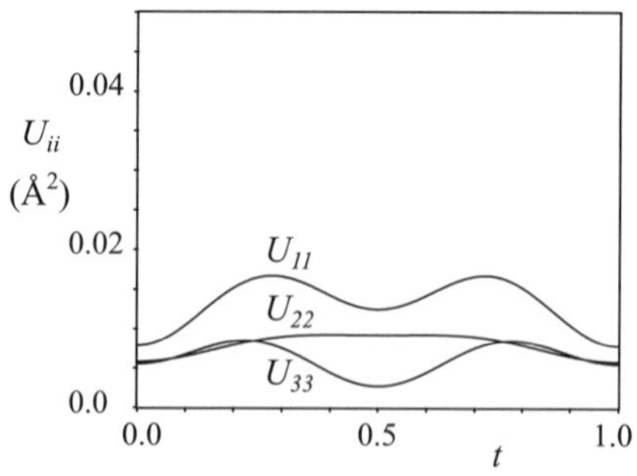

FIG. 4.12. t-Plot of the modulated temperature parameters U_{ii} of La in $[LaS]_{1.14}[NbS_2]$ ($i = 1, 2, 3$). Reprinted with permission from Jobst and van Smaalen (2002), copyright (2002) IUCr.

contact thus represents the tightest environment of La, that then corresponds to the smallest amplitude of vibration (Jobst and van Smaalen, 2002).

5

SUPERSTRUCTURES AND THE COMMENSURATE APPROXIMATION

5.1 Commensurate modulations

An essential property of an incommensurately modulated structure is the irrational character of at least one component of the modulation wave vector. Commensurate modulations are described by modulation wave vectors with only rational-valued components. In that case, an integer, N_{super}, exists, for which $N_{\mathrm{super}}\,\mathbf{q}$ is equal to a reciprocal lattice vector of the basic structure,

$$N_{\mathrm{super}}\,\mathbf{q} = \mathbf{G}_0\,. \tag{5.1}$$

N_{super} is defined as the smallest positive integer for which eqn (5.1) is fulfilled. Obviously, $N_{\mathrm{super}} > 1$ and integer multiples of N_{super} again are solutions to eqn (5.1).

Commensurately modulated crystals possess three-dimensional translational symmetry according to a superlattice with a N_{super}-fold supercell of the basic-structure unit cell. Crystallographic methods for analysing three-dimensionally periodic structures can thus be used to study commensurately modulated crystals. Nevertheless, it is known that structure determination and structure refinement of superstructures may be difficult to impossible, if the deviations from the basic structures are small. The superspace approach is an elegant way out of these problems. It allows us to concentrate on the most important parameters, like low-order harmonics in the Fourier expansions of the modulation functions, while less important parameters can be introduced at later stages. However, scattering information is often insufficient to fix the values of the higher-order harmonics of the modulation functions, and these parameters are kept zero throughout the analysis, thus avoiding correlated parameters in the structure refinements, as they would be present in supercell refinements.

A commensurate modulation wave vector can be expressed as a fraction of a reciprocal lattice vector of the basic structure [eqn (5.1)],

$$\mathbf{q} = \frac{1}{N_{\mathrm{super}}}\,\mathbf{G}_0\,. \tag{5.2}$$

The property of reciprocal and direct lattice vectors, that

$$\mathbf{G}\cdot\mathbf{L} = l = \text{integer}$$

then implies that periodic modulation functions are sampled at exactly N_{super} different values of their arguments $\bar{x}_{s4}$. These values are given by [eqn (1.5)]

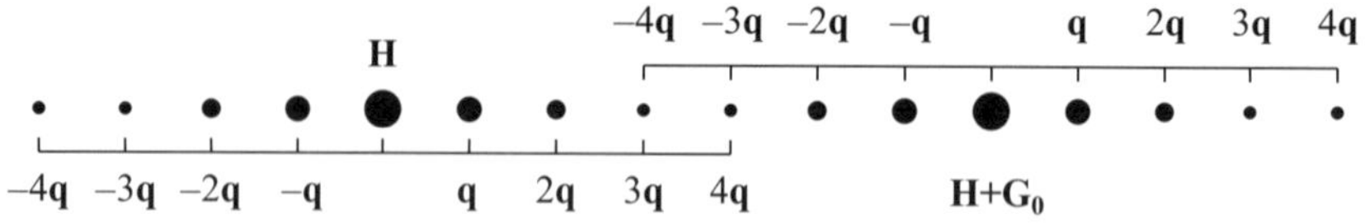

(a)

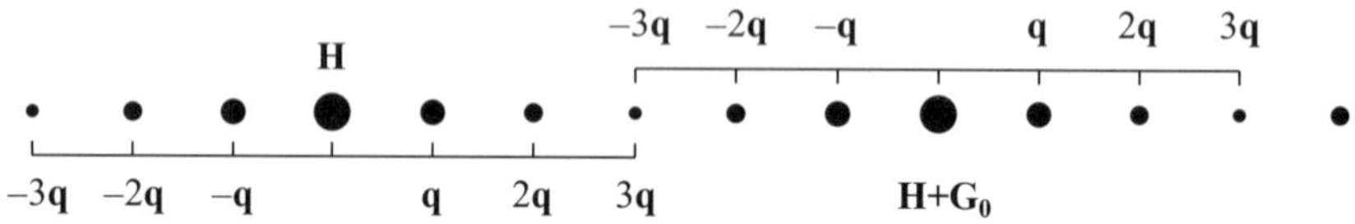

(b)

FIG. 5.1. One row of satellite reflections along the reciprocal lattice direction $\mathbf{G}_0$ [eqn (5.2)]. (a) $N_{\rm super} = 7$, and (b) $N_{\rm super} = 6$. $\mathbf{H}$ is a reciprocal lattice vector of the basic structure. Compare with Fig. 2.1(a).

$$\bar{x}_{s4} = t + \mathbf{q}\cdot\mathbf{x}^0(\mu) + \frac{l}{N_{\rm super}} \qquad \text{for} \qquad l = 0, \cdots, (N_{\rm super} - 1)\,. \tag{5.3}$$

The modulated structure is completely characterized by $N_{\rm super}$ values of each modulation function. The most general form of such a function is a finite Fourier series containing $N_{\rm super}$ independent parameters, *i.e.* containing $n_{\rm max} = \frac{1}{2}(N_{\rm super} - 1)$ harmonics in the case of $N_{\rm super}$ = odd [eqn (1.10)],

$$u_i^\mu(\bar{x}_{s4}) = \sum_{n=1}^{n_{\rm max}} A_i^n(\mu)\,\sin(2\pi n\bar{x}_{s4}) + B_i^n(\mu)\,\cos(2\pi n\bar{x}_{s4})\,, \tag{5.4}$$

for $i = 1, 2, 3$. For $N_{\rm super}$ = even, only one of the sine and cosine parameters should be used for the highest-order harmonic, now given by $n_{\rm max} = \frac{1}{2}N_{\rm super}$. Complications may arise in the presence of non-trivial point symmetries (Section 5.4).

The diffraction pattern is characterized by $(N_{\rm super} - 1)$ superlattice reflections on lines between Bragg reflections $\mathbf{H}$ and $\mathbf{H} + \mathbf{G}_0$ of the basic structure (Fig. 5.1). A unique indexing with four integers is obtained by restricting the satellite index h_4 to [eqn (1.37)]

$$|h_4| = m \leqslant m_{\rm max}\,. \tag{5.5}$$

Odd and even values of $N_{\rm super}$ need to be distinguished again. For $N_{\rm super}$ = odd, $m_{\rm max} = \frac{1}{2}(N_{\rm super} - 1)$, and satellites $\pm(m_{\rm max} + 1)\,\mathbf{q}$ and $\mp m_{\rm max}\,\mathbf{q}$ indicate the same reflections [Fig. 5.1(a)]. For $N_{\rm super}$ = even, $m_{\rm max} = \frac{1}{2}N_{\rm super}$, and $\pm m_{\rm max}\,\mathbf{q}$ point towards the same satellite reflections [Fig. 5.1(b)]. The indexing is employed with positive integers as satellite indices for these reflections.

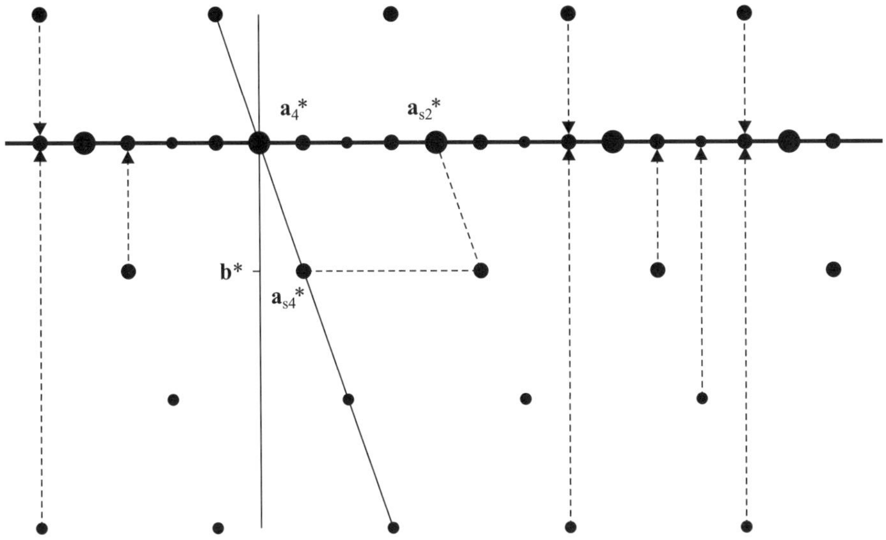

FIG. 5.2. $(0h_2 0h_4)$ Section of reciprocal superspace for $\mathbf{q} = (0,\ \frac{1}{4},\ 0)$. Reciprocal lattice points $(h_1\ h_2\ h_3\ 3)$ project onto the same points in physical space as $(h_1\ h_2{+}1\ h_3\ \bar{1})$ do. Compare to Fig. 2.1(b).

5.2 Commensurately modulated structures in superspace

Construction of a superspace model for commensurately modulated crystal structures closely follows the construction of superspace for incommensurately modulated structures (Chapter 2). The reciprocal basis vectors Λ^* and $\mathbf{a}_4^* = \mathbf{q}$ are considered to be projections of basis vectors of a reciprocal lattice Σ^* in (3+1)-dimensional superspace (Section 2.2). Similarly, Bragg reflections $(h_1\,h_2\,h_3\,h_4)$ in three-dimensional space are projections of reciprocal lattice points $(h_1\ h_2\ h_3\ h_4)$ in superspace [eqn (2.1)]. However, a unique indexing of all Bragg reflections is obtained with indices restricted to $|h_4| \leqslant m_{\max}$ [eqn (5.5)], and non-zero structure factors will only be assigned to reciprocal lattice vectors in superspace with $|h_{s4}| = |h_4| \leqslant m_{\max}$. Zero intensity is assigned to reciprocal lattice points with $|h_4| > m_{\max}$, in accordance with the property of commensurate modulations that these reciprocal lattice points do not represent Bragg reflections on their own. Instead, reciprocal lattice points $(h_1\,h_2\,h_3\,h_4)$ project onto the same positions in three-dimensional space as the points with indices $(h_1\ h_2\ h_3\ h_4^0)$ do, where $h_4^0 = h_4 + 2\,l\,m_{\max}$ and l is an integer such that $|h_4^0| \leqslant m_{\max}$ (Fig. 5.2).

The generalized electron density, $\rho_s(\mathbf{x}_s)$, is obtained as the inverse Fourier transform of the structure factors $F(h_1\ h_2\ h_3\ h_4)$ [eqn (2.12)], and it possesses translational symmetry according to the direct lattice Σ in superspace [eqn (2.9)]. The generalized electron density exhibits continuous strings of density, that are on the average parallel to $\mathbf{a}_{s4}$, and that have shapes according to the finite Fourier series of the modulation functions [eqn (5.4)]. However, physically relevant values

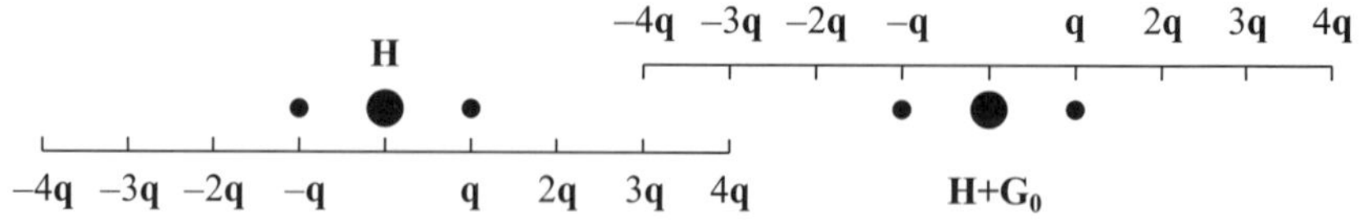

FIG. 5.3. A commensurately modulated structure with $\mathbf{q} = \frac{1}{7}\,\mathbf{G}_0$ and with first-order satellites only [compare to Fig. 5.1(a)].

of the modulation functions are only obtained for the N_{super} different values of their arguments $\bar{x}_{s4}$ [eqn (5.3)]. Equation (5.3) shows that sections of superspace

$$t = t_0 + \frac{l}{N_{\text{super}}} \qquad \text{with} \qquad l = \text{ integer} \tag{5.6}$$

provide equivalent descriptions of the electron density in three-dimensional space. Sections with different t_0 define different structures ($0 \leqslant t_0 \leqslant \frac{1}{N_{\text{super}}}$), unlike the case of incommensurate modulations, for which all values of t provide equivalent descriptions of three-dimensional space. Different sections can even give different space groups for the superstructures (Section 5.3), and finding the true value of t_0 is one of the tasks of the structure determination of superstructures.

Structure factors of incommensurately modulated crystals have been obtained as the Fourier transform in superspace of $\rho_s(\mathbf{x}_s)$ (Section 2.5). The integral over t needs to be replaced by a sum over the N_{super} equidistant values of t for commensurately modulated structures [eqns (2.33) and (5.6)],

$$\begin{aligned} F(\mathbf{S}_s) = \sum_{\mu=1}^{N} f_\mu(\mathbf{S}) \exp\left[2\pi i(\mathbf{S} - S_4\mathbf{q})\cdot\mathbf{x}^0(\mu)\right] \\ \times \sum_{l=0}^{N_{\text{super}}-1} \exp[2\pi i\,(\mathbf{S}\cdot\mathbf{u}^\mu[\tau_l] + S_4\,\tau_l)\,]\,, \end{aligned} \tag{5.7a}$$

with

$$\tau_l = t_0 + \frac{l}{N_{\text{super}}} + \mathbf{q}\cdot\mathbf{x}^0(\mu)\,. \tag{5.7b}$$

Nevertheless, the differences between the diffraction from commensurately and incommensurately modulated crystals are smaller than they seem to be. In one approach the integral over t is numerically evaluated on the basis of the values of the integrand at a discrete set of points, thus transforming integration into summation for incommensurate modulations too. The diffraction of a commensurately modulated crystal resembles the diffraction of an incommensurately modulated structure, if higher-order satellites have zero intensities (Fig. 5.3). For example, the presence of first-order satellites and main reflections allows the refinement of first-order harmonics only. The summation over N_{super} values of $\bar{x}_{s4}$—as imposed by the commensurability—leads to almost equal results as they would follow from a proper evaluation of the integral in case of an incommensurate modulation [eqn (2.34b)]. A consequence is that diffracted intensities cannot distinguish between incommensurate and commensurate modulations, and they cannot distinguish between structures with different values of t_0. In this way, determination of the true structure may be difficult to impossible, even if the superspace analysis is used. An example is provided by the fourfold superstructure of NaV_2O_5 (Lüdecke *et al.*, 1999),

5.3 Superspace symmetry

5.3.1 *Translations*

Symmetries of superstructures are given by space groups (G) based on the lattice translations of the supercell in three-dimensional space. The symmetry can also be characterized by a superspace group, if the superstructure is described as a commensurately modulated structure. The space group of the supercell then follows in a unique way from the superspace group.

First, consider lattice translations in superspace [eqn (3.20)]. Application of a superspace translation to the basic-structure coordinates and modulation functions gives [eqn (3.37)],

$$\{E, 1|l_{s1},\, l_{s2},\, l_{s3},\, l_{s4}\} : \bar{\mathbf{x}}(1) \longrightarrow \bar{\mathbf{x}}(2) = \bar{\mathbf{x}}(1) + \mathbf{L} \tag{5.8a}$$

$$\{E, 1|l_{s1},\, l_{s2},\, l_{s3},\, l_{s4}\} : \mathbf{u}^1[\bar{x}_{s4}(1)] \longrightarrow \mathbf{u}^2[\bar{x}_{s4}(2)] = \mathbf{u}^1[\bar{x}_{s4}(2) - l_{s4}]\,, \tag{5.8b}$$

with

$$\bar{x}_{s4}(2) = \mathbf{q}\cdot\bar{\mathbf{x}}(2) = \bar{x}_{s4}(1) + \mathbf{q}\cdot\mathbf{L}\,. \tag{5.8c}$$

The superspace translation defines translational symmetry in three-dimensional space if the modulation functions of atoms 1 and 2 have equal values. This condition is fulfilled if [eqn (5.8b)]

$$\bar{x}_{s4}(1) = \bar{x}_{s4}(2) - l_{s4}\,.$$

Employing the relation between basic-structure coordinates of atoms 1 and 2 [eqn (5.8a)], this leads to the condition on the translational operator of

$$\mathbf{q}\cdot\mathbf{L} = l_{s4}\,. \tag{5.9}$$

Any superspace translation $\mathbf{L}_s = (l_{s1},\, l_{s2},\, l_{s3},\, l_{s4})$ that obeys this condition represents translational symmetry in physical space, given by the operator $\mathbf{L} = (l_1,\, l_2,\, l_3) = (l_{s1},\, l_{s2},\, l_{s3})$. Substitution of the commensurate modulation wave vector [eqn (5.2)] leads to the alternative formulation of the condition [eqn (5.9)],

$$\mathbf{G}_0\cdot\mathbf{L} = l_{s4}\, N_{\mathrm{super}}\,. \tag{5.10}$$

While $\mathbf{G}_0\cdot\mathbf{L}$ is always equal to an integer, eqn (5.10) selects those translations $\mathbf{L}$ for which $\mathbf{G}_0\cdot\mathbf{L}$ is a N_{super}-fold integer. These translations together define the lattice of translations of the N_{super}-fold supercell.

For an incommensurate modulation the condition eqn (5.9) shows that these structures do not possess three-dimensional translation symmetry. For example, for $\mathbf{q} = (0,\, 0,\, \sigma_3)$ with σ_3 an irrational number, three-dimensional symmetry is restricted to a plane of lattice translations $\mathbf{L} = (l_1,\, l_2,\, 0)$. For $\mathbf{q} = (\sigma_1,\, \sigma_2,\, \sigma_3)$ translational symmetry is completely missing in three-dimensional space.

Complications arise, because basic-structure lattices and superlattices may have different centrings. If a centred unit cell is chosen for the superspace lattice (Table 3.9), the condition for a symmetry operator to define symmetry in three-dimensional space needs to be modified to include centring translations. Let $\{E|v_{s1}^r,\, v_{s2}^r,\, v_{s3}^r,\, v_{s4}^r\}$ be a centring translation of superspace with rational components v_{si}^r $(i = 1, \cdots, 4)$ (Table 3.9). Three-dimensional translational symmetry of the superstructure then follows from the modified condition [eqn (5.9)],

$$\mathbf{q}\cdot(\mathbf{v}^r + \mathbf{L}) = v_{s4}^r + l_{s4}\,, \tag{5.11}$$

where $\mathbf{v}^r = (v_1^r,\, v_2^r,\, v_3^r) = (v_{s1}^r,\, v_{s2}^r,\, v_{s3}^r)$. Notice that the superlattice may be centred even if eqn (5.11) has solutions with $\mathbf{v}^r = \mathbf{0}$ only. These features are illustrated by commensurately modulated structures with orthorhombic lattices $a_1 \times a_2 \times a_3$ of their basic structures:

1. First consider a basic structure with a primitive lattice and a modulation with wave vector $\mathbf{q} = (0,\, 0,\, \frac{1}{3})$. All lattice translations have $v_{si}^r = 0$. Superspace translations $\mathbf{L}_s = (l_{s1},\, l_{s2},\, 3l_{s4},\, l_{s4})$ define translational symmetry in physical space [eqn (5.9)], given by $\mathbf{L} = (l_{s1},\, l_{s2},\, 3l_{s4})$ with components with respect to the basic-structure unit cell. They define a set of primitive translations with respect to the $a_1 \times a_2 \times 3a_3$ supercell. The lattice of the threefold superstructure is found to be primitive.
2. A primitive orthorhombic lattice with $\mathbf{q} = (\frac{1}{2},\, 0,\, \frac{1}{4})$ has $v_{si}^r = 0$ for all translations. Superspace translations that obey eqn (5.9) are $\mathbf{L}_s = (2l'_{s4},\, l_{s2},\, 4l''_{s4},\, [l'_{s4} + l''_{s4}])$, and a $2a_1 \times a_2 \times 4a_3$ supercell is obtained. However, superspace translations $(2l'_{s4} + 1,\, l_{s2},\, 4l''_{s4} + 2,\, [l'_{s4} + l''_{s4} + 1])$ also give rise to translational symmetry in physical space. With respect to the supercell, the latter translations correspond to $(l_1 + \frac{1}{2},\, l_2,\, l_3 + \frac{1}{2})$. Accordingly, the $2a_1 \times a_2 \times 4a_3$ supercell is B-centred.
3. A C-centred lattice with $\mathbf{q} = (0,\, 0,\, \frac{1}{4})$ has $\mathbf{v}_s^r = (\frac{1}{2},\, \frac{1}{2},\, 0,\, 0)$. Translational symmetry in physical space is obtained for the superspace translations $\mathbf{L}_s = (l_{s1},\, l_{s2},\, 4l_{s4},\, l_{s4})$ and $\mathbf{L}_s = (\frac{1}{2} + l_{s1},\, \frac{1}{2} + l_{s2},\, 4l_{s4},\, l_{s4})$ [eqn (5.11)]. Accordingly, a C-centred $a_1 \times a_2 \times 4a_3$ supercell is obtained.
4. A C-centred lattice with $\mathbf{q} = (1,\, 0,\, \frac{1}{4})$ has solutions to eqn (5.11) given by $\mathbf{L}_s = (l_{s1},\, l_{s2},\, 4l_{s4},\, l_{s4})$ and $\mathbf{L}_s = (\frac{1}{2} + l_{s1},\, \frac{1}{2} + l_{s2},\, 2 + 4l_{s4},\, 1 + l_{s4})$. The $a_1 \times a_2 \times 4a_3$ supercell is I-centred in this case.
5. An I-centred lattice with modulation wave vector $\mathbf{q} = (\frac{1}{2},\, \frac{1}{2},\, \frac{1}{2})$ has $\mathbf{v}_s^r = (\frac{1}{2},\, \frac{1}{2},\, \frac{1}{2},\, 0)$. Solutions to eqn (5.11) involve those integer translations $(l_{s1},\, l_{s2},\, l_{s4},\, \frac{1}{2}[l_{s1}+l_{s2}+l_{s4}])$ for which $(l_{s1}+l_{s2}+l_{s4}) =$ even. This appears to define an F-centred, $2a_1 \times 2a_2 \times 2a_3$ supercell.

5.3.2 *The supercell space group*

Symmetry of modulated crystals is given by superspace groups. The most general form of an element of a superspace group is (Section 3.4)

$$\{R_s|\mathbf{v}_s\} = \{R_s|v_{s1},\, v_{s2},\, v_{s3},\, v_{s4}\}\,.$$

Symmetry $\{R_s|\mathbf{v}_s\}$ of the modulated structure corresponds to symmetry $\{R|\mathbf{v}\}$ in physical space, with $\{R|\mathbf{v}\} = \{R|v_1,\, v_2,\, v_3\} = \{R|v_{s1},\, v_{s2},\, v_{s3}\}$, if the condition,

$$\mathbf{q}\cdot\mathbf{v} = v_{s4} \tag{5.12}$$

is fulfilled. This is a direct generalization of the condition that superspace translation must fulfil, in order to define translational symmetry in physical space [eqns (5.9)–(5.11)]. Indeed, for lattice translations with $\mathbf{v}_s = \mathbf{v}_s^r + \mathbf{L}_s$, the condition eqn (5.12) reduces to the condition on translations [eqn (5.9) or eqn (5.11)].

Rotational symmetry in superspace is represented by operators R_s different from the unit operator. They are combined with lattice translations $\mathbf{v}_s$, with integer components $\mathbf{L}_s$ and rational components $\mathbf{v}_s^r$ of centring translations. Furthermore, glide planes and screw axes contain translation vectors with rational components $\mathbf{v}_s^r$ that do not represent translational symmetry themselves. In all these cases, the general condition for $\{R_s|\mathbf{v}_s\}$ providing symmetry of physical space [eqn (5.12)] reduces to the condition eqn (5.11). However, some components of $\mathbf{v}_s$ may be real numbers, with values that depend on the relative positions of the symmetry element and the origin of the coordinate system. Specifically, this applies to operators with $\epsilon = -1$, for which the value of v_{s4} defines the location of the symmetry element along the fourth coordinate axis in superspace. The condition eqn (5.12) will be fulfilled for specific values of v_{s4} only, and superspace operators with $\epsilon = -1$ will be part of the supercell space group for only these special choices of the origin in superspace.

As an example consider the superspace group $P2_1/m(00\frac{1}{4})0\bar{1}$ with symmetry operators,

$$\{E, 1|l_{s1},\, l_{s2},\, l_{s3},\, l_{s4}\} \qquad \{i, \bar{1}|v_{s1}^{\mathrm{O}}+l_{s1},\, v_{s2}^{\mathrm{O}}+l_{s2},\, v_{s3}^{\mathrm{O}}+l_{s3},\, v_{s4}^{\mathrm{O}}+l_{s4}\}$$

$$\{2^z, 1|v_{s1}^{\mathrm{O}}+l_{s1},\, v_{s2}^{\mathrm{O}}+l_{s2},\, \tfrac{1}{2}+l_{s3},\, l_{s4}\} \quad \{m_z, \bar{1}|l_{s1},\, l_{s2},\, \tfrac{1}{2}+v_{s3}^{\mathrm{O}}+l_{s3},\, v_{s4}^{\mathrm{O}}+l_{s4}\}\,.$$

Origin-dependent translational components of different symmetry operators are related to each other, because of the closeness condition of groups. $\mathbf{v}_s^{\mathrm{O}} = (v_{s1}^{\mathrm{O}},\, v_{s2}^{\mathrm{O}},\, v_{s3}^{\mathrm{O}},\, v_{s4}^{\mathrm{O}})$ represents the relative position of the origin of the coordinate system and the inversion point. Origin-dependent translational components of the screw axis and mirror plane are based on the same values v_{si}^{O} $(i = 1, \cdots, 4)$. However, the intrinsic translational component $(0,\, 0,\, \frac{1}{2},\, 0)$ of the screw axis implies that the translational components along $\mathbf{a}_{s3}$ of the inversion and the mirror differ from each other by exactly one half. This feature has consequences for the symmetry of the supercell, as is shown below.

Without loss of generality, origin-dependent components v_{si}^{O} ($i = 1, 2, 3$) can be set to zero, thus leading to symmetry operators,

$$\begin{array}{ll} \{E, 1|l_{s1},\, l_{s2},\, l_{s3},\, l_{s4}\} & \{i, \bar{1}|l_{s1},\, l_{s2},\, l_{s3},\, v_{s4}^{\mathrm{O}}{+}l_{s4}\} \\ \{2^z, 1|l_{s1},\, l_{s2},\, \tfrac{1}{2}{+}l_{s3},\, l_{s4}\} & \{m_z, \bar{1}|l_{s1},\, l_{s2},\, \tfrac{1}{2}{+}l_{s3},\, v_{s4}^{\mathrm{O}}{+}l_{s4}\}\,. \end{array}$$

Application of condition eqn (5.12) with $\mathbf{q} = (0,\, 0,\, \frac{1}{4})$ for any possible combination of values v_{s4}^{O} and l_{si} ($i = 1, \cdots, 4$) leads to the following supercell symmetries corresponding to the superspace group $P2_1/m(0\,0\,\frac{1}{4})0\bar{1}$:

1. Translations $\{E, 1|l_{s1},\, l_{s2},\, 4l_{s4},\, l_{s4}\}$ define a primitive, $a_1 \times a_2 \times 4a_4$ superlattice for all values v_{s4}^{O}.
2. The twofold screw axis never leads to symmetry in three-dimensional space, because $\frac{1}{4}(\frac{1}{2}{+}l_{s3})$ is different from an integer for all integers l_{s3}. In general, the presence or absence of symmetry in three-dimensional space does not depend on the location of the origin for operators with $\epsilon = 1$.
3. The condition $\frac{1}{4}\,l_{s3} = v_{s4}^{\mathrm{O}} + l_{s4}$ [eqn (5.12)] has solutions for $v_{s4}^{\mathrm{O}} = \frac{1}{4}l$ ($l =$ integer). Because translations in three-dimensional space do not explicitly depend on the value of l_{s4} or v_{s4}^{O}, different solutions are characterized by the value of v_{s4}^{O} in the interval $0 \leqslant v_{s4}^{\mathrm{O}} < 1$. Inversion symmetry in three-dimensional space thus follows as:
   (a) For $v_{s4}^{\mathrm{O}} = 0 \pmod 1$, $\{i, \bar{1}|l_{s1},\, l_{s2},\, 4l_{s4},\, l_{s4}\}$ defines an inversion at the origin of the supercell.
   (b) For $v_{s4}^{\mathrm{O}} = \frac{1}{4} \pmod 1$, $\{i, \bar{1}|l_{s1},\, l_{s2},\, 4l_{s4} + 1,\, \frac{1}{4} + l_{s4}\}$ defines an inversion centre with origin-dependent translational component $(0,\, 0,\, \frac{1}{4})$ with respect to the supercell.
   (c) For $v_{s4}^{\mathrm{O}} = \frac{1}{2} \pmod 1$, $\{i, \bar{1}|l_{s1},\, l_{s2},\, 4l_{s4} + 2,\, \frac{1}{2} + l_{s4}\}$ defines the inversion centre $\{i|0,\, 0,\, \frac{1}{2}\}$ with respect to the supercell.
   (d) For $v_{s4}^{\mathrm{O}} = \frac{3}{4} \pmod 1$, $\{i, \bar{1}|l_{s1},\, l_{s2},\, 4l_{s4} + 3,\, \frac{3}{4} + l_{s4}\}$ defines the inversion centre $\{i|0,\, 0,\, \frac{3}{4}\}$ with respect to the supercell.

   Any other value of v_{s4}^{O} implies the absence of inversion symmetry in three-dimensional space.
4. Mirror symmetry in three-dimensional space is obtained for $v_{s4}^{\mathrm{O}} = \frac{1}{8} + \frac{1}{4}l \pmod 1$ with $l = 0, 1, 2, 3$. The additional term of $\frac{1}{8}$ arises because the translational components along $\mathbf{a}_{s3}$ of i and m differ by one half.

Conditions 1–4 show that possible supercell space groups depend on the value of v_{s4}^{O}. For $P2_1/m(0\,0\,\frac{1}{4})0\bar{1}$, the supercell space group either contains the inversion centre or the mirror plane or neither of these two (Table 5.1). The mutual exclusion of inversion and mirror symmetry originates in the presence of screw axes or glide planes in the superspace group. In case of the symmorphic superspace group $P2/m(0\,0\,\frac{1}{4})0\bar{1}$, only two supercell space groups are possible: $P2/m$ for $v_{s4}^{\mathrm{O}} = 0 \pmod{\frac{1}{4}}$ and $P2$ for $v_{s4}^{\mathrm{O}} \neq 0 \pmod{\frac{1}{4}}$.

TABLE 5.1. Supercell space groups derived from $P2_1/m(0\,0\,\sigma_3)0\bar{1}$ for $\sigma_3 = \frac{1}{4}$ and for various choices of the origin, defined by the value of v_{s4}^{O} in $\{i, \bar{1}|0,\ 0,\ 0,\ v_{s4}^{\text{O}}\}$.

v_{s4}^{O}	$0 \pmod{\frac{1}{4}}$	$\frac{1}{8} \pmod{\frac{1}{4}}$	other
Supercell space group	$P\bar{1}$	Pm	$P1$

5.4 Modulation functions and supercell coordinates

The description of a crystal structure in terms of basic-structure parameters and modulation functions is equivalent to the description as a superstructure. In case of triclinic symmetry $P1$, the equivalence of these descriptions follows from the fact that each parameter x_i or U_{ij} $(i, j = 1, 2, 3)$ of the N_{super} atoms in the supercell is uniquely determined by the corresponding basic-structure parameter (x_i^0 or U_{ij}^0) together with a finite Fourier series containing $(N_{\text{super}} - 1)$ independent Fourier coefficients (Section 5.1). Non-trivial point symmetries lead to restrictions on the modulation functions that follow from the same relations as have been used for symmetry restrictions on modulation functions of incommensurate crystals (Section 3.6.2). Alternatively, interdependencies amongst the parameters of the atoms in the supercell follow from the supercell space group. The modulated-structure and supercell descriptions are equivalent in this case too, if the number of Fourier coefficients in the finite Fourier series is chosen to be equal to the number of parameters in the supercell description (Perez-Mato, 1991). It then may so happen that harmonics n need to be included into the Fourier series with $n > \frac{1}{2}N_{\text{super}}$ [eqn (5.4)].

The true symmetry of a superstructure is given by its supercell space group. The superspace group can therefore be nothing more than an approximation to the symmetry. The usefulness of the superspace description of the structure derives from the fact that structural distortions often are restricted to less harmonics than would be allowed by the supercell symmetry. The superspace group then provides the necessary restrictions between the Fourier components. However, superspace symmetry does not apply to other properties, if the superspace group and the supercell space group belong to different crystal classes.

An example is provided by the twofold superstructure of NbS_3, with superspace group $P2_1/m(0\,0\,\frac{1}{2})0\bar{1}$ and supercell space group $P\bar{1}$ (van Smaalen, 1988). The basic structure of NbS_3 (space group $P2_1/m$) contains four independent atoms, each one in a mirror plane on a twofold position (x_1^0, x_2^0, $\frac{1}{4}$). A single-harmonic distortion leads to $4 \times 5 = 20$ independent positional parameters in the modulated structure. The supercell contains eight independent atoms with a total of 24 parameters within $P\bar{1}$. The discrepancy of four parameters is made up by second-order Fourier components defining displacements along $\mathbf{a}_3$. Diffraction symmetry is $\bar{1}$. However, main reflections obey the higher point symmetry $2/m$, if the second-order harmonics are exactly zero (van Smaalen, 1988). This situation has been confirmed by the diffraction experiment. Internal R-values have been obtained for averaging reflections according to point symmetry $2/m$

of $R_{\text{int}} = 0.04$ for main reflections and of $R_{\text{int}} = 0.27$ for superstructure reflections. Furthermore, main reflections obey the reflection conditions implied by the superspace group $P2_1/m(0\,0\,\frac{1}{2})0\bar{1}$, while satellite reflection violate these conditions. Accordingly, all of the deviation from monoclinic symmetry originate in a distortion by a single harmonic, *i.e.* by a modulation corresponding to a single normal mode of the basic-structure space group.

The superspace group and supercell space group define different symmetry restrictions for other properties than atomic coordinates, including diffraction symmetry, monoclinic versus triclinic restrictions on the lattice parameters and symmetry restrictions on tensor properties like elasticity, thermal expansion and electrical conductivity. In all cases the correct symmetry is given by the supercell space group, but deviations from superspace-group symmetry are expected to be small, if the magnitude of the modulations is small. The superspace description thus explains pseudo-symmetries of superstructures.

Apart from its usefulness in structural analysis and in understanding pseudo-symmetries of superstructures, superspace groups are also of importance for understanding relations between different structures. For example, the basic structure of NbS_3 is isostructural to $ZrSe_3$, and the superspace analysis immediately makes clear that the crystal structure of NbS_3 can be obtained as a distortion of the $ZrSe_3$ structure type. Different structures of a single compound may occur at different temperatures, as they are related by phase transitions. A non-modulated structure may thus transform into a series of incommensurately and commensurately modulated structures, that differ only in the period of the modulation, but that have similar modulation parameters, and that are described by a single superspace group. An example of such an analysis is presented in Section 5.5 on the phase transitions in A_2BX_4 type compounds.

In an alternative approach, a supercell model can be used as an approximate representation of the atomic structure of an incommensurately modulated crystal. Translational symmetric structures are required if methods for the analysis of periodic structures should be applied to incommensurate crystals. This desire will exist in particular, if an 'incommensurate counterpart' of a particular method is not available, as it is the case for electronic band-structure calculations and computations involving the method of molecular dynamics. The first step towards a supercell is the choice of a commensurate modulation wave vector that provides a reasonably accurate approximation to the incommensurate wave vector. For example, the irrational number $\sigma_3 = 0.359\cdots$ can be approximated by the fractions $\frac{1}{3} \approx 0.333$, $\frac{4}{11} \approx 0.364$ and $\frac{5}{14} \approx 0.357$. Accordingly, an incommensurately modulated structure with a modulation wave vector $\mathbf{q} = (0,\, 0,\, \sigma_3)$ can be approximated by threefold, elevenfold and fourteenfold superstructures, whereby the size of the supercell increases concomitantly with the quality of the approximation. The supercell space group can be derived from the rules presented in Section 5.3. However, the supercell space group does not only depend on the locations of symmetry elements with $\epsilon = -1$ with respect to the origin of the coordinate system, but it also depends on the parity of the commensurate

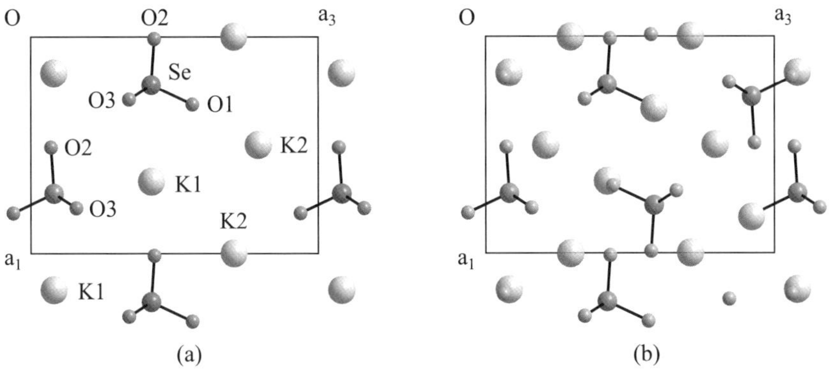

FIG. 5.4. Projection of the basic structure of K_2SeO_4 along $\mathbf{a}_2$. (a) Atoms about one mirror plane. (b) Atoms in one unit cell. Crystallographically independent atoms are indicated by atom names, with K1, K2, Se, O1 and O2 in the mirror plane and with O3 above and below the mirror plane. Coordinates from Yamada *et al.* (1984).

approximation to $\mathbf{q}$. In the example, the supercell space group will depend on $\mathbf{q} = r/s$ being the quotient of odd/odd, even/odd or odd/even integers r and s, and supercell approximations of increasing size will have different symmetries. This feature is again illustrated in Section 5.5 for the example of A_2BX_4 compounds

5.5 Symmetry of A_2BX_4 compounds

Compounds A_2BX_4 consist of monovalent cations A^+, with A = K, Rb, Cs, $NH_4{}^+$ or $N(CH_3)_4{}^+$, and complex anions $BX_4{}^{2-}$, like $SeO_4{}^{2-}$, $ZnCl_4{}^{2-}$, $HgCl_4{}^{2-}$ and others (Cummins, 1990). At room temperature they crystallize in the β-K_2SO_4 structure type, with space group $Pnma$ and four formula units in the unit cell (Fig. 5.4). Several of these compounds undergo a series of phase transitions on cooling, finally adopting an acentric superstructure that is ferroelectric.

K_2SeO_4 has an incommensurately modulated structure below $T_i = 129.5$ K with superspace group $Pnma(\sigma_1 0 0)\bar{1}s0$ and $\sigma_1 = \frac{2}{3} + \delta$. The incommensurability δ varies continuously with temperature from approximately 0.022 at T_i to 0.007 at $T_{\text{lock-in}} = 93$ K. $T_{\text{lock-in}}$ is the transition temperature of a first-order phase transition, at which δ becomes zero. The resulting threefold superstructure has space group $Pn2_1a$, but it can alternatively be described as a commensurately modulated structure with $\mathbf{q} = (\frac{2}{3}, 0, 0)$ and with the same superspace group as applies to the incommensurate phase (Janner and Janssen, 1980*a*).

Many A_2BX_4 compounds possess incommensurately modulated structures and threefold superstructures similar to those of K_2SeO_4, while other compounds have more complex phase diagrams. Superstructures include commensurately modulated structures with $\mathbf{q} = (\frac{3}{5}, 0, 0)$ and $\mathbf{q} = (\frac{1}{2}, 0, 0)$. It is thus of interest to consider the relations between supercell and superspace symmetries for $Pnma(\sigma_1 0 0)\bar{1}s0$.

TABLE 5.2. Supercell space groups derived from $Pnma(\sigma_1 00)\bar{1}s0$ for $\sigma_1 = \frac{r}{s}$ with integers r and s and for various choices of the origin of the coordinate system, defined by the value of v_{s4}^{O} in $\{i, \bar{1}|0,\, 0,\, 0,\, v_{s4}^{\mathrm{O}}\}$.

$\sigma_1 = r/s$	$v_{s4}^{\mathrm{O}} = 0 \pmod{\frac{1}{s}}$	$v_{s4}^{\mathrm{O}} = \frac{1}{2s} \pmod{\frac{1}{s}}$	$v_{s4}^{\mathrm{O}} \neq 0 \pmod{\frac{1}{2s}}$
$r=$ odd / $s=$ odd	$P2_1/n11$	$P2_12_12_1$	$P2_111$
$r=$ even / $s=$ odd	$P112_1/a$	$Pn2_1a$	$P11a$
$r=$ odd / $s=$ even	$P12_1/a1$	$Pna2_1$	$P1a1$

Superspace translations $\{E\, 1|l_{s1},\, l_{s2},\, l_{s3},\, l_{s4}\}$ (l_{si} = integer) imply translational symmetry in three-dimensional space if [eqn (5.9)],

$$\sigma_1\, l_{s1} = l_{s4} \qquad \Longleftrightarrow \qquad r\, l_{s1} = s\, l_{s4}\,,$$

where $\sigma_1 = r/s$ and r and s are integers. This condition can be fulfilled for a suitable choice of l_{s4} if l_{s1} is an s-fold integer. A commensurate modulation with $\mathbf{q} = (r/s,\, 0,\, 0)$ thus implies an s-fold supercell in physical space. Superspace operators $\{m_y 1|l_{s1},\, \frac{1}{2}+l_{s2},\, l_{s3},\, \frac{1}{2}+l_{s4}\}$ define symmetry in three-dimensional space if [eqn (5.12)],

$$\sigma_1\, l_{s1} = \tfrac{1}{2} + l_{s4} \qquad \Longleftrightarrow \qquad r\, l_{s1} = \tfrac{1}{2}s + s\, l_{s4}\,. \tag{5.13}$$

Solutions to eqn (5.13) exist only for modulation wave vectors with r = odd and s = even. Following procedures outlined in Section 5.3, this leads to nine different supercell space groups, depending on the parities of r and s as well as on the location of the origin of the coordinate system. The latter is expressed by the value of v_{s4}^{o} in $\{i\,\bar{1}|0,\, 0,\, 0,\, v_{s4}^{o}\}$ (Table 5.2). The threefold superstructure of K_2SeO_4 with $r/s = 2/3$ = even/odd is obtained as the special section $v_{s4}^{o} = \frac{1}{6} \pmod{\frac{1}{3}}$ of superspace. Supercell space groups of A_2BX_4 compounds with periods other than three can also be obtained from Table 5.2 as the appropriate section of the superspace group $Pnma(\sigma_1 00)\bar{1}s0$. Three space groups occur twice in Table 5.2, but they indicate only formal equivalence of symmetries. Monoclinic symmetries with unique axes a and c define different superstructures, if both supercell space groups are derived from a single setting of the superspace group.

6

QUANTITATIVE DESCRIPTION OF THE DIFFRACTION BY APERIODIC CRYSTALS

6.1 Diffracted intensities

The physical process of interaction between radiation and matter is the same for periodic crystals, aperiodic crystals and amorphous substances. The consequence is that periodic and aperiodic crystals both give rise to diffuse scattering, including incoherent scattering (Compton scattering), thermal diffuse scattering (TDS) and fluorescence. Further contributions to the diffuse scattering come from air scattering, scattering off the crystal mount and noise within the detector system. However, the methods of structural analysis discussed in this book apply to Bragg scattering. Diffuse scattering varies slowly with the scattering angle, and its intensity can be measured at scattering angles just next to the diffraction angles of Bragg reflections. Intensities of Bragg reflections are obtained by subtracting the intensity of the diffuse scattering from the scattered intensity measured at the positions of the Bragg reflections. This procedure is usually performed by the software for processing of diffraction data.

Both periodic crystals and aperiodic crystals diffract X rays in the form of Bragg reflections, as it has been shown in Chapter 1. The positions of the reflections are different for these two states of matter, but other properties follow the same rules. The width of a Bragg reflection is initially governed by the size of a coherently scattering patch of material, but other effects usually give larger contributions, like the contributions from the mosaic spread of the crystal and the spectral distribution, divergence and limited coherence of the radiation. Main reflections and satellite reflections can be very narrow and of comparable widths, when measured with synchrotron radiation on good-quality crystals (Fig. 6.1). Alternatively, incommensurately modulated structures are often the result of phase transitions within the solid phase, resulting in a limited coherence of the modulation wave and concomitantly reflection widths that are larger for satellite reflections than for main reflections.

In all cases, the integrated intensities are the appropriate measure for the relative scattering powers of the Bragg reflections. They are proportional to the squares of the structure factors. The constant of proportionality incorporates the same effects as apply to periodic crystals, including Lorentz and polarization factors, absorption correction, extinction correction and the scale factor. Usually, the measured intensity is corrected by the Lorentz and polarization factors and for the effect of absorption, arriving at the so-called observed intensity, $I_{\text{obs}}(\mathbf{H}_s)$, and observed structure factor amplitude, $|F_{\text{obs}}(\mathbf{H}_s)|$, with,

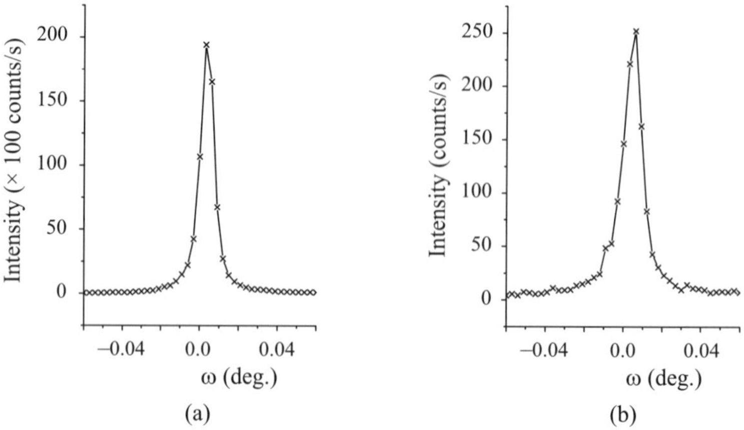

FIG. 6.1. Scattered intensity as function of the crystal orientation (ω scan) showing narrow Bragg reflections. (a) Main reflection ($\bar{3}\ \bar{5}\ 5\ 0\ 0$). (b) First-order satellite reflection ($\bar{3}\ \bar{5}\ 5\ \bar{1}\ 0$). Measurement with synchrotron radiation ($\lambda = 0.5608$ Å) on $P_4W_8O_{32}$ with $\mathbf{q}^1 = (0.330, 0.292, 0)$ and $\mathbf{q}^2 = (0.330, -0.292, 0)$. Reprinted with permission from Lüdecke *et al.* (2000), copyright (2000) EDP Sciences.

$$I_{\mathrm{obs}}(\mathbf{H}_s) = |F_{\mathrm{obs}}(\mathbf{H}_s)|^2 \,. \tag{6.1}$$

Secondary extinction (k_E) and the scale factor (k_s) are described by one parameter each, that are optimized in the refinement procedure.

The expression for the structure factor of an incommensurately modulated structure has been derived in Sections 1.3 and 2.5. This structure factor is hereafter denoted as the calculated structure factor,

$$F_{\mathrm{cal}}(\mathbf{H}_s) \,. \tag{6.2}$$

Calculated structure factors of periodic crystals are obtained as sums over the atoms in the unit cell. For incommensurately modulated crystals the sum extends over all atoms in the basic-structure unit cell, while the modulation is accounted for by an additional atomic scattering factor, $g_\mu(\mathbf{H}_s)$, that multiplies the atomic scattering factor $f_\mu(|\mathbf{H}|)$ [eqn (2.34a)]. Equation (3.49) gives an expression for $F_{\mathrm{cal}}(\mathbf{H}_s)$ that explicitly depends on the superspace symmetry. Modifications to these expressions are required in order to account for anomalous diffraction and thermal vibrations of the atoms ('lattice' vibrations). The atomic scattering factor is replaced by a modified atomic scattering factor [eqn (3.49)]:

$$f_\mu(|\mathbf{H}|) \longrightarrow f_\mu^0(|\mathbf{H}|) + \Delta f'_\mu + i\, f''_\mu \,. \tag{6.3}$$

$f_\mu^0(|\mathbf{H}|)$ represents the Fourier transform of the electron density of atom μ [eqn (1.19)], while $\Delta f'_\mu$ and f''_μ are the real and imaginary parts of the anomalous scattering factor. Atomic vibrations can be described by the conventional

Debye–Waller factor, but an additional phason Debye–Waller factor or modulated temperature parameters are sometimes required (Section 6.4).

With all these modifications, we find that $|F_{\rm obs}(\mathbf{H}_s)| = k_s\, k_E\, |F_{\rm cal}(\mathbf{H}_s)|$ for the ideal experiment and the correct structure model. This relation forms the basis of structural analysis.

6.2 Overlapping reflections

Goal of the diffraction experiment is to measure the integrated intensities of all Bragg reflections. Overlap of reflections prevents this goal being reached. Either the sum of intensities of a few reflections can be measured or computational procedures to separate neighbouring reflections lead to highly inaccurate values for the integrated intensities of individual reflections. Lack of accurate intensities of reflections hampers the application of methods of structural analysis, and it deteriorates the accuracy of the structure model as it is obtained from structure refinements.

Overlap of reflections with nearly equal scattering angles 2θ is inherent to powder diffraction. The problems are solved by application of the Rietveld method to the measured diffraction profiles $I(2\theta)$, but the fraction of Bragg reflections that can be resolved individually determines the quality of the refined structure model and the probability of success of crystallographic methods for structure determination.

Overlap of reflections occurs in single-crystal diffraction, if the scattering vectors differ by an amount that is smaller than the widths of the reflections. This situation typically occurs for twinned crystals (Section 6.3) and for crystals with large lattice parameters. The resolving power of an experiment is defined by the width of the reflections. The divergence and spectral width of Mo-K_α radiation prevents reflections $(h_1\, h_2\, h_3)$ and $(h_1\, h_2\, h_3{+}1)$ being resolved, if the lattice parameter a_3 is larger than ~ 50 Å. Synchrotron radiation together with large detector-to-crystal distances or with carefully selected detector slits has much higher resolving power. Reflections have been measured for protein crystals with lattice parameters of up to ~ 500 Å. For many crystals, however, the crystal quality—as defined by the mosaic spread, lattice strain and other materials properties—is the limiting factor determining the widths of the reflections.

Different reflections with nearly equal scattering vectors are inherent to aperiodic crystals. Modulation wave vectors $\mathbf{q} = (0,\, 0,\, \sigma_3)$ with $0.2 < \sigma_3 < 0.4$ correspond to superlattice periods Na_3 that vary between $N = 3$ and $N = 5$. Even for moderately sized basic-structure unit cells, the distances between main reflections and first-order satellite reflections (differences between their scattering vectors) will be similar to the distances between reflections of a periodic crystal with a large lattice parameter, and special procedures may be necessary to resolve these reflections.

The incommensurability of aperiodic crystals determines that any reflection has other reflections at arbitrary small distances from it. However, decreasing distances correspond to increasing differences between the satellite orders of the central reflection and putative neighbouring reflections. Resolved main reflections and resolved low-order satellite reflections can be observed, because the intensities of higher-order satellite reflections are zero. The latter property follows from the finite size of the modulation displacements. Nevertheless, incommensurate substances exist for which satellite reflections up to order 9 have been observed (Marmeggi *et al.*, 1990). The possibility to measure the integrated intensities of the reflections of incommensurate crystals thus depends on the components of the modulation wave vector as well as on the maximum order of satellite reflections with non-zero intensities. It is then a matter of chance rather than a fundamental property of modulations, whether or not all reflections can be resolved.

Overlap need not be a problem, as is illustrated by Na_2CO_3 (van Aalst *et al.*, 1976). Satellite reflections up to order $m = 4$ lie along lines $\pm m\mathbf{q}$ centred on the main reflections $(h_1\, h_2\, h_3\, 0)$ (Fig. 1.11). These lines of satellite reflections are inclined with respect to the reciprocal lattice lines of the basic structure, because $\mathbf{q} = (0.182,\ 0,\ 0.318)$ possesses two non-zero components. As a result, reflections $(h_1\, h_2\, h_3\, 2)$ and $(h_1\, h_2\, h_3{+}1\, -1)$ are well resolved instead of overlapping as it would occur for $\mathbf{q} = (0,\ 0,\ 0.318)$.

Many members of the A_2BX_4 class of compounds have modulation wave vectors $\mathbf{q} = (\frac{1}{3}{-}\delta,\ 0,\ 0)$ (Cummins, 1990). δ is a small irrational number with a value that depends on temperature. Main reflections and first-order satellite reflections are well resolved, but second-order satellite reflections $(h_1\, h_2\, h_3\, 2)$ are at small distances $3\delta a_1^*$ off the first-order satellite reflections $(h_1{+}1\, h_2\, h_3\, -1)$. In many experiments only first-order satellite reflections have been measured, that are then well resolved. However, several experiments have found higher-order satellites at temperatures close to the transition towards the lock-in phase. Measuring the integrated intensities of all these reflections is a major challenge or it can be impossible, depending on the particular value of δ. An example is provided by Rb_2ZnCl_4, where reflections are well resolved at $T = 198.05$ K with $\delta = 0.0189$, but they overlap at $T = 193.4$ K with $\delta = 0.0112$ (Fig. 6.2).

Overlap of reflections is a fundamental problem for incommensurate composite crystals, even if only low-order satellite reflections can be observed. Main reflections are found at the nodes of the reciprocal lattices of both subsystems, and non-zero intensities are expected for reflections $(h_1\, h_2\, h_3\, h_4)$ whereby any of the indices h_k can have a large value (Section 4.2). If h_1 and h_4 represent the indices along the mutually incommensurate reciprocal basis vectors $\mathbf{a}_{11}^* = \mathbf{a}_1^*$ and $\mathbf{a}_{21}^* = \mathbf{a}_4^*$, main reflections $(h_1\, h_2\, h_3\, 0)$ of subsystem 1 can be found close to main reflections $(0\, h_2\, h_3\, h_4)$ of subsystem 2, if sufficiently large values of h_1 and h_4 are allowed. The incommensurate composite crystal $[LaS]_{1.14}[NbS_2]$ has been introduced in Fig. 1.4. For this compound it has been found that $\mathbf{a}_{11}^* = \sigma_1\, \mathbf{a}_{21}^*$ with $\sigma_1 = 0.5667\,(1) = \frac{4}{7} - 0.0047\,(1)$ (Jobst and van Smaalen, 2002). The ratio of lattice parameters close to $\frac{4}{7}$ is responsible for the fact that reflections $(h_1\, h_2\, h_3\, h_4)$ are close to the two reflections $(h_1{\pm}4\, h_2\, h_3\, h_4{\mp}7)$. However, the non-standard centring F' of the superspace lattice (Table 3.9) implies that one of

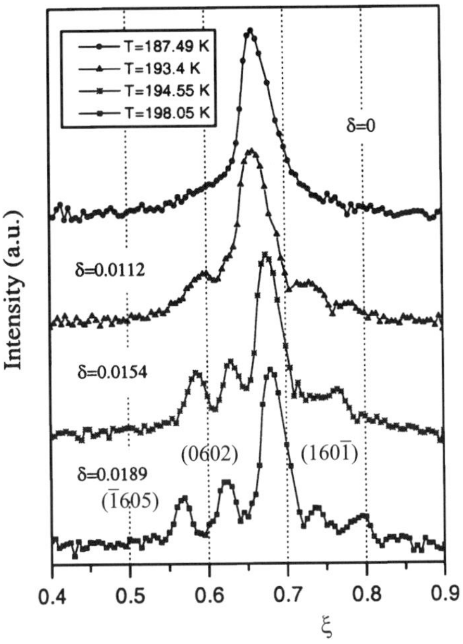

FIG. 6.2. Diffracted intensity as a function of the scattering vector (ξ 6 0) of Rb_2ZnCl_4 at four different temperatures. The incommensurabilities δ are defined by the positions of the highest maxima at ($\frac{2}{3}$+δ 6 0) = (1 6 0 $\bar{1}$). Reprinted with permission from Babkevich and Cowley (1999), copyright (1999) Institute of Physics.

the two low-index reflections is extinct, while the third reflection will be a high-order satellite. For example, main reflections ($4\,h_2\,h_3\,0$) are close to extinct main reflections ($0\ h_2\ h_3\ 7$), while intensities of the seventh-order satellite reflections ($8\,h_2\,h_3\,-7$) can be neglected. Overlap of reflections does not exist at this level. A reflection ($h_1\,h_2\,h_3\,h_4$) is also close to the two reflections ($h_1{\pm}8\,h_2\,h_3\,h_4{\mp}14$), none of which are extinct. Taking into account that reflections with non-zero intensities have only been observed for main reflections and satellite reflections up to order $m = 2$, the lowest-angle reflections for which overlap is a problem are main reflections ($6\,h_2\,h_3\,0$) with second-order satellite reflections ($-2\,h_2\,h_3\,14$) (Jobst and van Smaalen, 2002). These reflections are high-angle reflections with $2\theta \gtrsim 80°$ for Mo-K_α radiation. Other composite crystals are less benign, and the problem of overlap can occur at much smaller scattering angles.

6.3 Twinning

Twinned crystals (twins) are composed of grains of a crystalline substance that occur in a few different orientations of the lattice of this substance (Giacovazzo *et al.*, 2002). Growth twins are obtained if a crystal grows with different orientations of its lattice out of a single nucleus. This type of twinning can be avoided or facilitated by carefully selected conditions of growth. Grains are often of macroscopic sizes, and they can be mechanically separated, thus resulting in untwinned single crystals. Transformation twins are the result of a phase transition from a crystalline state with higher point symmetry towards a state with lower point symmetry. Phase transitions may independently commence at many points within the sample. Accordingly, a microtwinned crystal is formed, with sizes of the twin domains (grain sizes) down to the nanometer range. Transformation twins are unavoidable, except in cases where an external field can be used to promote the formation of domains of one particular orientation. Examples of external fields include the application of uniaxial stress and the application of an electrical field to ferroelectric materials. The boundaries between twin domains usually are low-index lattice planes, and a definite relation exists between the orientations of the different domains. For transformation twins, these orientations are related to each other by the symmetry operators that have been lost at the phase transition.

The diffraction pattern of a twinned crystal is the superposition of the diffraction pattern of the compound in various orientations, as they correspond to the orientations of the twin domains. Observed diffraction maxima may be due to individual Bragg reflections from single domains or they may be the superposition of Bragg reflections from different domains. Merohedral twinning refers to the situation of perfect coincidence of the lattices of the twin domains. Two domains with orientations related by the inversion centre do not pose a problem for structural analysis, because Friedel's law allows intensities of individual reflections to be extracted from the measured intensities. Other examples of merohedral twinning include periodic structures with point group $4/m$, that occur in two orientations on the tetragonal lattice. Each observed integrated intensity is the sum of intensities of two inequivalent reflections. Standard methods of structural analysis do not work and special procedures need to be invoked, as discussed in detail by Giacovazzo *et al.* (2002).

Incommensurately modulated structures are often obtained as a low-temperature phase of some periodic structure. The symmetry of the high-temperature structure is preserved as symmetry of the basic structure of the modulated phase, but loss of rotational symmetry cannot be excluded, then resulting in a twinned crystal. The possibility of twinning is a major problem in the structural analysis of modulated crystals, because structure refinements cannot discriminate between an interpretation of the diffraction data as coming from an untwinned crystal with high point symmetry and an interpretation of the data as coming from a twinned crystal with low symmetry of the individual domains. These problems are illustrated by the example of $(TaSe_4)_2I$.

$(TaSe_4)_2I$ crystallizes in space group $I422$ with lattice parameters $a_{t1} = a_{t2} =$ 9.519 Å and $a_{t3} = 12.762$ Å. (The subscript t indicates tetragonal lattice parameters.) An incommensurately modulated structure develops below $T_{\mathrm{CDW}} = 263$ K (Fujishita *et al.*, 1984). Eight satellite reflections are found at equal distances around the main Bragg reflections. They can be indexed as first-order satellite reflections on the basis of four, symmetry-related modulation wave vectors,

$$\begin{aligned} \mathbf{q}_t^1 &= (0.0641,\, 0.0641,\, 0.1510) & \mathbf{q}_t^3 &= (0.0641,\, -0.0641,\, 0.1510) \\ \mathbf{q}_t^2 &= (-0.0641,\, -0.0641,\, 0.1510) & \mathbf{q}_t^4 &= (-0.0641,\, 0.0641,\, 0.1510)\,. \end{aligned} \tag{6.4}$$

Because $\mathbf{q}_t^4 = \mathbf{q}_t^1 + \mathbf{q}_t^2 - \mathbf{q}_t^3$, three modulation wave vectors are sufficient for an integer indexing of the diffraction pattern, and $(TaSe_4)_2I$ would possess a three-dimensionally modulated structure that is characterized by a tetragonal (3+3)-dimensional superspace group with the tentative symbol $I422(\sigma_1\, \sigma_1\, \sigma_3)$.

However, a reduction of point symmetry towards orthorhombic, monoclinic or triclinic will result in a twinned crystal. Orthorhombic symmetry $I222$ of the basic structure results in two twin domains, but each domain is modulated by all four wave vectors of eqn (6.4). An incentive does not seem to exist for this reduction of symmetry, and $I222$ has been discarded as possible symmetry of the modulated structure. Interesting possibilities pertain to structures with one-dimensional or two-dimensional modulations. They are obtained if diagonal twofold axes of the tetragonal symmetry are preserved as superspace symmetry of the modulated structure. A classification of all possibilities is then most easily obtained from an F-centred setting of the lattice, obtained by the transformation,

$$\mathbf{a}_1 = -\mathbf{a}_{t1} + \mathbf{a}_{t2} \qquad \mathbf{a}_2 = \mathbf{a}_{t1} + \mathbf{a}_{t2} \qquad \mathbf{a}_3 = \mathbf{a}_{t3}\,. \tag{6.5}$$

Modulation wave vectors with respect to the transformed basis are,

$$\begin{aligned} \mathbf{q}^1 &= (0,\, \sigma_1,\, \sigma_3) & \mathbf{q}^3 &= (-\sigma_1,\, 0,\, \sigma_3) \\ \mathbf{q}^2 &= (0,\, -\sigma_1,\, \sigma_3) & \mathbf{q}^4 &= (\sigma_1,\, 0,\, \sigma_3)\,, \end{aligned} \tag{6.6}$$

with $\sigma_1 = 2\sigma_{t1} = 0.1282$ and $\sigma_3 = 0.1510$ [eqn (6.4)]. The (3+3)-dimensional superspace group of the tetragonal structure is described by the tentative symbol $F422(0\sigma_1\sigma_3)$ with respect to the F-centred lattice. It is a simple exercise to show that the symmetry operators of the orthorhombic point group 222 transform $\mathbf{q}^1$ into $\mathbf{q}^1$ and $\mathbf{q}^2$, and $\mathbf{q}^2$ into $\mathbf{q}^2$ and $\mathbf{q}^1$. A structure with orthorhombic symmetry thus is two-dimensionally modulated by $\mathbf{q}^1$ and $\mathbf{q}^2$. The other domain of the twinned crystal is obtained by application of the fourfold rotation to the first domain, and it follows that its modulation is given by $\mathbf{q}^3$ and $\mathbf{q}^4$. The five non-trivial possibilities for the modulated structure are summarized in Table 6.1. Notice that lower symmetry of the crystal structure corresponds to a reduced dimensionality of the modulation and an increased number of domains.

A complete data set of integrated intensities of main reflections ($\mathbf{G}$) and first-order satellite reflections ($\mathbf{G} \pm \mathbf{q}^j$, $j = 1, \cdots, 4$) has been measured with synchrotron radiation by van Smaalen *et al.* (2001). The diffraction pattern has tetragonal symmetry that can be explained by all models in Table 6.1, if equal volume fractions of the domains are assumed. Main reflections have contributions from all domains, while satellite reflections originate exclusively in a single domain. For example, for monoclinic symmetry each pair of satellite reflections

TABLE 6.1. Five possible symmetries of the modulated structure of $(TaSe_4)_2I$. Given are the $(3+d)$-dimensional superspace group, the number of domains, the dimension of the modulation (d), the modulation wave vectors of the modulation in the first domain, and the partial R values for main reflections and for satellite reflections of the best refinement in each symmetry.

Superspace group	Domains	d	Wave vectors	R_F(main)	R_F(sat)
$F422(0\,\sigma_1\,\sigma_3)000$	1	3	$\mathbf{q}^1$, $\mathbf{q}^2$, $\mathbf{q}^3$, $\mathbf{q}^4$	–	–
$F222(0\,\sigma_1\,\sigma_3)000$	2	2	$\mathbf{q}^1$, $\mathbf{q}^2$	0.085	0.099
$F222(0\,\sigma_1\,\sigma_3)00s$	2	2	$\mathbf{q}^1$, $\mathbf{q}^2$	0.085	0.099
$F211(0\,\sigma_1\,\sigma_3)000$	4	1	$\mathbf{q}^1$	0.090	0.082
$F1(0\,\sigma_1\,\sigma_3)0$	8	1	$\mathbf{q}^1$	0.091	0.098

$\mathbf{G} \pm \mathbf{q}^j$ originates in a different domain, and the four modulation wave vectors $\mathbf{q}^j$ thus define the four domains in the sample. Structure refinements have led to fits to the data of comparable quality for the different models. The similarity between the different refinements extends towards the values of the modulation parameters, that are equal to each other up to a scale factor provided by the volume fractions of the domains (0.5 for orthorhombic and 0.25 for monoclinic symmetries). The monoclinic structure model can be characterized by the presence of a single wave with wave vector $\mathbf{q}^1$ in the structure, while the orthorhombic model has two symmetry-equivalent waves, $\mathbf{q}^1$ and $\mathbf{q}^2$, that are simultaneously present in a single domain. Despite these relations, models of different symmetries describe entirely different structures. The problem is that structure refinements cannot distinguish between them, and more detailed diffraction experiments are required, in order to pin down the correct crystal structure.

The reduction of symmetry towards orthorhombic or monoclinic allows for different lattice distortions. In the diffraction this should lead to split maxima at the positions of the main reflections. However, lattice distortions and splittings of reflections are expected to be small, and the failure of observing splitting may not be taken as evidence against the structures of low symmetry. Favre-Nicolin (1999) has measured X-ray diffraction of $(TaSe_4)_2I$ employing synchrotron radiation and imaging plates. Main reflections are found to possess increased widths below T_{CDW}, while several of them are composed of three, partially overlapping Bragg reflections (Fig. 6.3). Accordingly, the modulated crystal must have contained at least three domains, and these observations support the monoclinic structure model.

The different symmetries can also be distinguished by consideration of higher-order satellite reflections. For all models satellite reflections $m\,\mathbf{q}^j$ can occur in the diffraction. In addition, structures with a two-dimensional modulation will give rise to mixed second-order satellite reflections $\mathbf{q}^1 \pm \mathbf{q}^2$ and $\mathbf{q}^3 \pm \mathbf{q}^4$. Favre-Nicolin (1999) has observed reflections $\pm 2\mathbf{q}^j$, but he failed to find any of the mixed second-order satellite reflections (Fig. 6.3), in accordance with the fail-

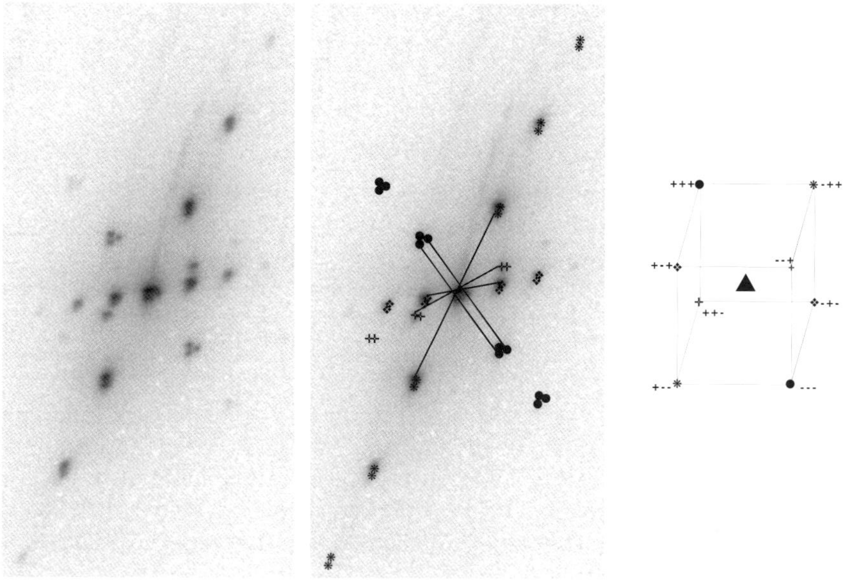

(a) (b) (c)

FIG. 6.3. Diffraction by $(TaSe_4)_2I$ recorded on an imaging plate by the rotating crystal technique. (a) (16 4 4) main reflection surrounded by satellites $\pm m\mathbf{q}^j$ $(j = 1, \cdots, 4)$ [eqn (6.4)]. Maxima consist of two or three overlapping reflections. (b) Indexing of the reflections overlayed onto the experimental image. (c) Projection of reciprocal lattice onto the plane of the imaging plate. $++-$ indicates the satellite $-\mathbf{q}_t^2 = (+0.0641, +0.0641, -0.1510)$ [eqn (6.4)]. Reprinted from Favre-Nicolin (1999) by courtesy of V. Favre-Nicolin.

ure to observe mixed second-order satellite reflections by a point detector in ω-scans at the positions $\mathbf{q}^1 + \mathbf{q}^2 = (0, 0, 2\sigma_3)$ and $\mathbf{q}^1 - \mathbf{q}^2 = (0, 2\sigma_1, 0)$ (van Smaalen *et al.*, 2001). Again these observations support the monoclinic model. Based on all experimental evidence, it has been concluded that the incommensurately modulated structure of $(TaSe_4)_2I$ has a one-dimensional modulation with monoclinic symmetry $F211(0\,\sigma_2\,\sigma_3)000$.

The possibility of pseudo-merohedral twinning, similar to the pseudo-merohedral twinning of $(TaSe_4)_2I$, is a major problem in the structural analysis of modulated crystals. Split reflections or the observation of mixed higher-order satellite reflections may be taken as evidence for a particular symmetry. However, the failure to observe splitting of reflections and the failure to observe higher-order satellite reflections can only be meaningful if X-ray diffraction with synchrotron radiation has been performed on good-quality single crystals. Other methods to discriminate between mono-domain and multi-domain structures include dark-field electron microscopy. Generating an image of the crystal from a single satellite reflection will show whether this satellite originates in all of the sample or only in one of its domains.

6.4 Atomic vibrations in incommensurate structures

6.4.1 *Phonons and the Debye–Waller factor*

Atoms in solid substances vibrate about average positions. Vibrations in periodic crystals can be decomposed into phonons, with the displacement $\delta\mathbf{u}(\mu, \mathbf{L}, t)$ of atom μ in unit cell $\mathbf{L}$ at time t given by (Maradudin *et al.*, 1963),

$$\delta\mathbf{u}(\mu, \mathbf{L}, t) = \frac{1}{M_\mu} \frac{1}{N_{\text{cell}}} \sum_{\mathbf{k}\in BZ}^{N_{\text{cell}}} \sum_{\nu=1}^{3N} \mathbf{Q}(\mathbf{k}, \nu, \mu) \exp[2\pi i \mathbf{k}\cdot\mathbf{L}(\mu) + i\omega_\nu(\mathbf{k})t] \,. \quad (6.7)$$

A phonon is a harmonic wave with wave vector $\mathbf{k}$, frequency $\omega_\nu(\mathbf{k})$ and amplitudes $\mathbf{Q}(\mathbf{k}, \nu, \mu)$ for all atoms in the unit cell. A crystal with N atoms in each of its N_{cell} unit cells contains $3N \times N_{\text{cell}}$ independent phonon modes, that split into $3N$ branches (indexed by ν). The sum over the wave vectors $\mathbf{k}$ includes N_{cell} points selected from the first Brillouin zone (BZ) of the reciprocal lattice. The mass of atom μ is given by M_μ.

Each branch of phonons is characterized by a dispersion relation that gives the phonon frequency as a function of the wave vector, $\omega_\nu = \omega_\nu(\mathbf{k})$. Acoustic phonons have linear dispersions in the long wavelength limit ($k \to 0$):

$$\omega_\nu(\mathbf{k}) = v_{s\nu} k \,, \quad (6.8)$$

where $v_{s\nu}$ is the sound velocity. Three branches of acoustic phonons correspond to one branch of longitudinal waves (LA) and two branches of transversal waves (TA) [Fig. 6.4(a)]. Acoustic phonons have zero energy ($\hbar\omega \to 0$) in the limit $k \to 0$. This is explained by the fact that all atoms in the unit cell have equal amplitudes for acoustic phonons. In the long wavelength limit this implies nearly equal displacements of neighbouring atoms, that become more equal for smaller k. Because displacements of the sample do not change its energy, the corresponding phonons have zero energies. Three independent directions of displacements in space define three branches of acoustic phonons.

The amplitudes of phonons depend on their energy as well as on the temperature. Lower energies and higher temperatures correspond to larger amplitudes and, consequently, larger displacements of the atoms. The effect of lattice vibrations on the intensities of Bragg reflections has been found to depend on the mean-square displacements of the atoms,

$$B^\mu = 8\pi^2 \left\langle \delta\mathbf{u}(\mu, \mathbf{L}, t)\, \delta\mathbf{u}(\mu, \mathbf{L}, t) \right\rangle \,, \quad (6.9)$$

where $\langle \cdots \rangle$ denotes the time average. B^μ is the temperature parameter or displacement parameter of atom μ. The diffraction of a crystal with lattice vibrations is given by a structure factor in which each atomic scattering factor $f_\mu(H)$ is replaced by the product of $f_\mu(H)$ and the Debye–Waller factor,

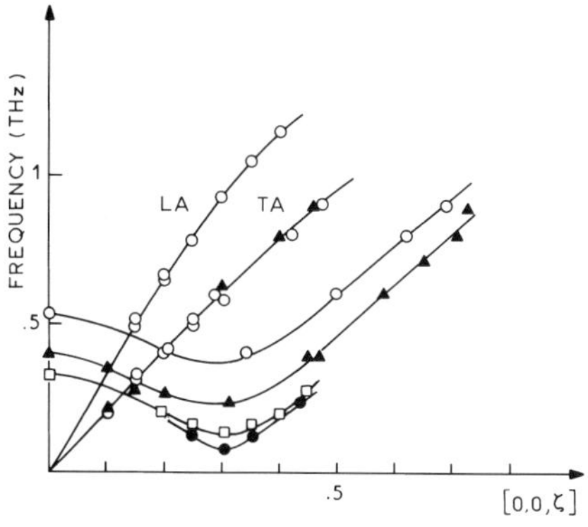

(a)

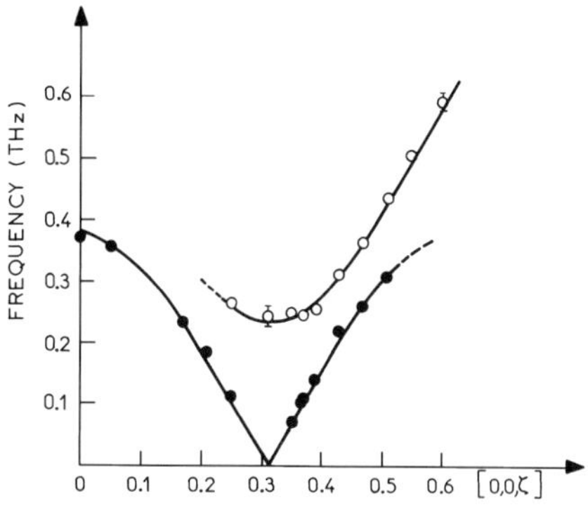

(b)

FIG. 6.4. Modes of atomic vibrations in $ThBr_4$ as measured with inelastic neutron scattering. (a) Phonon dispersion relations in the lattice-periodic state for **k** along $(0\,0\,\zeta)$, and at temperatures of 300 K (open circles), 150 K (triangles), 120 K (squares), and 101 K (dots). (b) Dispersion relations of the phason branch (dots) and amplitudon branch (open circles) at a temperature of 81 K [eqn (6.14)]. The normal to incommensurate phase transition occurs at $T_i = 95$ K. Reprinted with permission from Currat *et al.* (1986).

$$T^{\mu}_{\text{phonon}} = \exp\left[-\tfrac{1}{4}B^{\mu}H^2\right]. \tag{6.10}$$

In the general case of anisotropic vibrations, the Debye–Waller factor or temperature factor is,

$$\begin{aligned} T^{\mu}_{\text{phonon}} = \exp\big[-2\pi^2\,\big((h_1a_1^*)^2U^{\mu}_{11} + (h_2a_2^*)^2U^{\mu}_{22} + (h_3a_3^*)^2U^{\mu}_{33} \\ + 2h_1h_2a_1^*a_2^*U^{\mu}_{12} + 2h_1h_3a_1^*a_3^*U^{\mu}_{13} + 2h_2h_3a_2^*a_3^*U^{\mu}_{23}\big)\big], \end{aligned} \tag{6.11}$$

with

$$U^{\mu}_{ij} = \langle \delta u_i(\mu, \mathbf{L}, t)\, \delta u_j(\mu, \mathbf{L}, t)\rangle . \tag{6.12}$$

The Debye–Waller factor depends on the mean-square amplitudes of displacement of the atoms, but not on the existence of phonons. Consequence is that the effect of thermal vibrations on the diffraction of aperiodic crystals is in first approximation given by a Debye–Waller factor of the form of eqn (6.10) or eqn (6.11). Furthermore, time-dependent displacements and uncorrelated static displacements of atoms out of their lattice-periodic positions have similar effects on the intensities of Bragg reflections, and both are reflected by contributions to the temperature parameters [eqn (6.9) or eqn (6.12)].

6.4.2 *Phasons and the sliding mode*

Mean-square displacements of the atoms in any solid material are finite and bounded. The material melts if $\langle \delta\mathbf{u}(\mu, \mathbf{L}, t)\, \delta\mathbf{u}(\mu, \mathbf{L}, t)\rangle$ tends to become too large.

Nevertheless, the mechanisms of atomic vibrations need not be the same in periodic and aperiodic crystals. In particular, the energy of an incommensurately modulated crystal does not depend on the phase of the modulation. Noticing the relation between acoustic phonons and displacements of the crystal at zero cost of energy, Overhauser (1971) has proposed a fourth branch of acoustic-like crystal vibrations with atomic displacements that correspond to a shift of the phase of the modulation wave: the so-called phasons.

Consider an incommensurately modulated structure with atomic displacements,

$$u = A\sin[2\pi(\bar{x}_4 + \phi)]\,. \tag{6.13}$$

A time-dependent phase ϕ is introduced, that has average $\langle\phi\rangle = 0$ and finite mean-square value $\langle\phi^2\rangle \neq 0$. Employing the zero-energy effect of phase fluctuations allows one to derive two special branches of atomic vibrations, with dispersion relations,

$$\omega_\phi^2 = v_\phi^2\,(\delta q)^2 \tag{6.14a}$$

$$\omega_A^2 = \omega_0^2 + v_\phi^2\,(\delta q)^2\,, \tag{6.14b}$$

for

$$\mathbf{k} = \mathbf{q} + \delta\mathbf{q}\,. \tag{6.14c}$$

$\mathbf{q}$ is the modulation wave vector of the incommensurate structure, and $\mathbf{k}$ denotes a vector in reciprocal space of the basic structure. Equation (6.14a) describes the linear dispersion of the phason branch of vibrations. It is characterized by the phason sound velocity v_ϕ. The dispersion of eqn (6.14b) is based on the same velocity v_ϕ, but the energy of these modes is always finite, with a minimum of $\hbar\omega_0$ at $\mathbf{k} = \mathbf{q}$. These modes are called amplitudons, because the atomic displacements can be derived from a variation of the amplitude of the modulation wave. Phasons and amplitudons have been observed by inelastic neutron scattering [Fig. 6.4(b)].

Phasons are fundamentally different from acoustic phonons, because relative displacements of neighbouring atoms do not vanish in the limit $\delta q \to 0$. Impurities will therefore pin the modulation wave, thus raising the energies of phason modes to finite values at all δq. Even without impurities, requirements of stability of aperiodic substances lead to an intrinsic gap of energy $\hbar\eta$ in the phason branch of atomic vibrations (Zeyher and Finger, 1982). The modified dispersion relation is,

$$\omega_\phi^2 = \eta^2 + v_\phi^2\,(\delta q)^2\,. \tag{6.15}$$

An alternative characterization of the phason gap is provided by the damping of phasons (Γ_ϕ) being finite at all wavelengths, such that modes with $\omega_\phi < \Gamma_\phi$ are diffusive instead of propagating.

The notion of a phason gap and the diffusive character of low-energy phason modes is in agreement with all experimental observations, with estimated gaps of the order of 0.2 meV and damping constants that are slightly larger than this value (Currat *et al.*, 1986; Ollivier *et al.*, 1998; Ravy *et al.*, 2004). Overdamped modes might still affect the line shapes in Nuclear Magnetic Resonance (NMR) experiments. Many reports exist that claim to have observed the signature of phasons in NMR (Blinc *et al.*, 1994; Taye *et al.*, 2004).

Phase changes in incommensurate composite crystals define relative displacements of the subsystems along their mutually incommensurate direction (Section 4.4). The corresponding dynamical mode is now called the sliding mode. At high frequencies and small wavelengths one may expect independent acoustic-phonon-like modes on the two subsystems (Heilmann *et al.*, 1979). At low frequencies one mode becomes the acoustic phonon with equal displacement amplitudes for all atoms, while the other mode (the sliding mode) has displacements in opposite directions for atoms of different subsystems. Impurity pinning and the criterion of stability make the sliding mode overdamped for low frequencies (Schmicker and van Smaalen, 1996). The observation of sliding modes at very low frequencies has been attributed to lack of long-range order in one of the subsystems [Fig. 6.5; Heilmann *et al.* (1979), Schmicker *et al.* (1995)].

6.4.3 *The Debye–Waller factor for aperiodic crystals*

Low-energy acoustic phonons have large amplitudes and constitute a major contribution to the Debye–Waller factor in X-ray diffraction. Phasons without damping or gaps [eqn (6.14)] also have large amplitudes, and they should have a major effect on the diffraction by aperiodic crystals. In one approach Overhauser (1971) has proposed that the effect of phasons on the intensities of Bragg reflections is described by an additional multiplicative factor,

$$T_{\text{phason}} = \exp\left[-B_\phi {h_4}^2\right] , \tag{6.16}$$

resulting in a Debye–Waller factor [compare eqns (6.10) and (6.11)],

$$T^\mu_{\text{DW}} = T_{\text{phason}} \times T^\mu_{\text{phonon}} \, . \tag{6.17}$$

In a second approach Axe (1980) has shown that phasons lead to renormalized modulation amplitudes [eqn (6.13)],

$$\langle u \rangle = A \langle \cos(2\pi\phi) \rangle \, , \tag{6.18}$$

as well as a phason contibution to the Debye–Waller factor of,

$$T_{\text{phason}} = \exp[-B_\phi h_4 (h_4 - 1)] \, . \tag{6.19}$$

Both forms of the phason Debye–Waller factor only modify the intensities of satellite reflections and not of the main reflections. The effect may be large and this has led Overhauser (1971) to propose that satellite reflections might be too weak to be observable. Yet, satellite reflections have been observed in the diffraction of many incommensurate crystals. The reason obviously is that

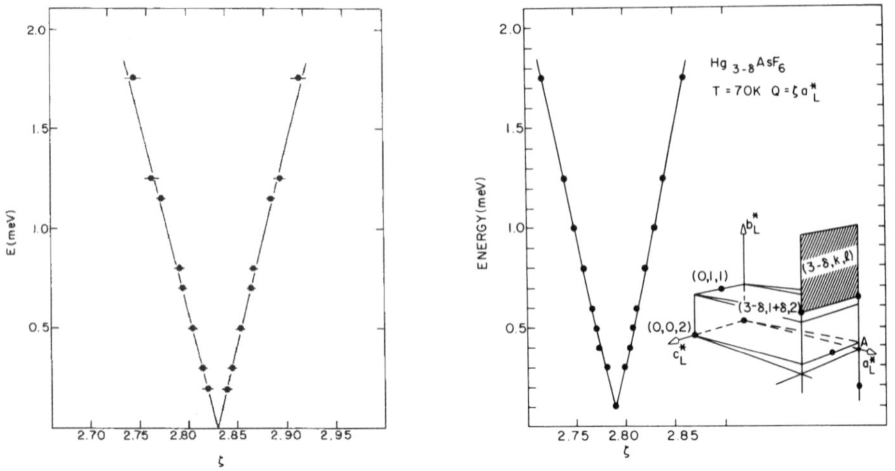

(a) (b)

FIG. 6.5. Dispersion relations of the atomic vibrations in the incommensurate composite crystal $[\mathrm{Hg}]_{2.83}[\mathrm{AsF}_6]$, as determined by inelastic neutron scattering along $(\zeta\ 0\ 0)$. (a) At $T = 295$ K without long-range order of the Hg subsystems. Straight lines are a fit of eqn (6.14a) to the data, with $v_\phi = 23.8$ meVÅ. (b) At $T = 70$ K with completely ordered subsystems. Lines are a fit to eqn (6.15) with $\eta = 0.103$ meV and $v_\phi = 29$ meVÅ. The main reflection (1 0 0) of the Hg subsystem correspond to $\zeta = 2.83$ as indexed with respect to the AsF_6 subsystem. Notice that scattering at $(\zeta\ 0\ 0)$ selects longitudinal modes of the Hg subsystem with displacements along $\mathbf{a}_1$ only. (Compare to Fig. 4.8.) Reprinted with permission from Heilmann *et al.* (1979), copyright (1979) by the American Physical Society.

low-energy phasons are diffusive modes and that the phason branch of atomic vibrations is gapped, such that eqns (6.16) and (6.19) do not apply, or that the phason temperature factor B_ϕ is a small quantity.

In a different approach, Perez-Mato *et al.* (1991) have shown that phase and amplitude fluctuations affect the diffraction through Debye–Waller factors that contain modulated temperature parameters $B^\mu = B^\mu(\bar{x}_{s4})$ [eqn (6.10)]:

$$B^\mu(\bar{x}_{s4}) = B^0(\mu) + \sum_{n=1}^{\infty} B^{sn}(\mu)\sin(2n\pi\bar{x}_{s4}) + B^{cn}(\mu)\cos(2n\pi\bar{x}_{s4})\,. \qquad (6.20)$$

Each harmonic n is described by two modulation parameters for the isotropic temperature parameter. Anisotropic temperature parameters contain 12 independent parameters for each harmonic. The effect of phasons and amplitudons is represented by second-order harmonic coefficients B^{s2} and B^{c2} in eqn (6.20). However, the presence of other harmonics in $B^\mu(\bar{x}_{s4})$—as it would result from other mechanisms—cannot be excluded. In any case, structure refinements should include second-harmonic modulation parameters for the temperature parameters, if phase and amplitude fluctuations are suspected to be present in the crystal.

Evidence for a phason Debye–Waller factor [eqns (6.16) or (6.19)] has been claimed in several diffraction studies, but all of them are inconclusive. More recently, the temperature dependence of the intensities of satellite reflections of Rb_2ZnCl_4 up to the seventh order has been explained by a variation of the modulation functions rather than by a phason Debye–Waller factor (Aramburu *et al.*, 1997). Structure refinements of $[NH_3(C_3H_7)]_2MnCl_4$ have shown that modulated temperature parameters better explain the diffraction data than either of the two phason Debye–Waller factors, although all three forms of extended Debye–Waller factors improve the fit to the data as compared to the simple form of eqn (6.11) (Meyer *et al.*, 1994). The effect of phasons on the diffracted intensities generally is small. Significant values for the modulated temperature parameters are only obtained with extensive data sets that include high-angle reflections and at least second-order satellite reflections. The fit to limited data sets can sometimes be improved by the phason Debye–Waller factor, but this factor then mimics the lack of higher-order harmonics in the structure model (Aramburu *et al.*, 1996).

While phase and amplitude fluctuations affect diffracted intensities through second-harmonic modulation coefficients of the temperature parameters, the reverse need not be true. An alternative interpretation relates the modulation of the temperature parameters of a composite crystal to the variations of the environments of the atoms with the variation of the incommensurate phase t (Section 4.8). Tighter environments correspond to smaller U_{ij} and less compact environments correspond to larger values of U_{ij} (Jobst and van Smaalen, 2002). This interpretation of modulated temperature parameters is closely related to the interpretation of temperature parameters of commensurately modulated structures. The N_{super} different values of $B^{\mu}(\bar{x}_{s4})$ in a N_{super}-fold superstructure reflect the N_{super} different environments in the superstructure for each independent atom of the basic structure.

6.5 Statistical properties of diffracted intensities

6.5.1 *Preliminary considerations*

Diffraction experiments provide magnitudes of the structure factors through the measured intensities of Bragg reflections [eqn (6.1)], while information is not obtained on the reflection phases, $\varphi(\mathbf{H}_s)$, defined by,

$$F(\mathbf{H}_s) = |F(\mathbf{H}_s)|\ \exp[i\varphi(\mathbf{H}_s)]\,. \tag{6.21}$$

Different crystals have different sets of structure factor amplitudes and phases, but possible combinations of amplitudes and phases are severely restricted by two properties of the electron density, that are valid for both periodic and aperiodic crystals. The electron density is positive at all points in space, and the electron density has atomic character, *i.e.* it can be separated into a sum of independent atomic electron densities in good approximation [eqn (1.17)]. These properties lead to probability relations between phases and amplitudes of different reflections, that form the basis of direct methods of structure solution (Giacovazzo, 1998; Shmueli and Weiss, 1995). The generalization towards incommensurately modulated crystals has been made for statistical properties of structure factor amplitudes. They are presented in Sections 6.5.2 and 6.5.3. Structure invariants and direct methods are discussed in Section 10.4.

6.5.2 *The average diffracted intensity*

The Wilson plot for periodic crystals requires knowledge of the contents of the unit cell. The combined scattering power of the atoms in the unit cell is,

$$\mathcal{S}_2 = \sum_{\mu=1}^{N} [f_\mu(H)]^2 . \tag{6.22}$$

It is found to depend on the length of the scattering vector. In the literature this quantity is often denoted by Σ_s. Averaging over many reflections with the same length of scattering vector leads to an averaged intensity of Bragg reflections that is a simple function of $\mathcal{S}_2$ and the average temperature parameter B (Giacovazzo, 1998),

$$\frac{1}{\mathcal{S}_2} \left\langle |F_{\text{cal}}(\mathbf{H})|^2 \right\rangle_H = \exp\left[-\tfrac{1}{2} B H^2\right] , \tag{6.23}$$

where $\langle\cdots\rangle_H$ denotes the average over all reflections with the same length H of scattering vectors $\mathbf{H}$. Values of $\left\langle |F_{\text{obs}}(\mathbf{H})|^2 \right\rangle_H$ are obtained from the experiment as the averages over all reflections with lengths of their scattering vectors falling within a suitably chosen interval of H. The result is the Wilson plot,

$$\ln\left[\frac{1}{\mathcal{S}_2} \left\langle |F_{\text{obs}}(\mathbf{H})|^2 \right\rangle_H\right] = \ln\left[K^2\right] - \frac{1}{2} B H^2 , \tag{6.24}$$

where K is the scale factor.

The extension of the Wilson plot towards incommensurately modulated structures requires the explicit consideration of the effects of the modulation on the structure factors (Lam *et al.*, 1992). First, consider a single-harmonic displacive modulation [eqn (1.10)],

$$\mathbf{u}^\mu(\bar{x}_{s4}) = \mathbf{A}(\mu) \sin[2\pi \bar{x}_{s4}] + \mathbf{B}(\mu) \cos[2\pi \bar{x}_{s4}] . \tag{6.25}$$

The atomic modulation scattering factor is [eqn (2.34b)],

$$g_\mu(\mathbf{H}_s) = \int_0^1 \exp[2\pi i \, (\mathbf{H} \cdot \mathbf{u}^\mu(\tau) + h_4 \, \tau)] \, d\tau , \tag{6.26}$$

where $\mathbf{H}$ is the scattering vector in physical space, that is obtained as the projection of $\mathbf{H}_s$ onto physical space (Section 2.2). The integral can be evaluated if we define the quantities $C(\mu)$ and $\eta(\mu)$ through the relation,

$$\begin{aligned}\mathbf{H}\cdot\mathbf{u}^{\mu}(\bar{x}_{s4}) &= \mathbf{H}\cdot\mathbf{A}(\mu)\,\sin[2\pi\bar{x}_{s4}] + \mathbf{H}\cdot\mathbf{B}(\mu)\,\cos[2\pi\bar{x}_{s4}] \\ &= C(\mu)\sin(2\pi[\bar{x}_{s4} - \eta(\mu)])\,, \end{aligned} \tag{6.27}$$

from which follows,

$$[C(\mu)]^2 = [\mathbf{H}\cdot\mathbf{A}(\mu)]^2 + [\mathbf{H}\cdot\mathbf{B}(\mu)]^2\,. \tag{6.28}$$

Notice that both $C(\mu)$ and $\eta(\mu)$ depend on the scattering vector $\mathbf{H}$. Employing the Jacobi–Auger expansion of Bessel functions gives for the atomic modulation scattering factor,

$$g_\mu(\mathbf{H}_s) = J_{-h_4}[2\pi C(\mu)]\,\exp[2\pi i\,h_4\,\eta(\mu)]\,, \tag{6.29}$$

where $J_{h_4}[x] = (-1)^{h_4}\,J_{-h_4}[x]$ is a Bessel function of the first kind of order h_4 (Section 2.5.2). Substitution into the expression for the structure factor results in [eqns (2.34a) and (6.10)],

$$\begin{aligned}F_{\text{cal}}(\mathbf{H}_s) = \sum_{\mu=1}^{N} & f_\mu(\mathbf{S})\exp\left[-\tfrac{1}{4}B(\mu)H^2\right]J_{-h_4}[2\pi C(\mu)] \\ & \times\,\exp[2\pi i\,h_4\,\eta(\mu)]\,\exp\left[2\pi i(\mathbf{H}-h_4\mathbf{q})\cdot\mathbf{x}^0(\mu)\right].\end{aligned} \tag{6.30}$$

The quantity of interest is the average diffracted intensity for averaging over the direction of the scattering vector. It turns out that different averages need to be considered for each satellite order $m = |h_4|$,

$$\left\langle|F_{\text{cal}}(\mathbf{H}_s)|^2\right\rangle_{H,m} = \left\langle\sum_{\mu=1}^{N} f_\mu^2(\mathbf{H})\exp\left[-\tfrac{1}{2}B(\mu)H^2\right](J_m[2\pi C(\mu)])^2\right\rangle_{H,m}\,, \tag{6.31}$$

where $\langle\cdots\rangle_{H,m}$ denotes the average over the scattering vector $\mathbf{H}$ at constant length H and for fixed satellite order m. It is assumed that cross terms cancel each other in the average of the product $|F_{\text{cal}}(\mathbf{H}_s)|^2$. This approximation is better for larger structures, because the number of cross terms with different phase factors $\exp\left[2\pi i(\mathbf{H}-h_4\mathbf{q})\cdot\left(\mathbf{x}^0[\mu]-\mathbf{x}^0[\mu']\right)\right]$ is proportional to N^2. Secondly, structures with larger unit cells have more reflections that contribute to the average $\langle\cdots\rangle_{H,m}$ for a suitably chosen interval of the length of the scattering vector.

Further approximations include the introduction of an overall temperature parameter, B, and the notion that atomic scattering factors follow a universal dependence on the length of the scattering vector in good approximation. Denote the atomic number of atom μ by Z_μ. Then

$$f_\mu(\mathbf{H}) \approx Z_\mu\,f(\mathbf{H})\,,$$

and,

$$\langle |F_{\text{cal}}(\mathbf{H}_s)|^2 \rangle_{H,m} = \mathcal{S}_2 \exp\left[-\tfrac{1}{2}BH^2\right] \times \left\langle \frac{1}{\sum_{\mu=1}^{N} Z_\mu^2} \sum_{\mu=1}^{N} Z_\mu^2 \left(J_m[2\pi C(\mu)]\right)^2 \right\rangle_{H,m} . \tag{6.32}$$

Assuming equal modulation amplitudes for all atoms, and assuming a special form of the modulation functions [eqn (6.25)],

$$\mathbf{u}^\mu(\bar{x}_{s4}) = \mathbf{U}\,\sin[2\pi(\bar{x}_{s4} - \eta)] , \tag{6.33}$$

allows the average $\langle \cdots \rangle_{H,m}$ in eqn (6.32) to be evaluated towards,

$$Z(H; m, U) = \int_0^1 [J_m(2\pi HUx)]^2 \, dx , \tag{6.34}$$

where U is the magnitude of the modulation amplitude $\mathbf{U}$. Series expansions of the Bessel functions in eqns (6.32) and (6.34) are equal up to lowest orders in their arguments, provided U is defined as the weighted average modulation amplitude,

$$U^2 = \frac{1}{\sum_{\mu=1}^{N} Z_\mu^2} \sum_{\mu=1}^{N} Z_\mu^2 \left([A(\mu)]^2 + [B(\mu)]^2\right) . \tag{6.35}$$

The function $Z(H; m, U)$ describes the influence of the modulation on the average diffracted intensities through [compare eqn (6.23)],

$$\frac{1}{\mathcal{S}_2} \langle |F_{\text{cal}}(\mathbf{H}_s)|^2 \rangle_{H,m} = \exp\left[-\tfrac{1}{2}BH^2\right] Z(H; m, U) . \tag{6.36}$$

The modulated-structure counterpart of the Wilson plot follows as [eqn (6.24)],

$$\ln\left[\frac{1}{\mathcal{S}_2} \langle |F_{\text{obs}}(\mathbf{H}_s)|^2 \rangle_{H,m}\right] = \ln[G(H; m)] = \ln\left[K^2\right] - \frac{1}{2}BH^2 + \ln[Z(H; m, U)] . \tag{6.37}$$

It is noticed that different curves $G(H; m)$ are obtained for different satellite orders m, but all curves are described by a common set of three parameters: the scale factor K, the overall displacement parameter B, and the overall modulation amplitude U. The function $G(H; m)$ gives a good representation of the average diffracted intensities, both for compounds with small and large modulation amplitudes (Fig. 6.6). Furthermore, fitted values of B and U are in reasonable agreement with values computed from the known structural parameters (Table 6.2). Even better agreements are obtained for simulated data of artificial structures with 50 carbon atoms placed randomly in the unit cell. A2 is an acentric structure with superspace group $P1(\sigma_1\, \sigma_2\, \sigma_3)0$ and with randomly distributed modulation amplitudes for different atoms with a prescribed average magnitude

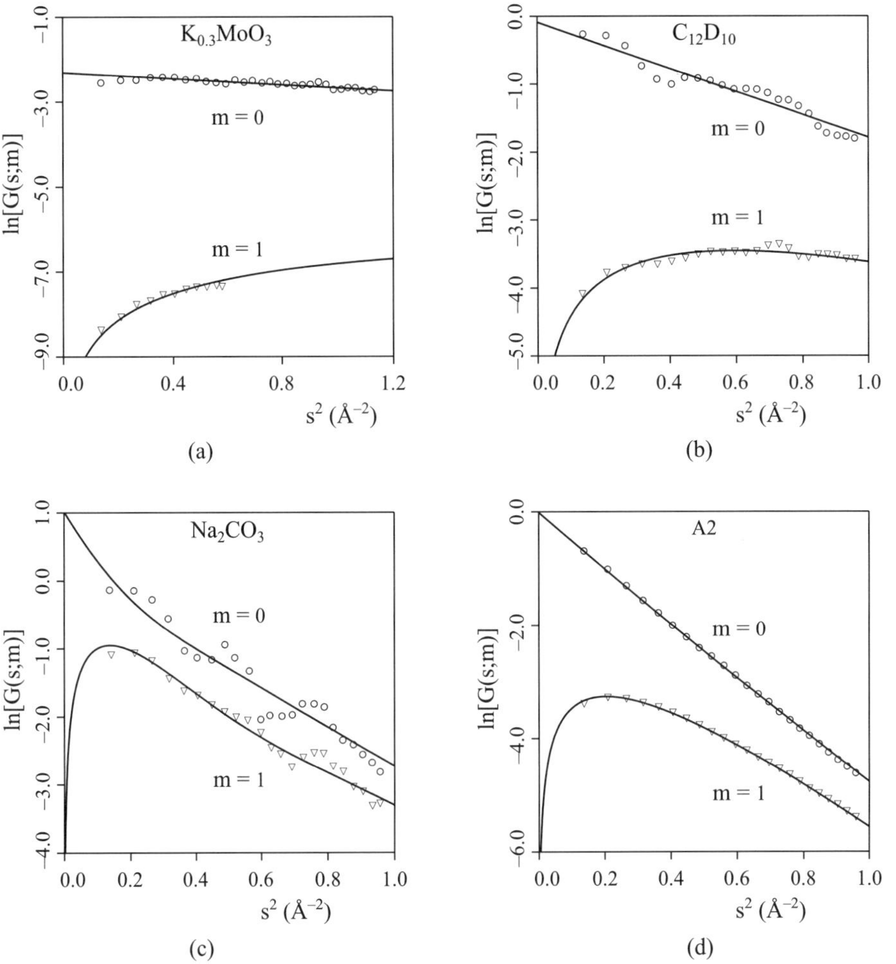

FIG. 6.6. Dependence of the average diffracted intensity on the length of the scattering vector for selected compounds with one-dimensional displacive modulations. (a) $K_{0.3}MoO_3$. (b) Deuterated biphenyl $C_{12}D_{10}$. (c) Na_2CO_3. (d) Artificial structure A2 (see text). s is defined as $\frac{1}{2}H$. Lines represent a fit of eqn (6.37) to the data. Adapted from Lam *et al.* (1993; 1994), copyright (1993, 1994) by the IUCr.

of $U = 0.200$ Å; C2 is the centrosymmetric counterpart of A2, with superspace group $P\bar{1}(\sigma_1\,\sigma_2\,\sigma_3)\bar{1}$.

Average diffracted intensities have also been considered for structures with one-dimensional modulations involving arbitrary numbers of harmonic amplitudes for both displacive and occupational modulations (Lam *et al.*, 1994). The

TABLE 6.2. Parameters of functions $G(H;m)$ for selected compounds. Given are the overall temperature parameter (B) and the overall modulation parameter (U) as obtained by a fit of $G(H;m)$ to diffraction data [eqn (6.37)], as well as the corresponding values computed from structure models [eqn (6.35)]. Artificial structures A2 and C2 are defined in the text. Values have been taken from Lam *et al.* (1993; 1994).

	Computed from model		Fitted to data	
Compound	B (Å^2)	U (Å)	B (Å^2)	U (Å)
$K_{0.3}MoO_3$	0.22	0.033	0.17	0.035
$C_{12}D_{10}$	0.99	0.112	0.70	0.110
Na_2CO_3	1.42	0.386	1.16	0.461
A2	2.00	0.200	1.98	0.198
C2	2.00	0.200	1.98	0.203

result of the lowest-order approximation is summarized here. For each harmonic n_u of the displacement modulation of eqn (1.10), an average modulation amplitude U^{n_u} is defined similar to the average modulation amplitude U of the single-harmonic modulation [eqn (6.35)]:

$$[U^{n_u}]^2 = \frac{1}{\sum_{\mu=1}^{N} Z_\mu^2} \sum_{\mu=1}^{N} Z_\mu^2 \left([A^{n_u}(\mu)]^2 + [B^{n_u}(\mu)]^2\right). \tag{6.38}$$

The Fourier series of the occupational modulation function is [eqn (2.37)]

$$p^\mu(\bar{x}_{s4}) = \sum_{n_p=-\infty}^{\infty} P^{n_p}(\mu) \exp[2\pi i n_p \bar{x}_{s4}], \tag{6.39}$$

with $P^0(\mu)$ equal to the average occupancy of site μ. The occupational modulation is defined by complex Fourier coefficients $P^{n_p}(\mu) = [P^{-n_p}(\mu)]^*$, where P^* indicates the complex conjugate of P. Each harmonic $n_p \geqslant 0$ leads to one overall occupational modulation amplitude,

$$[P^{n_p}]^2 = \frac{1}{\sum_{\mu=1}^{N} Z_\mu^2} \sum_{\mu=1}^{N} Z_\mu^2 \, |P^{n_p}(\mu)|^2 . \tag{6.40}$$

The average diffracted intensity is [eqn (6.37)],

$$G(H;m) = K^2 \exp\left[-\tfrac{1}{2}BH^2\right] Y(H;m). \tag{6.41}$$

The function $Y(H;m)$ incorporates the effect of the modulation on the average diffracted intensities of main reflections ($m = 0$) and of satellite reflections of order $m = |h_4|$. It depends on the average modulation amplitudes U^{n_u} and P^{n_p}:

$$Y(H;0) = \left(P^0\right)^2 \prod_{n_u=1}^{\infty} Z(H;0,U^{n_u}) \tag{6.42a}$$

$$Y(H;m) = \sum_{n_p=(1-m)}^{\infty} \left(P^{n_p}\right)^2 Z(H;1,U^{m+n_p}) . \tag{6.42b}$$

$\prod$ denotes a product in analogy with $\sum$ for summation. The average intensity of satellite reflections of order m is described by an infinite number of parameters. However, a good fit to the diffraction data can be expected, if the orders of harmonics, n_u and $|n_p|$, are restricted to values less than or equal to the maximum order of observed satellite reflections. In many applications the maximum values of n_u or $|n_p|$ will be even smaller. For example, if indications for an occupational modulation are absent, $P^0 = P^0(\mu) = 1$ and $P^{n_p} = P^{n_p}(\mu) = 0$ for $n_p \neq 0$, and the function $Y(H;m)$ reduces to,

$$Y(H;0) = \prod_{n_u=1}^{\infty} Z(H;0,U^{n_u}) \tag{6.43a}$$

$$Y(H;m) = Z(H;1,U^m) . \tag{6.43b}$$

The average diffracted intensities of reflections of satellite order m are found to depend on the average modulation amplitudes of mth-order, in first approximation. Satellite reflections have average intensities equal to those of a single-harmonic modulation of corresponding order m [compare eqn (6.43b) with eqn (6.36)], while average intensities of main reflections are affected by the cumulative effect of the various harmonics.

The quality of the fit of $G(H;m)$ [eqn (6.41)] to simulated diffraction data has been demonstrated for an artificial structure with 50 atoms placed randomly in the unit cell (DP). Up to second-order harmonic occupational and displacive modulation parameters were introduced into the structure model, with random values and prescribed averages of $P^0 = 0.900$, $P^1 = 0.250$, $P^2 = 0.125$, $U^1 = 0.200$ Å and $U^2 = 0.080$ Å. The fit with eqn (6.41) resulted in values $P^0 = 0.914$, $P^1 = 0.243$, $P^2 = 0.123$, $U^1 = 0.199$ Å and $U^2 = 0.084$ Å (Fig. 6.7). An excellent agreement between data and fit is thus obtained.

The function $G(H;m)$ provides a less good description of the average diffracted intensities of real crystals (Lam *et al.*, 1993), because crystal symmetry and other features lead to violations of the assumption of randomness—in agreement with observations for periodic crystals.

6.5.3 *Probability distributions of the structure factor amplitudes*

Disregarding prior information about the atomic structures of periodic crystals, all points in the unit cell have the same probability to be an atomic site. For

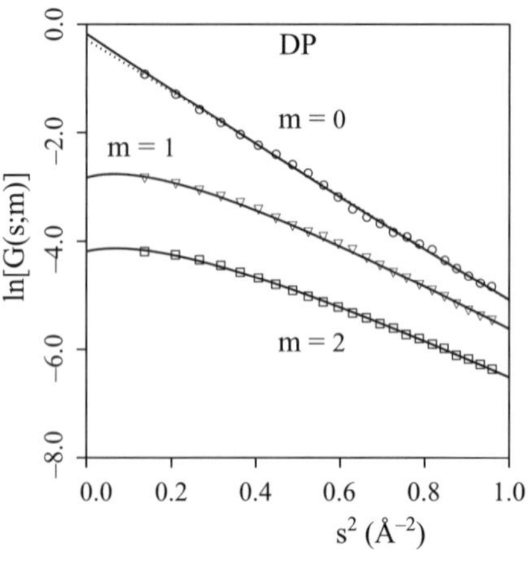

FIG. 6.7. Dependence of the average diffracted intensity on the length of the scattering vector for artificial structure DP. s is defined as $\frac{1}{2}H$. Lines represent a fit of eqn (6.41) to the diffraction data. Adapted from Lam *et al.* (1994), copyright (1994) by the IUCr.

a sufficiently large structure the position of one atom will be independent of the positions of the other atoms. The coordinates $\mathbf{x}^0(\mu)$ of the N atoms in the unit cell can then be considered as independent random variables. With these assumptions the probability distribution of the structure factor phases and amplitudes ($|F|$) has been found to be a function of (Giacovazzo, 1998)

$$\frac{|F|^2}{\mathcal{S}_2 \exp\left[-\frac{1}{2}BH^2\right]} \, . \tag{6.44}$$

The denominator is recognized as the average diffracted intensity of reflections with length of their scattering vectors equal to H [eqn (6.23)].

Modulated crystal structures are characterized by basic-structure coordinates, $\mathbf{x}^0(\mu)$, and modulation functions. The choice of independent random variables thus needs to be reconsidered (Lam *et al.*, 1993). A suitable set of random variables is provided by the basic-structure coordinates of the N atoms in the unit cell of the basic structure. However, values of the modulation functions depend on $\mathbf{x}^0(\mu)$ through their arguments $\bar{x}_{s4} = t + \mathbf{q}\cdot[\mathbf{L} + \mathbf{x}^0(\mu)]$ [eqn (2.20)], because modulation functions employ the basic-structure coordinates as phase reference point [compare eqn (1.10) and Fig. 6.8(a)]. Modulation functions and basic-structure coordinates can be decoupled, if the origin of the unit cell is used as alternative phase reference point, leading to modulation functions [Fig. 6.8(b)],

$$\mathbf{u}'^{\mu}[\bar{x}_{s4}] = \mathbf{u}^{\mu}\left[\bar{x}_{s4} - \mathbf{q}\cdot\mathbf{x}^0(\mu)\right] \, . \tag{6.45}$$

Computation of the structure factor based on modulation functions $\mathbf{u}'^{\mu}(\bar{x}_{s4})$, and subsequent substitution of $\mathbf{u}^{\mu}[\bar{x}_{s4}]$ for $\mathbf{u}'^{\mu}[\bar{x}_{s4}]$ gives [eqn (2.33)],

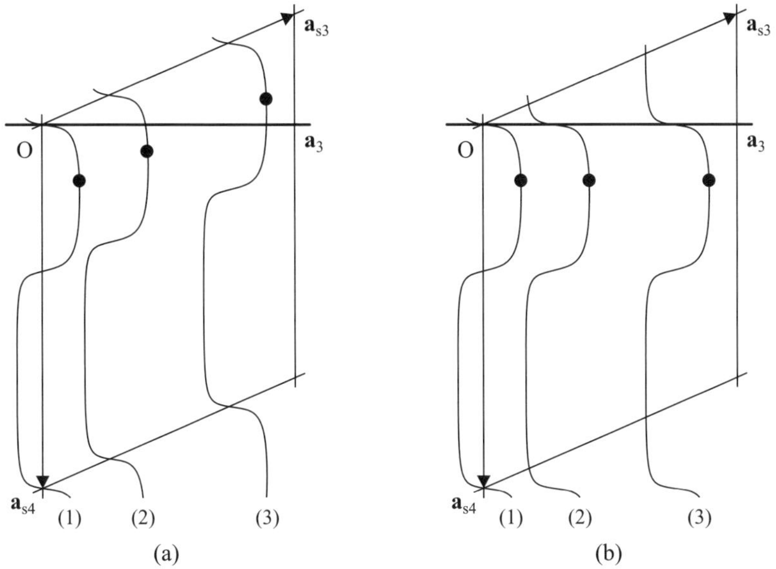

FIG. 6.8. An atom with modulation function $\mathbf{u}(\bar{x}_{s4})$ at three different positions $\mathbf{x}^0$ in the superspace unit cell. (a) $\mathbf{x}^0$ as phase reference point. (b) The origin as phase reference point [eqn (6.45)]. Dots indicate points of equal phases of the different modulation functions.

$$F(\mathbf{H}_s) = \sum_{\mu=1}^{N} g'_\mu(\mathbf{H}_s)\, f_\mu(H)\, \exp\!\left[-\tfrac{1}{4}B(\mu)H^2\right]\, \exp\!\left[2\pi i\,\mathbf{H}\cdot\mathbf{x}^0(\mu)\right], \qquad (6.46a)$$

with,

$$g'_\mu(\mathbf{H}_s) = \int_0^1 \exp\!\left[2\pi i\left(\mathbf{H}\cdot\mathbf{u}^\mu\left[\tau + \mathbf{q}\cdot\mathbf{x}^0(\mu)\right] + h_4\,\tau\right)\right] \mathrm{d}\tau\,. \qquad (6.46b)$$

The explicit dependence on $\mathbf{x}^0(\mu)$ is compensated by an implicit dependence of $\mathbf{u}^\mu[\bar{x}_{s4}]$ on $\mathbf{x}^0(\mu)$, such that the displacement of atom μ and concomitantly the alternate modulation scattering factor $g'_\mu(\mathbf{H}_s)$ do not depend on $\mathbf{x}^0(\mu)$ [compare Fig. 6.9 and eqn (6.45)].

The contribution of atom μ to the structure factor can be written as [eqn (6.46a)],

$$\psi_\mu + i\,\eta_\mu = |g_\mu(\mathbf{H}_s)|\, f_\mu(H)\, \exp\!\left[-\tfrac{1}{4}B(\mu)H^2\right]\, \exp\!\left[2\pi i\,\mathbf{H}\cdot\mathbf{x}^0(\mu) + i\Theta(\mu)\right], \qquad (6.47)$$

with $|g_\mu(\mathbf{H}_s)| = |g'_\mu(\mathbf{H}_s)|$ and $\Theta(\mu)$ is equal to the phase of $g'_\mu(\mathbf{H}_s)$. From the previous discussion it is clear that $\psi_\mu + i\,\eta_\mu$ depends on $\mathbf{x}^0(\mu)$ only through the

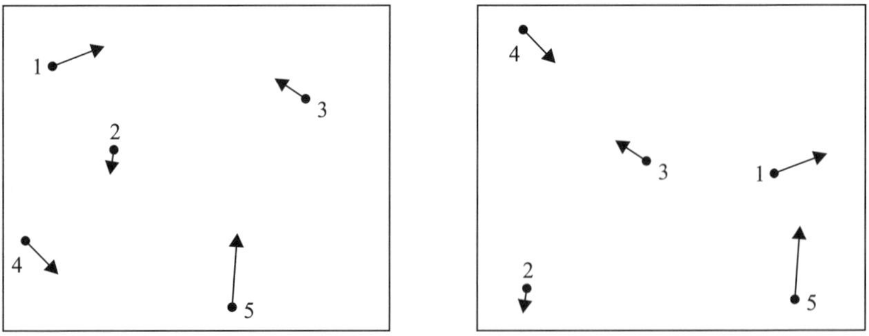

(a) (b)

FIG. 6.9. Basic-structure unit cell of a modulated crystal. (a) and (b) depict two possibilities of placing five atoms in the unit cell. Values of modulation functions (arrows) are independent of $\mathbf{x}^0(\mu)$ for each atom $\mu = 1, \cdots, 5$, as it is achieved through the alternate definition of modulation functions according to eqn (6.45).

factor $\exp\left[2\pi i\, \mathbf{H}\cdot\mathbf{x}^0(\mu)\right]$. Expectation values $\langle\cdots\rangle_{\mathbf{x}^0}$ with respect to the average over $\mathbf{x}^0(\mu)$ as independent random variables can thus be obtained by the same procedures as for periodic crystals. As the only difference, atomic scattering factors $f_\mu(H)$ are replaced by $|g_\mu(\mathbf{H}_s)|\, f_\mu(H)$ in all formulas. Following arguments by Giacovazzo (1998) then leads to the probability distribution, ${}_1P_{|F|}(|F|)\,\mathrm{d}\,|F|$, for the structure factor amplitude $|F|$ of an acentric structure to have values between $|F|$ and $|F| + \mathrm{d}|F|$ of,

$$
\begin{aligned}
{}_1P_{|F|}(|F|)\,\mathrm{d}|F| = & \frac{2|F|}{\pi \sum_{\mu=1}^{N} |g_\mu(\mathbf{H}_s)|^2\,[f_\mu(H)]^2 \exp\left[-\frac{1}{2}B(\mu)H^2\right]} \\
& \times \exp\left[-\frac{|F|^2}{\sum_{\mu=1}^{N} |g_\mu(\mathbf{H}_s)|^2\,[f_\mu(H)]^2 \exp\left[-\frac{1}{2}B(\mu)H^2\right]}\right] \mathrm{d}|F|\,. \qquad (6.48)
\end{aligned}
$$

Similar to periodic crystals, all values for the phase of the structure factor have equal probabilities.

The distribution ${}_1P_{|F|}(|F|)\,\mathrm{d}\,|F|$ differs from the acentric distribution of structure factor amplitudes for periodic crystals only through the dependence on the average diffracted intensities. For periodic crystals the latter do not depend on the structural parameters, but for modulated crystals they are found to depend on the modulation functions. Suitable independent random variables for modulations are the harmonic coefficients of the modulation functions, with Gaussian distributions about mean values that are equal to the overall modulation amplitudes U^n and P^n as defined in Section 6.5.2. Expectation values evaluate towards [eqns (6.37) and (6.41)],

$$\left\langle |F_{\text{obs}}(\mathbf{H}_s)|^2 \right\rangle_{H,m} = \left\langle \sum_{\mu=1}^{N} |g_\mu(\mathbf{H}_s)| \, [f_\mu(H)]^2 \, \exp\left[-\tfrac{1}{2}B(\mu)H^2\right] \right\rangle_{H,m}$$
$$= \mathcal{S}_2 \, G(\mathbf{H}; m). \tag{6.49}$$

As discussed in Section 6.5.2, the average intensity is a function of H and different averages need to be considered for each satellite order m. Accordingly, normalized structure factors $E(\mathbf{H}_s)$ for modulated crystals can be defined as,

$$E(\mathbf{H}_s) = \frac{F(\mathbf{H}_s)}{\sqrt{\left\langle |F_{\text{obs}}(\mathbf{H}_s)|^2 \right\rangle_{H,m}}}\,. \tag{6.50}$$

The acentric distribution of normalized structure factor amplitudes follows as,

$${}_1P_{|E|}(|E|)\, \mathrm{d}|E| = 2|E| \, \exp\left[-|E|^2\right] \, \mathrm{d}|E|\,. \tag{6.51}$$

The distribution of normalized structure factors is found to be independent of all structural parameters and of the filling of the unit cell of the basic structure. The functional form of ${}_1P_{|E|}(|E|)$ is identical to the acentric distribution function for periodic crystals (Giacovazzo, 1998), any differences between periodic and aperiodic crystals being incorporated into the definition of normalized structure factors. For the artificial structure A2, defined in Section 6.5.2, the distributions of normalized structure factors closely follow the theoretical expression (Fig. 6.10).

Deviations from the acentric distribution in eqn (6.51) will occur when correlations between structural parameters exist. These correlations may pertain to rigid-body modulations as well as to the presence of non-trivial superspace symmetries. In particular, the presence of an inversion centre determines that the phase of each structure factor is restricted to one out of two possible values, instead of being a continuous variable. Following the analysis for periodic crystals, the centric probability distribution for normalized structure factors is obtained as (Lam *et al.*, 1993; Giacovazzo, 1998),

$${}_{\bar{1}}P_{|E|}(|E|)\, \mathrm{d}|E| = \sqrt{\frac{2}{\pi}} \, \exp\left[-\frac{|E|^2}{2}\right] \, \mathrm{d}|E|\,. \tag{6.52}$$

Again, this distribution is closely followed by the diffracted intensities of an artificial structure with superspace symmetry $P\bar{1}(\sigma_1\,\sigma_2\,\sigma_3)\bar{1}$ (Fig. 6.10).

A superspace group based on non-trivial point symmetry leads to reflections with different multiplicities. Accordingly, a symmetry enhancement factor, $\varepsilon(\mathbf{H}_s)$, needs to be introduced, that has the same meaning as for periodic crystals. The modified definition of normalized structure factors is [eqn (6.50)],

$$E(\mathbf{H}_s) = \frac{F(\mathbf{H}_s)}{\sqrt{\varepsilon(\mathbf{H}_s)\, \mathcal{S}_2 \, G(\mathbf{H}; m)}}\,. \tag{6.53}$$

Even with $\varepsilon(\mathbf{H}_s)$ taken into account, the distributions ${}_1P_{|E|}(|E|)$ and ${}_{\bar{1}}P_{|E|}(|E|)$ do not provide exact descriptions of the diffracted intensities of real crystals,

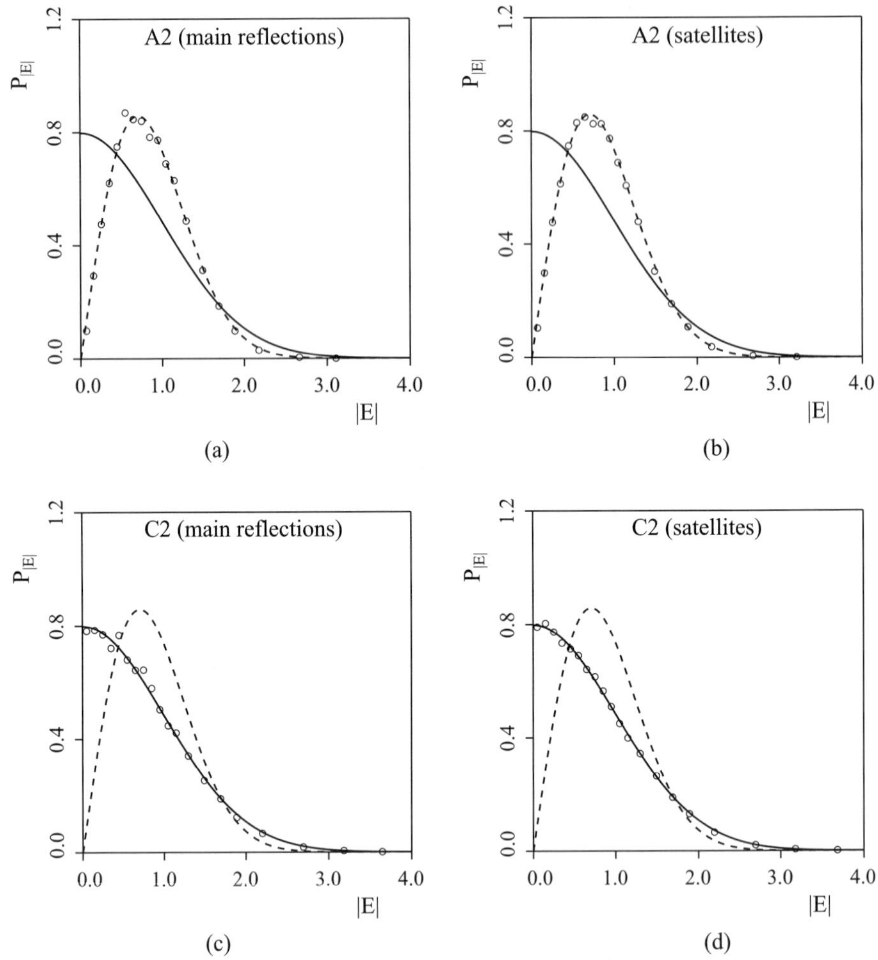

FIG. 6.10. Probability distributions $P_{|E|}(|E|)$ of the normalized structure factor amplitudes $|E|$. (a) Main reflections of the acentric artificial structure A2. (b) Satellite reflections. (c) Main reflections of the centrosymmetric artificial structure C2. (d) Satellite reflections. Solid and dashed lines represent the theoretical distribution functions for acentric and centrosymmetric structures, respectively [eqns (6.51) and (6.52)]. Adapted from Lam *et al.* (1993), copyright (1993) IUCr.

because the true distributions will depend on the precise superspace symmetry (Giacovazzo, 1998).

7

STRUCTURE REFINEMENTS

7.1 Principle of refinement

The goal of a structure refinement is to find those values for the parameters of the structure model that provide the best fit to the diffraction data. Usually, the least-squares criterion is employed. Those values of the parameters are searched for, that minimize the quantity χ^2, defined by

$$\chi^2 = \sum_{i=1}^{N_{\text{ref}}} w_i \left(|F_{\text{obs}}(\mathbf{H}_i)| - k_s\, k_E\, |F_{\text{cal}}(\mathbf{H}_i)|\right)^2 . \tag{7.1}$$

The sum is over all (N_{ref}) Bragg reflections for which structure factor amplitudes $|F_{\text{obs}}(\mathbf{H}_i)|$ have been measured (Section 6.1). $F_{\text{cal}}(\mathbf{H}_i)$ depends on the parameters of the structure model [eqn (2.34a)]. k_s is the scale factor and k_E is a factor accounting for secondary extinction. The weights w_i are related to the standard uncertainties of the measured structure factor amplitudes, according to

$$w_i = \frac{1}{[\sigma(\mathbf{H}_i)]^2} . \tag{7.2}$$

Alternative weights are used too, involving functions of $|F_{\text{obs}}|$ and $|F_{\text{cal}}|$.

The quality of the fit to the diffraction data is evaluated by consideration of the R index (residual)

$$R_F = \frac{\sum_{i=1}^{N_{\text{ref}}} \left||F_{\text{obs}}(\mathbf{H}_i)| - k_s\, k_E\, |F_{\text{cal}}(\mathbf{H}_i)|\right|}{\sum_{i=1}^{N_{\text{ref}}} |F_{\text{obs}}(\mathbf{H}_i)|} . \tag{7.3}$$

A good fit to excellent diffraction data may have R_F as low as 0.01, while the failure of the model to describe advanced structural features (like anharmonic temperature movements) or less accurate data (*e.g.* many weak reflections) may result in values of R_F as high as 0.1 for the best-fitting model. The weighted R index is

$$wR_{F^2} = \left(\frac{\sum_{i=1}^{N_{\text{ref}}} w_i \left(|F_{\text{obs}}(\mathbf{H}_i)| - k_s\, k_E\, |F_{\text{cal}}(\mathbf{H}_i)|\right)^2}{\sum_{i=1}^{N_{\text{ref}}} w_i\, |F_{\text{obs}}(\mathbf{H}_i)|^2} \right)^{\frac{1}{2}} . \tag{7.4}$$

wR_{F^2} is often used to monitor the progress of the refinement, because it is proportional to $\sqrt{\chi^2}$. However, values of wR_{F^2} depend on the weights w_i. Accordingly, it is not a strict measure of the misfit between $F_{\rm obs}(\mathbf{H}_i)$ and $F_{\rm cal}(\mathbf{H}_i)$, and the index R_F is considered to be a better measure of the quality of the fit to the data. The advantage of R indices over χ^2 is that R indices define the relative deviation between observed and calculated structure factors. Consequently, R_F will be independent of the number of reflections that is used to compute it, and a good fit to the data always corresponds to $R_F \lesssim 0.1$ (but see below).

Refinements of periodic and aperiodic structures follow from the same mathematical principles and they employ the same numerical procedures. For a discussion of these procedures I refer to the extensive literature on the crystallography of periodic crystals (Giacovazzo *et al.*, 2002).

Equations (7.1)–(7.4) pertain to all measured reflections, that include satellite reflections in the case of incommensurate crystals. Minimisation of χ^2 based on all reflections is the correct procedure. However, in the typical situation of satellite reflections that are much weaker than main reflections, low values of R_F may be attained by a good fit to the main reflections, while satellite reflections are only poorly described by the model. It is thus necessary to consider partial R indices, whereby the sums in eqn (7.3) and eqn (7.4) are restricted to either main reflections [$m = 0$; $R_F(0)$] or satellite reflections of order m [$R_F(m)$]. A reliable model of the modulation must have small values for the partial R indices of the satellite reflections, but it should be taken into account that measured intensities are less accurate for weak reflections than they are for strong reflections. Thus the expected values for R indices,

$$R_{\rm exp} = \frac{\sum_{i=1}^{N_{\rm ref}} \sigma(\mathbf{H}_i)}{\sum_{i=1}^{N_{\rm ref}} |F_{\rm obs}(\mathbf{H}_i)|}, \tag{7.5}$$

will be higher for partial R indices of groups of weak reflections than they are for partial R indices of groups of strong reflections. Accordingly, the correct model for the modulation may lead to values for $R_F(m)$ ($m \geqslant 1$) that will be in the range 0.05–0.15.

The same considerations apply to periodic structures that can be described as superstructures or commensurately modulated structures. In order to establish that the deviations from the basic structure have been correctly described, it is imperative to have low values for the partial R indices of the superlattice reflections.

7.2 Modulation parameters

Modulation functions need to be parameterized, if refinement techniques should be applied to incommensurate structures. Most common are truncated Fourier series for the functions describing displacive modulations [eqn (1.10)] or occupational modulations (Section 2.5.3). Each harmonic involves up to six parameters

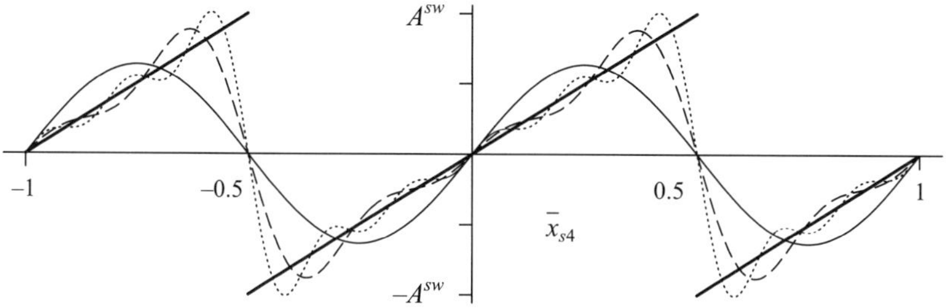

FIG. 7.1. Saw-tooth function of amplitude A^{sw} and centre $x^{sw} = 0$ (heavy line) and approximations by truncated Fourier series of a single harmonic (full line), three harmonics (dashed line) and five harmonics (dotted line) [eqns (7.6) and (7.7)].

of displacive modulation and up to two parameters of occupational modulation for each atom in the basic-structure unit cell. Superspace symmetry reduces the number of independent parameters (Section 3.6). Nevertheless, interdependencies are commonly observed between modulation parameters, thus leading to a failure of the refinement. Correlated parameters can only be avoided, if the maximum harmonic of modulation parameters is less or equal to the maximum order of satellite reflections for which significant non-zero intensity has been observed.

On the other hand, many compounds have been found for which modulation functions deviate from a simple harmonic shape, while higher-order satellite reflections are too weak to be measured. This problem is illustrated by the saw-tooth function. Each of the three components of the displacive modulation function is characterized by an amplitude A_i^{sw} and a centre $\bar{x}_i^{sw}$ according to (Fig. 7.1)

$$u_i^{sw}(\bar{x}_{s4}) = 2A_i^{sw}\,(\bar{x}_{s4} - x_i^{sw}) \qquad \text{for} \qquad -\tfrac{1}{2} < (\bar{x}_{s4} - x_i^{sw}) < \tfrac{1}{2}\,. \tag{7.6}$$

The Fourier series of the periodic saw-tooth function is

$$u_i^{sw}(\bar{x}_{s4}) = \frac{2A_i^{sw}}{\pi}\sum_{n=1}^{\infty}\frac{(-1)^{n+1}}{n}\,\sin[2\pi n\,(\bar{x}_{s4} - x_i^{sw})]\,. \tag{7.7}$$

Figure 7.1 shows that a single harmonic is but a poor approximation to the saw-tooth function, while both functions contain two independent parameters. Three harmonics provide a reasonable approximation for $|\bar{x}_{s4} - x_i^{sw}| \lesssim 0.3$, but five harmonics still give a poor description of the saw-tooth function for $0.4 \lesssim |\bar{x}_{s4} - x_i^{sw}| \lesssim 0.6$. To a first approximation, intensities of satellite reflections of order m are proportional to the square of the amplitude of the mth-order harmonic. For the saw-tooth function they amount to

$$I(h_1\, h_2\, h_3 \pm m) \propto \frac{1}{m^2}\,.$$

Intensities of satellite reflections of a structure with saw-tooth-shaped modulation functions will rapidly decrease with the satellite order of the reflections, such that significant diffracted intensity can be expected for a few satellite orders at most. Low-order harmonics hardly contribute to high-order satellite reflections, but the reverse is not true. Intensities of low-order satellite reflections contain information on higher harmonics in the modulation functions, although correlations between parameters prevent the higher-order harmonics being determined directly in a refinement procedure. Instead, refinement of the few parameters of the saw-tooth function may provide a better fit to the diffraction data than the fit provided by a truncated Fourier series. It can then be concluded that the true modulation functions are better approximated by saw-tooth functions than by a Fourier series with several harmonics. The high-temperature superconductor $Bi_2Sr_2CaCu_2O_{8+\delta}$ ($\delta \approx 0.1$) is an example of a compound with saw-tooth-shaped functions for the modulations of the oxygen atoms (Petricek *et al.*, 1990).

Experience has shown that an *ab initio* determination of the shapes of modulation functions cannot be achieved through refinements of the parameters in truncated Fourier series for these functions. Instead, special functions need to be considered that are highly anharmonic but contain a few refinable parameters only. Alternative approaches include the analysis of Fourier maps and electron densities obtained by the maximum entropy method (Chapter 8), as well as the use of constraints and restraints between parameters (see below).

Structure models for incommensurate crystals include three basic-structure positional parameters and six anisotropic temperature parameters for each independent atom in the unit cell of the basic structure. Modulation functions for temperature parameters can be introduced into the model as far as it is necessary to find a good fit to the data. Particularly interesting are the second-order harmonic coefficients, that might represent the major contribution to the modulation of temperature parameters, because they define the phason degree of freedom (Section 6.4). Usually, this is only possible if extensive data sets are available. Advanced structure models include anharmonic temperature parameters or multipole parameters for the atomic scattering factors and their modulations (Gourdon *et al.*, 1999; Palatinus *et al.*, 2004).

7.3 Constraints

Correlations in structure refinements can be diminished by a reduction of the number of parameters that are independently varied. Values of the dependent parameters are uniquely defined by the independent parameters through mathematical equations: the constraint relations or simply constraints. Computationally most easy to handle are linear constraints.

A special type of linear constraint is represented by the restrictions imposed by space group or superspace group symmetry (Section 3.6). For example, two atoms ($\mu = 1, 2$) might be equivalent by the superspace inversion centre $(-x_{s1}, -x_{s2}, -x_{s3}, -x_{s4})$. Symmetry restrictions on the basic-structure coordinates, temperature parameters and modulation parameters of atom 2 follow as [eqn (1.1), eqn (1.10) and eqn (6.12)]

$$\begin{aligned} x_i^0(2) &= -x_i^0(1); \quad U_{ij}(2) = U_{ij}(1) \\ A_i^n(2) &= A_i^n(1); \quad B_i^n(2) = -B_i^n(1) \end{aligned} \tag{7.8}$$

for $i, j = 1, 2, 3$ and for each harmonic n of modulation parameters. Alternatively, an atom may be located on a symmetry element in the basic-structure unit cell. Symmetry restrictions then follow as linear relations between the parameters of this atom, often involving zero values for certain parameters. For example, an atom μ on the twofold axis $(-x_{s1}, -x_{s2}, x_{s3}, x_{s4})$ adheres to the symmetry restrictions

$$\begin{aligned} &x_1^0(\mu) = 0; \quad x_2^0(\mu) = 0; \quad U_{13}(\mu) = 0; \quad U_{23}(\mu) = 0 \\ &A_1^0(\mu) = 0; \quad B_1^n(\mu) = 0; \quad A_2^n(\mu) = 0; \quad -B_2^n(\mu) = 0\,. \end{aligned} \tag{7.9}$$

An atom on the diagonal twofold axis $(x_{s2}, x_{s1}, -x_{s3}, -x_{s4})$ in a structure with tetragonal symmetry leads to the more complicated restrictions

$$\begin{aligned} &x_2^0(\mu) = x_1^0(\mu); \;\; x_3^0(\mu) = 0; \quad U_{22}(\mu) = U_{11}(\mu); \quad U_{23}(\mu) = -U_{13}(\mu) \\ &A_2^n(\mu) = -A_1^n(\mu); \qquad\qquad B_2^n(\mu) = -B_1^n(\mu); \quad B_3^n(\mu) = 0\,. \end{aligned} \tag{7.10}$$

It is easily checked that both kinds of twofold axes give rise to the same number of independent parameters for atoms located on these symmetry elements.

Linear constraints beyond symmetry restrictions are commonly applied to refinements of periodic structures. In particular, the re-distribution of electron density due to chemical bonding and anharmonicity of thermal vibrations—however small—are part of any crystal. Nevertheless, most refinements are restricted to structure models involving the positions of spherical atoms and their anisotropic, harmonic temperature parameters. In one point of view, these structure models can be considered to involve implicit constraints of the form

$$\text{parameter} = 0$$

for the multipole parameters and the anharmonic temperature parameters. Explicit constraints that are regularly employed in refinements of periodic structures include restrictions applied to temperature parameters. Correlations can often be removed from refinements, if anisotropic temperature parameters of different atoms are set equal to each other or if isotropic instead of anisotropic temperature parameters are used.

The need for constraints beyond symmetry restrictions is more profound in incommensurate crystals. Allowance for arbitrary shapes of modulation functions requires an infinite number of parameters, although only a few modulation parameters can be determined for each atom in the basic-structure unit cell (Section 7.2). The use of a few low-order harmonic coefficients for each modulation function corresponds to the implicit constraints

$$A_i^n(\mu) = 0; \quad B_i^n(\mu) = 0 \qquad \text{for} \qquad n > n_{\max}. \tag{7.11}$$

Alternatively, special functions like the saw-tooth function correspond to constraints between all harmonic coefficients as they are defined by the Fourier series of the special function [eqn (7.7)].

Satellite reflections are much weaker than main reflections for many incommensurate crystals. The limited scattering information that can be obtained on these satellite reflections then implies that modulation functions cannot be determined for temperature factors. Again, this represents an implicit constraint on the modulation functions.

A different type of constraint might be required, if the modulated structure is the result of a second-order phase transition on cooling of a periodic crystal. Structural distortions involve a single normal mode of the high-temperature space group. This wave can often be represented by a single harmonic displacement modulation function that obeys the symmetry of the superspace group of the modulated phase, while the basic structure might obey higher symmetry. For example, $NbSe_3$ is monoclinic $P2_1/m$ at room temperature (**b** unique). Below $T_c = 145$ K $NbSe_3$ develops an incommensurately modulated structure with superspace symmetry $P2_1/m(0\,\sigma_2\,0)0\bar{1}$, that is obeyed by both the basic-structure coordinates and the modulation functions. In other compounds, the phase transition will be accompanied by a reduction of point symmetry. Mo_2S_3 is monoclinic and incommensurately modulated with superspace group $P2_1/m(\frac{1}{2}, 0.441, 0)$ at high temperatures. Below $T_C = 390$ K a second, independent incommensurate modulation wave develops with $\mathbf{q}^2 = (-0.056, \frac{1}{2}, 0.229)$ at room temperature. The twofold screw axis is incompatible with the component $\frac{1}{2}$ along **b** of the second modulation wave vector, and the superspace symmetry of the modulated structure is triclinic $P\bar{1}(\sigma_1\,\sigma_2\,\sigma_3)\,(\sigma_1'\,\sigma_2'\,\sigma_3')(\bar{1},\bar{1})$ with the 'accidental' restrictions $\sigma_1 = \frac{1}{2}$, $\sigma_3 = 0$ and $\sigma_2' = \frac{1}{2}$. Diffraction symmetry is $\bar{1}$, but the main reflections and $\mathbf{q}^1$-type satellite reflections have been found to obey monoclinic symmetry in good approximation. Deviations from monoclinic symmetry are restricted to the second modulation wave. Consequently, symmetry restrictions of $P2_1/m(\frac{1}{2}, 0.441, 0)$ have been applied as non-symmetry constraints to the basic-structure coordinates, temperature parameters and modulation functions of the $\mathbf{q}^1$-type modulation wave, while $\mathbf{q}^2$-type modulation functions obey triclinic symmetry restrictions of the true symmetry only (Schutte *et al.*, 1993).

7.4 Restraints

Bond lengths are known to lie in narrow ranges, depending on the type of bond. For example the covalent C–C bond between sp^3 hybridized carbon atoms has a length of 1.54 ± 0.06 Å (Wilson, 1995). Constraints would make a bond length equal to a prescribed value, although it can have any value within the narrow range of values. This type of information can be incorporated into the refinement by the addition to χ^2 of a penalty that increases with increasing deviation from the optimal value of the parameter under consideration. Instead of χ^2 the quantity Q is minimized, as it is defined by

$$Q = \chi^2 + \sum_j g_j \left[d_j - d_j(\text{opt})\right]^2 . \tag{7.12}$$

The summation is over all restraints j. d_j is the value of the jth structural parameter as computed from the structure model, and $d_j(\text{opt})$ is the optimal (or standard) value for this parameter. g_j is the weight of the jth restraint.

Each pair of atoms in the basic-structure unit cell of a modulated crystal defines an infinite number of distances in dependence on the incommensurate parameter t (Section 2.4). The restraints of eqn (7.12) need to be replaced by (Yamamoto *et al.*, 1984)

$$\sum_j g_j \int_0^1 \left[d_j(t) - d_j(\text{opt})\right]^2 \mathrm{d}t \,, \tag{7.13}$$

where the integration extends over one period of the modulation. Other forms of restraints are possible too. For example, Yamamoto *et al.* (1984) employed the function

$$\sum_j g_j \int_0^1 \left[\frac{\Delta d_j(t)}{d_j(\text{max}) - d_j(\text{min})}\right]^2 \mathrm{d}t \,, \tag{7.14}$$

with

$$\Delta d_j(t) = \begin{cases} d_j(\text{min}) - d_j(t) & \text{for} \quad d_j(t) < d_j(\text{min}) \\ 0 & \text{for} \quad d_j(\text{min}) < d_j(t) < d_j(\text{max}) \\ d_j(t) - d_j(\text{max}) & \text{for} \quad d_j(t) > d_j(\text{max}) . \end{cases} \tag{7.15}$$

Restraints are often combined with constraints on the weights g_j. The latter can be set equal to each other, or g_j can be made proportional to $(1/\sigma_j)^2$, where σ_j is a measure for the spread of values around $d_j(\text{opt})$ of parameter d_j. Restraints can be imposed on any quantity that can be calculated from the structure model, including non-bonded interatomic distances, bond angles, torsion angles, planarity of groups and site occupancy factors.

7.5 Rigid bodies and non-crystallographic site symmetries

A special type of constraint is the rigid body. The idea originates in the study of molecular crystals. Covalent chemical bonds ensure that the structure of a molecule is independent of its environment to a first approximation. Crystal structure analysis then amounts to finding the orientations and positions of the molecules in the unit cell. For incommensurate crystals modulation functions will apply to these orientations and positions rather than to the individual atoms. Molecular structures are obtained from crystal structures of similar compounds or from quantum chemical calculations. Molecules, like benzene, can be rigid bodies, but more often the concept of rigid bodies is applied to parts of molecules, like phenyl groups.

The atomic coordinates of a rigid group of atoms with prescribed geometry are most usefully given with respect to a cartesian coordinate system with an origin that coincides with the origin of the crystal coordinates. The orientation of rigid body κ is defined by three parameters, $\mathbf{R}(\kappa) = [R_1(\kappa), R_2(\kappa), R_3(\kappa)]$, that can be identified with Euler angles describing three consecutive rotations about the coordinate axes. Alternatively, the orientation of a rigid body follows as the rotation about an axis $\hat{\mathbf{R}}$ over an angle Φ. Two parameters are required to specify a unit vector $\hat{\mathbf{R}}(\kappa)$ that defines the direction of the rotation axis. The position of rigid body κ is defined by a translation, $\mathbf{T}(\kappa) = [T_1(\kappa), T_2(\kappa), T_3(\kappa)]$, to be applied after the rigid body has been rotated into the desired orientation. The coordinates of all atoms $\mu = 1, \cdots, N_\kappa$ of rigid body κ follow as (Goldstein, 1980)

$$\mathbf{x}^0(\mu) = \mathbf{T}(\kappa) + \mathbf{x}^0_{rb}(\kappa;\mu)\cos(\Phi) + \hat{\mathbf{R}}(\kappa)[\hat{\mathbf{R}}(\kappa)\cdot\mathbf{x}^0_{rb}(\kappa;\mu)][1-\cos(\Phi)] + [\hat{\mathbf{R}}(\kappa)\times\mathbf{x}^0_{rb}(\kappa;\mu)]\sin(\Phi)\,. \quad (7.16)$$

Reference coordinates $\mathbf{x}^0_{rb}(\kappa;\mu)$ ($\mu = 1, \cdots, N_\kappa$) define rigid body κ and they are not varied in structure refinements. Coordinates $\mathbf{x}^0(\mu)$ of atom μ of rigid body κ depend on six refinable parameters, $\mathbf{T}(\kappa)$, $\hat{\mathbf{R}}(\kappa)$ and Φ, independent of the number of atoms in the rigid body.

Rigid-body constraints are linear in the three translational parameters, but they are non-linear constraints for the rotational parameters. The latter property might pose a problem for the evaluation of modulation functions of rigid bodies. Therefore, a linearised form of eqn (7.16) has been used in computer programs for structure refinements:

$$\mathbf{x}^0(\mu) = \mathbf{T}(\kappa) + [\mathbf{R}(\kappa)\times\mathbf{x}^0_{rb}(\kappa;\mu)] \quad (7.17a)$$

with

$$\mathbf{R}(\kappa) = \hat{\mathbf{R}}(\kappa)\sin(\Phi) = [R_1(\kappa), R_2(\kappa), R_3(\kappa)]\,. \quad (7.17b)$$

This form is exact in the limit of infinitesimal rotations and it is a good approximation for rotations up to a few degrees in magnitude.

Modulations of molecular crystals principally affect entire rigid bodies, without 'modulating' their internal structure. Accordingly, modulation functions are desired that apply to rigid bodies rather than to individual atoms. To this end coordinates, $\mathbf{x}^0(rb\,\kappa)$, are introduced for the centre or reference point of rigid body κ. Because rotations have been defined about the origin of the coordinate system, the reference point is equal to the translation,

$$\mathbf{x}^0(rb\,\kappa) = \mathbf{T}(\kappa)\,. \quad (7.18)$$

Modulation functions of rigid body κ are a function of [eqn (2.20)]

$$\bar{x}_{s4}(rb\,\kappa) = t + \mathbf{q}\cdot\left[\mathbf{L} + \mathbf{x}^0(rb\,\kappa)\right] . \tag{7.19}$$

Displacive modulations are described by modulation functions for the translations and rotations. With $\mathbf{u}_T^{\kappa}[\bar{x}_{s4}(rb\,\kappa)]$ a vector function describing modulations of the translation and with $\mathbf{u}_R^{\kappa}[\bar{x}_{s4}(rb\,\kappa)]$ an axial vector describing modulations of the rotation, the displacement modulation of atom μ of rigid body κ is (Petricek *et al.*, 1985)

$$\mathbf{u}^{\mu}[\bar{x}_{s4}(\mu)] = \mathbf{u}_T^{\kappa}[\bar{x}_{s4}(rb\,\kappa)] + \left[\mathbf{u}_R^{\kappa}[\bar{x}_{s4}(rb\,\kappa)] \times \mathbf{x}_{rb}^0(\kappa;\mu)\right] . \tag{7.20}$$

Because modulation functions of atom μ are a function of

$$\bar{x}_{s4}(\mu) = t + \mathbf{q}\cdot\left[\mathbf{L} + \mathbf{x}^0(\mu)\right] ,$$

equation (7.19) does not lead to the equality of Fourier coefficients of $\mathbf{u}^{\mu}[\bar{x}_{s4}(\mu)]$ and $\mathbf{u}_T^{\kappa}[\bar{x}_{s4}(rb\,\kappa)]$ even in the case that $\mathbf{u}_R^{\kappa}[\bar{x}_{s4}(rb\,\kappa)]$ were zero. The appropriate phase shift

$$\bar{x}_{s4}(\mu) - \bar{x}_{s4}(rb\,\kappa) = \mathbf{q}\cdot\left[\mathbf{x}^0(\mu) - \mathbf{x}^0(rb\,\kappa)\right] ,$$

needs to be taken into account. Similar considerations apply to occupational modulations, that are described by modulation functions $p^{rb\,\kappa}[\bar{x}_{s4}(rb\,\kappa)]$.

Thermal motion does not affect the internal structure of a rigid body to a first approximation. Atomic vibrations are given in terms of time-dependent translations (δT) and rotations (δR) of the rigid body. Evaluation of the mean-square displacements results in a temperature tensor with 20 independent components, that are arranged in contributions from pure translations (U_{T}), pure librations (U_{L}) and the combination of translations and librations (U_{S}), with [eqn (6.12)] (Schomaker and Trueblood, 1968)

$$U^{\mu} = U_{\mathrm{T}} + A_{\mathrm{TLS}}(\mu)\, U_{\mathrm{L}}\, A_{\mathrm{TLS}}(\mu)^t + A_{\mathrm{TLS}}(\mu)\, {U_{\mathrm{S}}}^t + U_{\mathrm{S}} A_{\mathrm{TLS}}(\mu)^t. \tag{7.21a}$$

The superscript t indicates the transpose and the 3×3 matrix $A_{\mathrm{TLS}}(\mu)$ is

$$A_{\mathrm{TLS}}(\mu) = \begin{pmatrix} 0 & \mathbf{x}_{rb,3}^0(\kappa;\mu) & -\mathbf{x}_{rb,2}^0(\kappa;\mu) \\ -\mathbf{x}_{rb,3}^0(\kappa;\mu) & 0 & \mathbf{x}_{rb,1}^0(\kappa;\mu) \\ \mathbf{x}_{rb,2}^0(\kappa;\mu) & -\mathbf{x}_{rb,1}^0(\kappa;\mu) & 0 \end{pmatrix} . \tag{7.21b}$$

Equation (7.21a) shows the Translation-Libration-Screw formalism (TLS formalism) to be a particular kind of linear constraints on the temperature parameters U^{μ}. The TLS parameters can be employed as independent parameters instead of the temperature parameters of individual atoms. Modulated temperature parameters can be taken into account through modulations of the independent quantities within the TLS formalism.

The concept of rigid bodies can be generalized towards groups of atoms for which the internal structure is also varied in the refinements. Initially the same number of parameters is required for a description of the structure in terms of coordinates of individual atoms located somewhere in the unit cell and for a description in terms of a group of atoms centred on the origin of the coordinate system together with a translation $\mathbf{T}(\kappa)$ and a rotation $\mathbf{R}(\kappa)$. However, the description of the structure with the aid of non-rigid groups offers a variety of ways to reduce the number of independent parameters:

1. Several crystallographically independent copies of a molecular fragment can be described by independent rotations and translations for these copies, while a single internal structure of the fragment is optimized in the refinements.
2. All crystallographically independent atomic coordinates are refined, but temperature parameters are restricted through the TLS formalism, or modulation functions are described by modulations of the fragment rather than by modulations of individual atoms [eqn (7.20)].
3. A non-crystallographic site symmetry can be introduced, *e.g.* $6/mmm$ point symmetry for the benzene molecule, while the space group is of lower symmetry. This site symmetry can then be applied to all parameters, including basic-structure coordinates, temperature parameters and modulation functions (Dusek *et al.*, 2001).

7.6 Occupational modulations

Occupational modulations occur in two variants. In the first kind of compound, a single site of the basic-structure unit cell is alternatively occupied by atoms of different chemical elements. The chemical composition is defined by the average occupational probabilities $P^0(\mu)$ of atom types $\mu = 1, \cdots, N_{\text{site}}$. Full site occupancies correspond to the constraints

$$\sum_{\mu=1}^{N_{\text{site}}} P^0(\mu) = 1 \qquad \text{and} \qquad P^0(\mu) \geqslant 0\,. \tag{7.22}$$

Site occupancies less than one can be accommodated within this formalism by allowing one of the 'chemical types' to be a vacancy. Modulations of the occupational probabilities are described by modulation functions [eqn (2.37)]

$$p^{\mu}(\bar{x}_{s4}) = P^0(\mu)\left[1 + \sum_{n=1}^{\infty} P^{s\,n}(\mu)\sin(2\pi n\bar{x}_{s4}) + P^{c\,n}(\mu)\cos(2\pi n\bar{x}_{s4})\right]. \tag{7.23}$$

They define partial chemical order, except in the limiting case of block-wave functions with values of 0 and 1, when complete order is achieved. Full site occupancies are required for all values of t in $\bar{x}_{s4} = t + \mathbf{q}\cdot\mathbf{x}^0(\mu)$, leading to the condition [eqn (7.22)]

$$\sum_{\mu=1}^{N_{\text{site}}} p^{\mu}[t + \mathbf{q}\cdot\mathbf{x}^0(\mu)] = 1 \tag{7.24}$$

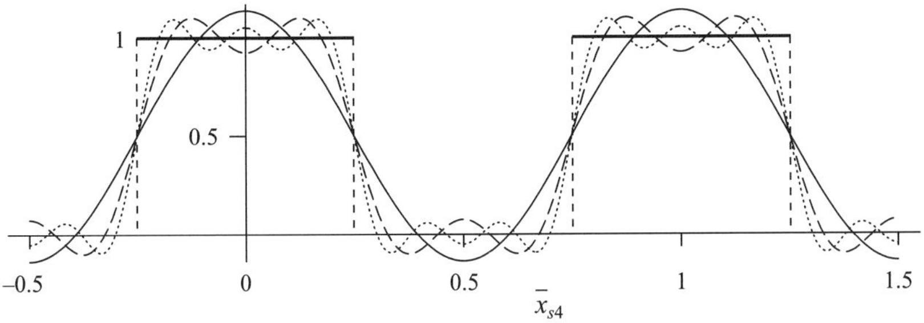

FIG. 7.2. Block-wave function of amplitude $\frac{1}{2}$, average value $\frac{1}{2}$ and centre $x^{bw} = 0$ (heavy line) and approximations by truncated Fourier series of a single harmonic (full line), two harmonics (dashed line) and three harmonics (dotted line) [eqn (7.26)].

for all values of t. In the special case of two chemical types, modulation functions are found to be complementary with $p^2(\bar{x}_{s4}) = 1 - p^1(\bar{x}_{s4})$. Site occupancies must be larger than 0 and smaller than 1, suggesting for each site μ a condition

$$p^{\mu}[t + \mathbf{q}\cdot\mathbf{x}^0(\mu)] \geqslant 0\,. \tag{7.25}$$

while the combination of eqn (7.24) and eqn (7.25) ensures occupational probabilities to be less than one. However, a block wave with average value $\frac{1}{2}$ and amplitude $\frac{1}{2}$ possesses a first-order harmonic coefficient that is larger than $\frac{1}{2}$,

$$p^{bw}(\bar{x}_{s4}) = \frac{1}{2} + \frac{2}{\pi}\sum_{n=1}^{\infty}\frac{(-1)^{n-1}}{2n-1}\cos\left[2\pi(2n-1)\left(\bar{x}_{s4} - x^{bw}\right)\right]. \tag{7.26}$$

A structure model restricted to a first-order harmonic modulation then involves site occupancies larger than 1 and smaller than 0 for certain values of t (Fig. 7.2). If refinements lead to such values, they provide strong evidence for the true modulation function being a block wave. Too narrow restraints would lead to an erroneous result for the modulation functions.

The second kind of occupational modulations pertains to structures where one atom can occupy several sites in the basic-structure unit cell, while for stereochemical reasons only one of these sites can be occupied in each unit cell. The same mathematical formalism can be used to describe these structures as it is given above for the first kind of compounds. However, the interpretation of the occupational probabilities is different. Average occupation probabilities $P^0(\mu)$ and modulation functions $p^{\mu}(\bar{x}_{s4})$ now refer to the occupational probabilities of different sites by a single atomic species. The chemical composition is given by the sum of $P^0(\mu)$ over the mutually exclusive sites μ.

Within a unified mathematical description sites μ are defined with independent structural parameters for position, occupation by a single chemical element, temperature parameters and their modulation functions. The various kinds of

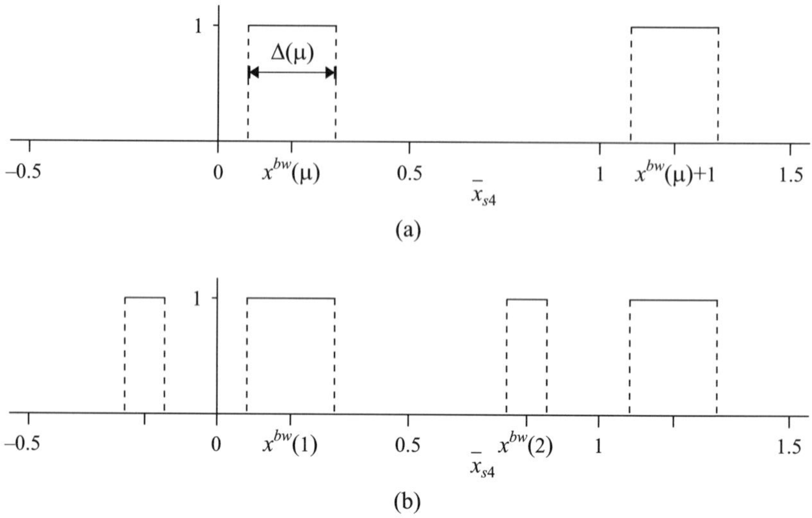

FIG. 7.3. (a) Block wave of width $\Delta(\mu)$ centred on $x^{bw}(\mu)$. (b) Complex block wave with two regions of function value one.

modulations are then described by different types of constraints between the parameters of related sites.

Occupational modulations can be used to describe complex superstructures that are based on simple but disordered basic structures (Perez-Mato *et al.*, 1999). Complete chemical order is described by occupational modulation functions that assume but two values: 0 and 1. The most general form of a block wave is defined by value 1 for an interval $\Delta(\mu)$ centred on $x^{bw}(\mu)$ on the $\bar{x}_{s4}$ axis and value 0 for the complementary interval within one period of $\bar{x}_{s4}$,

$$p^{\mu}(\bar{x}_{s4}) = \begin{cases} 1 & \text{for} \quad |\bar{x}_{s4} - x^{bw}(\mu)| < \Delta(\mu) \\ 0 & \text{for} \quad |\bar{x}_{s4} - x^{bw}(\mu) - \frac{1}{2}| < 1 - \Delta(\mu)\,, \end{cases} \tag{7.27}$$

with periodic continuation [Fig. 7.3(a)]. The average occupancy of site μ is equal to $\Delta(\mu)$. More complicated modulation functions can be defined as combinations of several regions of value 1 within one period on the $\bar{x}_{s4}$ axis. Alternatively, each region of value 1 can be considered as an independent site with its own width and centre parameters [$\mu = 1, 2$ in Fig. 7.3(b)].

Occupational modulations are invariably accompanied by displacive modulations. Special attention is required for the description of modulation waves of the sites that are affected by a block-wave-type occupational modulation. First it is noticed that the success of truncated Fourier series in refinements of modulation functions originates in the property that a Fourier series is a complete series of orthogonal functions. This property implies that the values of Fourier coefficients are independent of the values of all other Fourier coefficients of a given modulation function, thus minimizing correlations between parameters. In a general approach, modulation functions $f(\bar{x}_{s4})$ can be expanded in any complete set of orthogonal periodic functions, $g_n(\bar{x}_{s4})$,

$$f(\bar{x}_{s4}) = A^0 + \sum_{n=1}^{\infty} A^n \, g_n(\bar{x}_{s4}) \,. \tag{7.28}$$

Orthogonality is defined by the scalar product of two functions, that is the integral over one period of the product of two of these periodic functions,

$$(g_{n_1}, g_{n_2}) = \int_0^1 g_{n_1}(\bar{x}_{s4}) \, g_{n_2}(\bar{x}_{s4}) \, \mathrm{d}\bar{x}_{s4} = \delta_{n_1,n_2} \,, \tag{7.29}$$

where δ_{n_1,n_2} is the Kronecker delta with value 1 for $n_1 = n_2$ and with value 0 for $n_1 \neq n_2$. Sine and cosine functions are orthogonal according to

$$\begin{aligned}
&\int_0^1 \sin(2\pi n_1 \bar{x}_{s4}) \, \cos(2\pi n_2 \bar{x}_{s4}) \, \mathrm{d}\bar{x}_{s4} = 0 \,, \\
&\int_0^1 \sin(2\pi n_1 \bar{x}_{s4}) \, \sin(2\pi n_2 \bar{x}_{s4}) \, \mathrm{d}\bar{x}_{s4} = \delta_{n_1,n_2} \,, \\
&\int_0^1 \cos(2\pi n_1 \bar{x}_{s4}) \, \cos(2\pi n_2 \bar{x}_{s4}) \, \mathrm{d}\bar{x}_{s4} = \delta_{n_1,n_2} \,.
\end{aligned} \tag{7.30}$$

In the presence of an occupational modulation wave $p^\mu(\bar{x}_{s4})$, an alternative definition applies to the orthogonality of displacive modulation functions,

$$\int_0^1 g_{n_1}(\bar{x}_{s4}) \, g_{n_2}(\bar{x}_{s4}) \, p^\mu(\bar{x}_{s4}) \, \mathrm{d}\bar{x}_{s4} = \delta_{n_1,n_2} \,. \tag{7.31}$$

In case of a block-wave occupational modulation function, the scalar product of eqn (7.31) reduces to [eqn (7.27)]

$$\int_{x^{bw}(\mu)-\frac{1}{2}\Delta(\mu)}^{x^{bw}(\mu)+\frac{1}{2}\Delta(\mu)} g_{n_1}(\bar{x}_{s4}) \, g_{n_2}(\bar{x}_{s4}) \, \mathrm{d}\bar{x}_{s4} = \delta_{n_1,n_2} \,. \tag{7.32}$$

It is easily checked that sine and cosine functions do not represent a set of orthogonal functions according to this definition. As a consequence high correlations occur between modulation parameters, if the harmonic coefficients $\mathbf{A}^n(\mu)$ and $\mathbf{B}^n(\mu)$ are employed as independent parameters in refinements [eqn (1.10)]. The origin of this problem is visualized for the displacive modulation of an atom with a block-wave occupational modulation function (eqn (7.27) and Fig. 7.4). The displacement modulation is only defined for those values of $\bar{x}_{s4}$ for which $p^\mu(\bar{x}_{s4}) = 1$. Values of modulation functions outside this interval are irrelevant and they may assume any value without affecting the structure model. For example, the modulation in Fig. 7.4 (thick line) may be reasonably well approximated

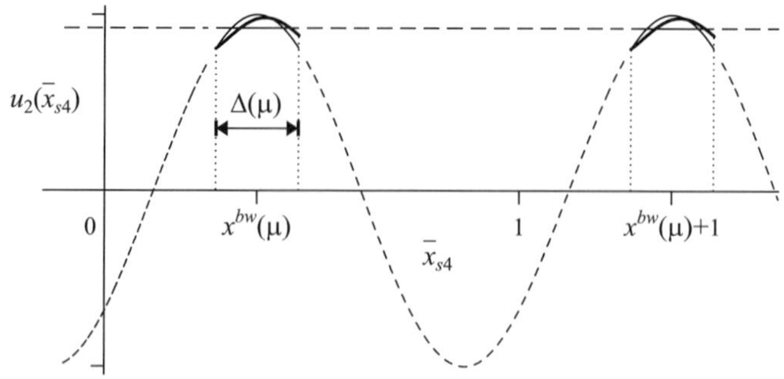

FIG. 7.4. Displacive modulation for a site μ with a block-wave occupational modulation of width $\Delta(\mu)$ centred on $x^{bw}(\mu)$ (compare Fig. 7.3). Shown are the displacive modulation function (thick line) and approximations by a constant function and by a single harmonic function. Dashed lines represent parts of the modulation functions that are irrelevant for the crystal structure.

by a constant function, *i.e.* the basic-structure coordinate will adapt itself to a value that is the average of the positions along the modulation curve if further modulation functions are not employed. However, the true modulation is better described by a single-harmonic function with a rather large amplitude, in order to correctly describe the curvature of the modulation function. Concomitantly, the basic-structure position is shifted to zero, thus representing a high correlation between first harmonic modulation coefficient and basic-structure position of this atom. As is shown in Fig. 7.4, the true modulation is reasonably well approximated by either the basic-structure position or a single-harmonic modulation function, but both functions possess grossly different values in the structurally irrelevant interval of $\bar{x}_{s4}$ (dashed portions of the curves).

The problem of correlated parameters can be resolved by the introduction of orthogonal basis functions that replace the harmonic functions (Petricek *et al.*, 1995). A Gram–Schmidt orthonormalization procedure is used to derive a set of orthonormal functions $g_n(\bar{x}_{s4})$ from the harmonic functions $\sin(2\pi n\bar{x}_{s4})$ and $\cos(2\pi n\bar{x}_{s4})$, employing the modified criterion of orthogonality of eqn (7.32) and starting with the constant function. In this way, the orthonormal basis functions $g_{2n-1}(\bar{x}_{s4})$ and $g_{2n}(\bar{x}_{s4})$ become linear combinations of the harmonic functions up to order n. For example, a crenel function—as the block-wave function has been named by Petricek *et al.* (1995)—with $\bar{x}^{bw} = 0$ and $\Delta = 0.5$ defines the following orthonormal functions for displacive modulations up to order 2:

$$\begin{aligned} g_0(\bar{x}_{s4}) &= 1.000 \\ g_1(\bar{x}_{s4}) &= 0.000 + 1.414\sin(2\pi\bar{x}_{s4}) \\ g_2(\bar{x}_{s4}) &= -2.069 + 3.249\cos(2\pi\bar{x}_{s4}) \\ g_3(\bar{x}_{s4}) &= 0.000 - 2.271\sin(2\pi\bar{x}_{s4}) + 2.675\sin(4\pi\bar{x}_{s4}) \\ g_4(\bar{x}_{s4}) &= 9.102 - 14.30\cos(2\pi\bar{x}_{s4}) + 6.381\cos(4\pi\bar{x}_{s4}) \,. \end{aligned} \tag{7.33}$$

It is now apparent that the introduction of a large amplitude for $\cos(2\pi\bar{x}_{s4})$ is accompanied by a corresponding shift of the constant term that is intrinsic to the function $g_2(\bar{x}_{s4})$. The block-wave function is symmetric around $\bar{x}_{s4} = 0$, so that odd functions (sine functions) do not mix with even functions (cosine functions and the constant function). The basic-structure position for the modulation in Fig. 7.4 will correspond to the average position of the atom, independent of the amplitudes of the modulation functions $g_1(\bar{x}_{s4})$ and $g_2(\bar{x}_{s4})$.

Maybe surprising, the use of orthonormal basis functions removes correlations between modulation parameters in most cases, although exactly the same number of independent parameters is used as it is contained in an expansion in sine and cosine functions. Correlations may remain for occupational modulations with small values of $\Delta(\mu)$, *i.e.* $\Delta(\mu) \ll 0.5$. These correlations can only be removed if certain basis function $g_n(\bar{x}_{s4})$ are discarded. To this end, the function $g_{n+1}(\bar{x}_{s4})$ is written as the linear combination of the functions $\{g_0(\bar{x}_{s4}), \cdots, g_n(\bar{x}_{s4})\}$ [together defining the parallel part, $g^{\parallel}_{n+1}(\bar{x}_{s4})$, of $g_{n+1}(\bar{x}_{s4})$] and a remainder $g^{\perp}_{n+1}(\bar{x}_{s4})$,

$$g_{n+1}(\bar{x}_{s4}) = g^{\parallel}_{n+1}(\bar{x}_{s4}) + g^{\perp}_{n+1}(\bar{x}_{s4}) \,. \tag{7.34}$$

Within the Gram–Schmidt orthonormalization procedure the function $g_{n+1}(\bar{x}_{s4})$ is added to the already selected basis functions $\{g_0(\bar{x}_{s4}), \cdots, g_n(\bar{x}_{s4})\}$ if the perpendicular component $g^{\perp}_{n+1}(\bar{x}_{s4})$ is sufficiently large. Petricek *et al.* (1995) have introduced the criterion that the cosine of the angle between the functions $g_{n+1}(\bar{x}_{s4})$ and $g^{\parallel}_{n+1}(\bar{x}_{s4})$,

$$\xi_{n+1} = \sqrt{\frac{\left(g^{\parallel}_{n+1}, g^{\parallel}_{n+1}\right)}{\left(g_{n+1}, g_{n+1}\right)}} \,,$$

must be smaller than some prescribed value λ. A value $\lambda = 1$ implies that all functions $g_{n+1}(\bar{x}_{s4})$ will be accepted into the set of orthonormal basis functions, but a value $\xi_{n+1} = 1$ implies that $g^{\perp}_{n+1}(\bar{x}_{s4}) = 0$ and that $g^{\parallel}_{n+1}(\bar{x}_{s4})$ is completely dependent on $\{g_0(\bar{x}_{s4}), \cdots, g_n(\bar{x}_{s4})\}$. Correlations between modulation parameters can be avoided by smaller values of λ, for example $\lambda = 0.95$.

7.7 Refinements against powder diffraction data

7.7.1 *Full-profile analysis for periodic crystals*

X-ray powder diffraction produces a profile of scattered intensity as a function of the scattering angle 2θ. The 2θ axis is discretized in most experiments and the profile is obtained as a table of scattered intensities y_i at scattering angles $2\theta_i$ for $i = 1, \cdots, N_{\text{pix}}$. Typical step widths are between 0.001° and 0.02°. A Bragg reflection $\mathbf{H}_s$ occurs at a scattering angle $2\theta(\mathbf{H}_s)$, as it is uniquely determined by Bragg's law. In principle this would allow to extract structure factor amplitudes from measured diffraction profiles. However, two properties of the diffraction profiles prevent the extraction of complete data sets in most cases.

Any Bragg reflection contributes diffracted intensity at a range of values 2θ about the ideal Bragg angle $2\theta(\mathbf{H}_s)$. For perfect microcrystalline material the peak profile is determined by instrumental parameters, including deviations from pure monochromaticity and divergence of the radiation and the finite size of the sample. The finite widths of Bragg reflections prevent structure factor amplitudes being determined, if Bragg angles $2\theta(\mathbf{H}_s)$ of the reflections differ from each other by an amount less than approximately the Full Width at Half Maximum (FWHM) of the peak profiles. This situation is found at high diffraction angles, because the density of Bragg reflections on the 2θ axis is proportional to the third power of the length of the diffraction vector, *i.e.* to $[2\sin(\theta)/\lambda]^3$. 'Accidental' overlap of Bragg reflections may be present at low scattering angles, depending on the values of the lattice parameters of the sample.

Exact overlap of Bragg reflections is present if two inequivalent Bragg reflections have identical Bragg angles. This situation is found for high-symmetry compounds, mostly compounds that belong to Laue classes $4/m$, $m\bar{3}$, $6/m$ or trigonal $\bar{3}m$.

Methods have been developed to extract structure factor amplitudes from powder diffraction profiles. Most widely used are the LeBail and Pawley methods (David *et al.*, 2002). All methods rely on an accurate description of the peak profiles, employing peak-profile functions, $\phi[2\theta - 2\theta(\mathbf{H}_s)]$, that depend on a few parameters only. The peak-profile parameters are characteristic for the instrument and the sample. They apply to all Bragg reflections, but they may be smooth functions of, for example, the scattering angle 2θ. A calculated profile is constructed of intensities y_{ci}, with

$$y_{ci} = y_{bi} + \sum_{\mathbf{H}_s} |F(\mathbf{H}_s)|^2 \, L(\mathbf{H}_s) \, \phi[2\theta_i - 2\theta(\mathbf{H}_s)]. \tag{7.35}$$

$L(\mathbf{H}_s)$ includes geometric corrections factors, like Lorentz and polarization factors, the scale factor, absorption correction and the multiplicity of reflection $\mathbf{H}_s$; y_{bi} is the background intensity at $2\theta_i$. The sum extends over all Bragg reflections for which $\phi[2\theta_i - 2\theta(\mathbf{H}_s)]$ is significantly different from zero at $2\theta_i$. This may include Bragg reflections with Bragg angles $2\theta(\mathbf{H}_s)$ within a range of about 1° around $2\theta_i$. In a full-profile analysis calculated intensities are matched with measured intensities according to the least-squares criterion

$$\chi^2 = \sum_{i=1}^{N_{\text{pix}}} w_i \, (y_i - y_{ci})^2 . \tag{7.36}$$

The weights can be defined as $w_i = 1/y_i$.

Refinable parameters in the LeBail or Pawley methods are profile parameters, lattice parameters, a few parameters describing the background as a smooth function of 2θ, and observed structure factor amplitudes $|F_{\mathrm{obs}}(\mathbf{H}_s)|^2$ for all Bragg reflections within the experimental range of 2θ. Values of $|F_{\mathrm{obs}}(\mathbf{H}_s)|^2$ of overlapping reflections are correlated to each other. The LeBail method resolves this problem by an iterative procedure in which the intensity of each point i is split into contributions of participating reflections. In this way values of $|F_{\mathrm{obs}}(\mathbf{H}_s)|$ can be extracted from the diffraction profiles, that may be used in methods of structure solution as well as for the evaluation of the Bragg R index (R_{Bragg}), defined in the same way as the index R_F according to [eqn (7.3)]

$$R_{\mathrm{Bragg}} = \frac{\sum\limits_{\mathbf{H}_s} ||F_{\mathrm{obs}}(\mathbf{H}_s)| - k_s\, k_E\, |F_{\mathrm{cal}}(\mathbf{H}_s)||}{\sum\limits_{\mathbf{H}_s} |F_{\mathrm{obs}}(\mathbf{H}_s)|} . \tag{7.37}$$

The sum extends over all extracted reflections.

The full-profile analysis provides accurate values for the lattice parameters and the profile parameters. Obviously, $|F_{\mathrm{obs}}(\mathbf{H}_s)|$ cannot be accurately determined for overlapping reflections. In the case of exact overlap any method cannot do better than to evenly split the observed intensity amongst the contributing reflections. As a consequence extracted structure factor amplitudes are not suitable for a classical structure refinement (Section 7.1). Instead a least-squares refinement is performed of observed and calculated profiles [eqn (7.36)].

In the Rietveld refinement (Young, 1995), calculated intensities, y_{ci}, are computed with calculated structure factors from the structure model [eqn (2.34a)], *i.e.* $|F(\mathbf{H}_s)| = |F_{\mathrm{cal}}(\mathbf{H}_s)|$ in eqn (7.35). The refinement determines structural parameters as well as profile and lattice parameters, but the latter may be kept fixed at values obtained from a preceding LeBail fit. The Rietveld refinement is considered to employ the maximum amount of information that is contained in powder diffraction data (David *et al.*, 2002). Nevertheless, this amount of information is smaller than it is contained in a complete set of observed structure factor amplitudes. Standard uncertainties and correlations between parameters are larger than in the case of refinements against single-crystal diffraction data. The use of constraints and restraints beyond those imposed by symmetry is unavoidable in Rietveld refinements.

7.7.2 *Incommensurate crystals*

Diffraction profiles of microcrystalline powders have equal appearances for periodic and aperiodic crystals. They comprise diffraction maxima at positions and with intensities that are characteristic for each compound. For incommensurately modulated crystals, diffraction profiles include contributions from main and satellite reflections (Fig. 7.5). Straightforward generalizations have been proposed of full-profile methods towards their applications to powder diffraction by incommensurate crystals (Yamamoto *et al.*, 1990; Dusek *et al.*, 2001): both main

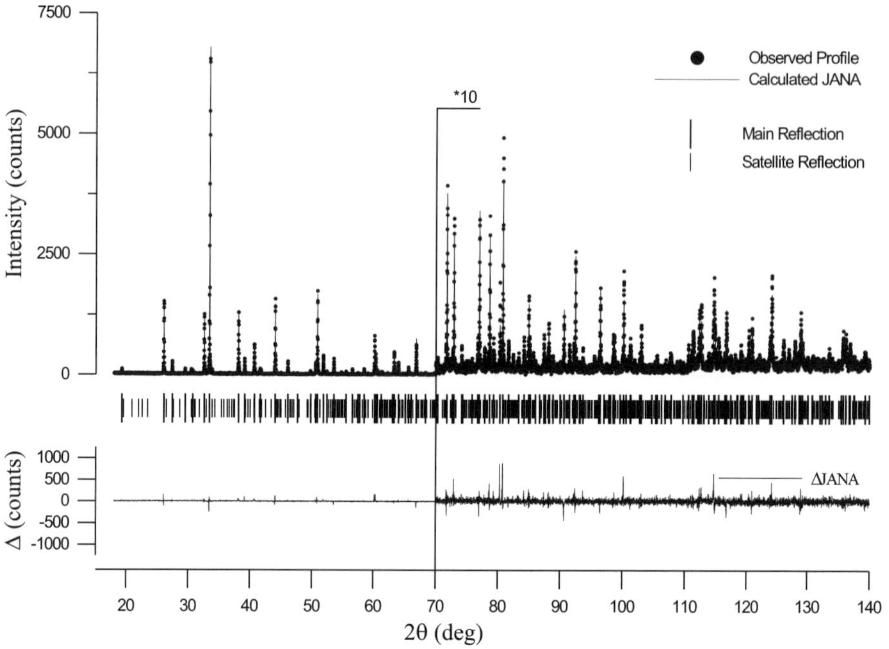

FIG. 7.5. Rietveld plot of incommensurately modulated $NbTe_4$. The top pane shows the experimental values y_i (dots), and the calculated profile (thin line). Reflection markers indicate the values $2\theta(\mathbf{H}_s)$ of main reflections (long bars) and satellite reflections up to order $m = 2$ (short bars). The lower pane displays the difference $\Delta = y_i - y_{ci}$. [eqn (7.35)]. Reprinted from Dusek *et al.* (2001), copyright (2001) by the IUCr.

and satellite reflections must be included in the construction of the calculated diffraction profile y_{ci} [eqn (7.35)].

Positions $2\theta(\mathbf{H}_s)$ of main reflections are determined by the reflection indices $(h_1\, h_2\, h_3\, 0)$ and the lattice parameters of the basic-structure unit cell. Matching observed and calculated reflection positions in a LeBail fit or Rietveld refinement leads to accurate values for the lattice parameters. Positions of satellite reflections $(h_1\, h_2\, h_3\, h_4)$ with $m = |h_4| \neq 0$, depend on the lattice parameters as well as the modulation wave vector through the length of the scattering vector in physical space (Section 2.2),

$$\mathbf{H} = (h_1 + \sigma_1\, h_4)\, \mathbf{a}_1^* + (h_2 + \sigma_2\, h_4)\mathbf{a}_2^* + (h_3 + \sigma_3\, h_4)\mathbf{a}_3^* \,. \tag{7.38}$$

Matching of observed and calculated positions of satellite reflections requires the correct values for the components of the modulation wave vector (compare Fig. 7.6). Successful LeBail fits and Rietveld refinements will thus produce accurate values for both lattice parameters and modulation wave vectors.

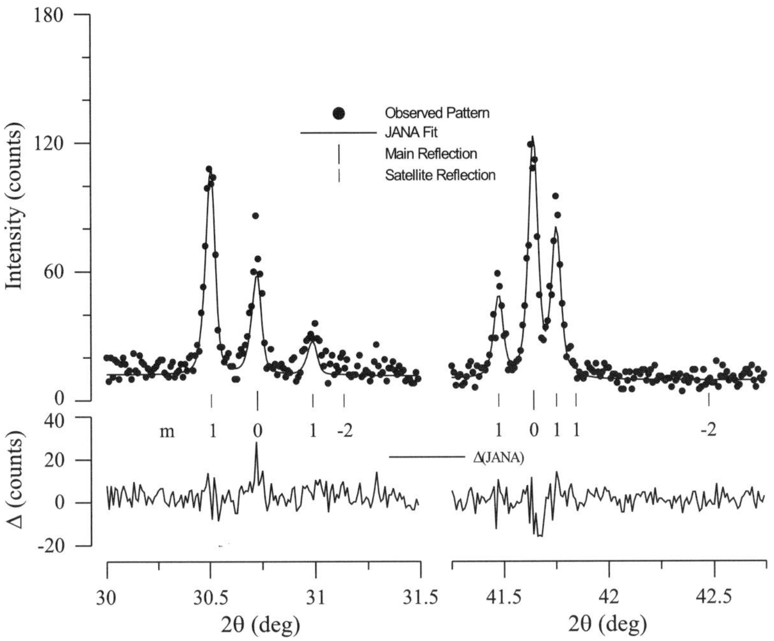

FIG. 7.6. Enlarged view of two regions of the Rietveld plot of incommensurately modulated $NbTe_4$. Reprinted from Dusek *et al.* (2001), copyright (2001) by the IUCr.

A complicating factor is that satellite reflections densely fill any interval on the 2θ axis. The observation of sharp diffraction maxima implies that reflection intensities are zero for satellite reflections beyond some order m (Section 1.5). Full-profile analyses thus require that the satellite order $m = |h_4|$ be restricted to some small value, usually 1 or 2. Composite crystals give rise to main reflections belonging to two, mutually incommensurate reciprocal lattices, and $|h_4|$ cannot be restricted to small values. In the simple example of $\mathbf{a}_{11}^* = \mathbf{a}_1^*$ and $\mathbf{a}_{21}^* = \mathbf{a}_4^*$, the satellite order m is defined as

$$m = \text{Minimum}\left(|h_1|, |h_4|\right).$$

m must be restricted to small values, in order to have but a finite number of reflections within the experimental range of 2θ values.

Diffraction profiles of modulated crystals are more complicated than those of periodic crystals, despite restrictions on the satellite order of reflections. Compounds with one-dimensionally modulated structures produce $(2m+1)$ as many reflections as there are produced by periodic crystals with unit cells equal to the basic-structure unit cells of the modulated compounds. On the average, diffraction by modulated crystals has the complexity of the diffraction by a $(2m+1)$-fold superstructure of the basic-structure unit cell. Structural analysis of periodic crystals is limited by overlap of reflections that increases with increasing size of the unit cell. Accordingly, this will limit the applicability or accuracy of Rietveld refinements of incommensurate crystal structures. The problem of overlap is aggravated if satellite reflections are not evenly spread between main reflections. For example, consider a compound with a modulation with $\mathbf{q} = (0, 0, 0.1173)$. Separations of reflections will reflect separations of reflections of a periodic compound with a ninefold supercell along $\mathbf{a}_3$, even if only first-order satellite reflections appear in the diffraction pattern.

The information content of powder diffraction profiles is much less than the information content of complete data sets of single-crystal diffraction data up to the same maximum diffraction angle. One reason is the high level of 'background' intensity at any value of 2θ in powder diffraction. The ratio between maximum and minimum intensity will not be better than ~ 100 in powder diffraction, while this ratio can be higher than 10^6 in a good single-crystal diffraction experiment. Weak satellite reflections will be drowned in the background, and powder diffraction is not suitable to study compounds with small modulation amplitudes, like the vast array of charge-density-wave systems (van Smaalen, 2005).

Rietveld refinements of incommensurate crystal structures proceed in the same way as Rietveld refinements of periodic crystals. Non-zero modulation functions are responsible for non-zero intensities of satellite reflections, and refinable parameters include both basic-structure and modulation parameters. Usually, the class of weak reflections contains many more satellite reflections than main reflections. Constraints and restraints are thus even more important for incommensurate structures than they already are for periodic crystals.

8

ELECTRON DENSITY IN SUPERSPACE

8.1 Fourier maps

Structure factors of Bragg reflections are the only non-zero Fourier coefficients of electron densities of incommensurate crystal structures. Neglecting anomalous scattering, the generalized electron density can thus be defined as the inverse Fourier transform in superspace of the structure factors [eqn (2.12)],

$$\rho_s(\mathbf{x}_s) = \frac{1}{V} \sum_{\mathbf{H}_s} F(\mathbf{H}_s) \exp[-2\pi i \, \mathbf{H}_s \cdot \mathbf{x}_s] \, . \tag{8.1}$$

Fourier maps are obtained if observed structure-factor amplitudes are combined with structure-factor phases,

$$F(\mathbf{H}_s) = |F_{\mathrm{obs}}(\mathbf{H}_s)| \exp[i \, \varphi(\mathbf{H}_s)] \, . \tag{8.2}$$

Depending on the application, reflection phases are phases of calculated structure factors of some structure model or they can be the result of direct methods or experimental methods of phase determination.

Fourier maps in superspace can be used to visualize modulation functions. Each atom in the basic structure of an incommensurately modulated crystal appears as an undulating string of high density in the Fourier maps. Atomic strings are parallel to $\bar{x}_{s4}$ on the average, and they represent the atomic strings defined in Chapter 2 (compare Fig. 2.3).

Several features of Fourier maps will be illustrated by the incommensurate composite crystal $[LaS]_{1.14}[NbS_2]$, employing diffraction data and the structure model by Jobst and van Smaalen (2002). The LaS and NbS_2 subsystems share the $(\mathbf{a}_{s2}^*, \mathbf{a}_{s3}^*)$ reciprocal lattice plane, and $\mathbf{a}_1 = \mathbf{a}_{11} = \sigma_1 \mathbf{a}_{21}$ defines the mutually incommensurate direction, as introduced in Fig. 1.4 and Section 4.7. Each subsystem has an incommensurately modulated structure. Accordingly, atomic strings are found to be parallel to $\bar{x}_{s4}$ for subsystem 1 and parallel to $\bar{x}_{s1}$ for subsystem 2 (compare Fig. 4.6). For LaS as first subsystem, the two-dimensional section of the (3+1)-dimensional Fourier map parallel to (x_{s1}, x_{s4}) and centred on La shows the atomic string corresponding to La (Fig. 8.1), in complete accordance with the definition of superspace atoms in Chapter 2. The location of the maximum density as a function of the physical-space section t = constant traces the x_{s1} component of the atomic position, and it thus defines the component u_1 of the modulation function of La. Indeed, the position derived from the refined structure model (dashed line in Fig. 8.1) follows quite well the trace of

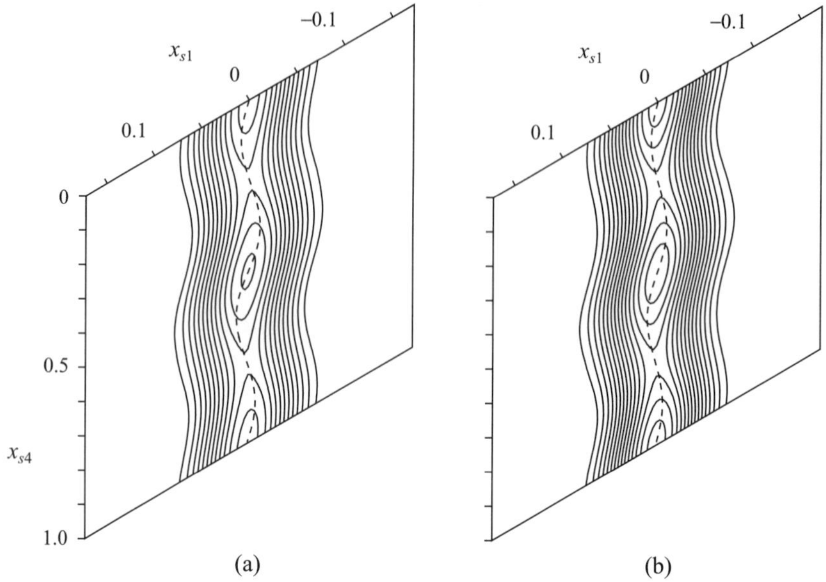

FIG. 8.1. (x_{s1}, x_{s4}) Sections of superspace Fourier maps of $[LaS]_{1.14}[NbS_2]$ centred on $\mathbf{x}^0(\text{La}) = (0,\, 0,\, 0.327)$. (a) Phases from a structure model with displacive modulation functions; $\rho^{\max} = 316.9$ electrons/Å^3. (b) Phases from a structure model with displacive modulation functions and modulated temperature parameters; $\rho^{\max} = 314.1$ electrons/Å^3. Contours of constant electron density at intervals of 24 electrons/Å^3. The dashed line represents the position of La. Coordinates from Jobst and van Smaalen (2002).

maximum density in the Fourier map. At this stage, gross differences between structure model and Fourier map would become visible. They can be used to develop an improved structure model. This method has been employed by Petricek *et al.* (1990) to show that the modulation functions of oxygen atoms in $Bi_2Sr_2CaCu_2O_{8.1}$ are better described by saw-tooth-shaped functions than by harmonic functions.

A section t = constant of the superspace Fourier map provides the Fourier map in physical space. The latter is not periodic, but it can be calculated by Fourier transform in physical space without reference to superspace, according to [eqn (2.13)],

$$\rho(\mathbf{x}) = \frac{1}{V} \sum_{\mathbf{G}} \sum_{h_4=-\infty}^{\infty} F(\mathbf{G},\, h_4)\, \exp[-2\pi i\, (\mathbf{G} + h_4\mathbf{q})\cdot\mathbf{x}]\,, \tag{8.3}$$

where $\mathbf{G}$ is a reciprocal lattice vector of the basic structure and $\mathbf{H} = \mathbf{G} + h_4\mathbf{q}$ is the scattering vector in physical space of Bragg reflections $(h_1\, h_2\, h_3\, h_4)$ [eqn (1.37)]. The modulation of an atom can be analysed by studying physical space

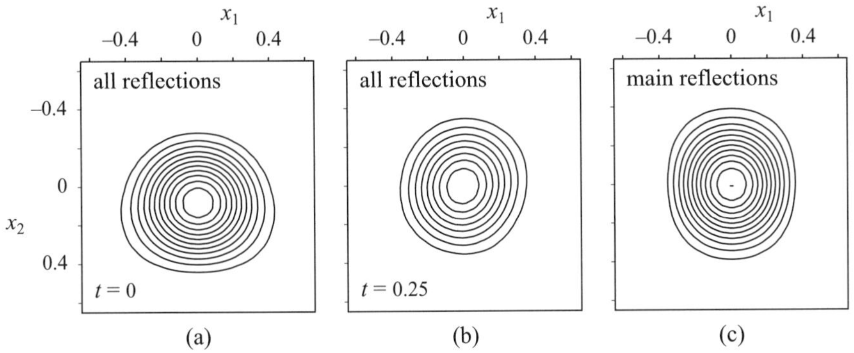

FIG. 8.2. (x_1, x_2) Physical-space sections of superspace Fourier maps of $[LaS]_{1.14}[NbS_2]$ centred on $\mathbf{x}^0(\mathrm{La}) = (0, 0, 0.327)$. (a) Section $t = 0$ and (b) section $t = 0.25$ of the map with all reflections; $\rho^{\max} = 325.0$ electrons/Å^3. (c) Section $t = 0$ of the map with main reflections of the LaS subsystem; $\rho^{\max} = 245.9$ electrons/Å^3. Phases from the best structure model. Contours of constant electron density at intervals of $\rho^{\max}/12$; coordinates in Å.

sections as a function of t in the neighbourhood of the basic-structure position of that atom. Figure 8.2 shows two such sections for La at $\mathbf{x}^0 = (0, 0, 0.327)$. The location of the maximum varies with t, representing the components $u_1(\bar{x}_{s4})$ and $u_2(\bar{x}_{s4})$ of the displacement modulation function of La. The distribution of density around the maximum depends on t too. This variation can be related to modulations of the concentration of about 5 % of vacancies on the La site (occupational modulation), and modulations of the temperature factors (Section 8.2).

Partial Fourier maps, employing subsets of reflections in the summation of eqn (8.1), define projections of the superspace density. Of particular interest is the partial Fourier map of main reflections. The resulting superspace density depends on the coordinates x_{s1}, x_{s2} and x_{s3} only, and all sections t = constant are identical. They represent the periodic average structure in physical space. Local maxima are centred on basic-structure positions of the atoms, but they are smeared according to variations defined by the modulation functions [Fig. 8.2(c)]. However, for most compounds modulation amplitudes are considerably smaller than atomic sizes and thermal vibration amplitudes, so that smearing is not obviously visible for an already spread-out density.

Fourier maps can be computed of many quantities. The checking Fourier map is obtained with $F(\mathbf{H}_s) = F_{\mathrm{cal}}(\mathbf{H}_s)$ in eqn (8.1), while the summation extends over those reflections for which an observed structure factor is available. Any differences between this Fourier map and the structure model are the result of series termination effects. The magnitude of these discrepancies sets a lower limit on differences between the corresponding Fourier map and the structure model, that can be interpreted as structural features missing in the model.

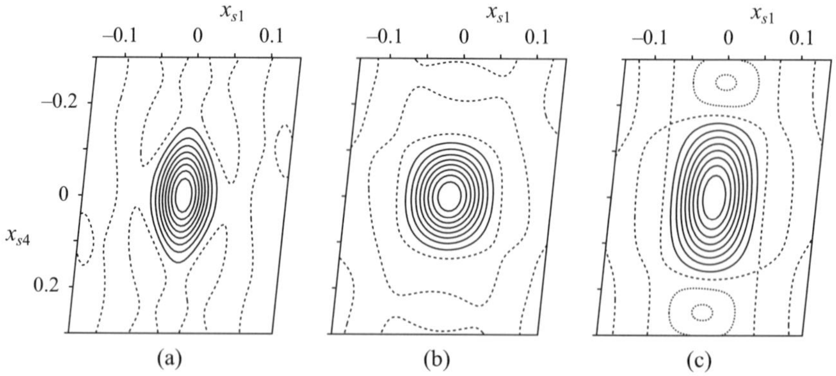

FIG. 8.3. (x_{s1}, x_{s4}) Sections of inverse Fourier transforms of shape functions for Na_2CO_3. (a) Complete data set. (b) Data restricted to $[\sin(\theta)/\lambda]_{\max} = 0.66$ Å^{-1}. (c) Data restricted to $[\sin(\theta)/\lambda]_{\max} = 0.66$ Å^{-1} and satellite index $m \leqslant 2$. Negative and zero densities are represented by dotted and dashed lines, respectively. Diffraction data from Dusek *et al.* (2003).

The inverse Fourier transform of the shape function—defined by $F(\mathbf{H}_s) = 1$ in eqn (8.1)—contains one narrow maximum at the origin of superspace, if reflections $\mathbf{H}_s$ are incorporated into the summation up to high values of $|h_i|$ for all reflection indices. For smaller data sets, the inverse Fourier transform of the shape function demonstrates the resolving power of these data. For example, the inverse Fourier transform of the shape function of Na_2CO_3 is characterized by a well localized maximum at the origin of the coordinate system for the extensive data set by Dusek *et al.* (2003). Restricting the data to $[\sin(\theta)/\lambda]_{\max} = 0.66$ Å^{-1} leads to a considerable broadening along x_{s1}, x_{s2} and x_{s3} [Fig. 8.3(b)]. Further limiting the data to satellite order $m \leqslant 2$ results in a broadening along x_{s4}, indicating considerably less information on the modulation in these limited data as compared to the complete data set up to $m = 4$ [Fig. 8.3(c)].

Other Fourier maps include inverse Fourier transforms of normalized structure factors and of integrated intensities of Bragg reflections (Chapter 10), and the difference Fourier map discussed in the next section.

8.2 Difference Fourier maps

The successful interpretation of Fourier maps relies on complete data sets. Any missing reflection $\mathbf{H}_s$ leads to noise in the Fourier map of a magnitude that is related to the value of the corresponding structure factor, $|F(\mathbf{H}_s)|$. Fourier maps become useless, if a few strong, low-order reflections are not available—a situation that is typical for many experiments. The Fourier transform of any electron density contains an infinite number of structure factors up to arbitrary large scattering vectors. Perfect Fourier maps require structure factors up to at least $[\sin(\theta)/\lambda]_{\max} = 5.0$ Å^{-1} (de Vries *et al.*, 1996). Accordingly Fourier maps suffer from series termination effects for any reasonable experiment. This noise is sufficiently large to conceal hydrogen atoms near heavy atoms, to obliterate any effects of chemical bonding on the electron density and to cover details of modulations. Specific to incommensurate crystals are series termination effects due to a limited availability of high-order satellite reflections: data sets often contain satellite reflections up to orders $m = 1$ or $m = 2$ only.

One way out of these problems is the difference Fourier map:

$$\begin{aligned}\Delta\rho_s(\mathbf{x}_s) &= \frac{1}{V}\sum_{\mathbf{H}_s} \Delta F(\mathbf{H}_s)\, \exp[-2\pi i\, \mathbf{H}_s \cdot \mathbf{x}_s] \\ \Delta F(\mathbf{H}_s) &= (|F_{\mathrm{obs}}(\mathbf{H}_s)| - |F_{\mathrm{cal}}(\mathbf{H}_s)|)\exp[\, i\, \varphi(\mathbf{H}_s)]\,,\end{aligned} \tag{8.4}$$

where reflection phases correspond to the phases of $F_{\mathrm{cal}}(\mathbf{H}_s)$. Difference Fourier maps suffer from series termination effects too, but the magnitude of the noise is related to $|\Delta F(\mathbf{H}_s)|$ rather than $|F(\mathbf{H}_s)|$ of the missing reflections. Difference Fourier maps are thus suitable to uncover deficiencies in structure models, that would not be apparent in the corresponding Fourier map. For example, for a structure model of $[LaS]_{1.14}[NbS_2]$ with displacive modulation only, the difference Fourier map exhibits clear features at the position of La [Fig. 8.4(a)]. An extended structure model gives a featureless difference Fourier map of which $|\Delta\rho^{\mathrm{max}}|$ is more than 10 times smaller than that of the original map of $[LaS]_{1.14}[NbS_2]$ [Fig. 8.4(b)]. Difference Fourier maps have thus been used to show that the modulation of La involves an occupational modulation of the 5% vacancies on the La site as well as a modulation of the temperature parameters by second-harmonic coefficients (Jobst and van Smaalen, 2002). The presence of the modulated temperature parameters has been related to variations of the atomic environments in this composite crystal, rather than to the presence of a 'sliding mode' (Section 6.4).

Difference Fourier maps for other incommensurate compounds have revealed anharmonic temperature parameters or modulated anharmonic temperature parameters (Gourdon *et al.*, 1999; Palatinus *et al.*, 2004). Especially the latter feature seems to be typical for modulated crystals. Variations of environments of atoms in dependence on the phase of the modulation wave will induce asymmetrical thermal vibrations with a skewness towards opposite directions for difference values of $\bar{x}_{s4}$. This feature can be modelled by modulations of anharmonic thermal parameters of odd orders (particularly third order), while their basic-structure values are zero.

8.3 Maximum entropy method in superspace

The maximum entropy method (MEM) is a versatile method in science, that can be used to extract the maximum amount of information from a given set of data (Buck and Macaulay, 1991). The starting point of any application of the MEM is an 'image' of strictly positive values. The goal is to find the most probable image amongst all possible images that are compatible with the data.

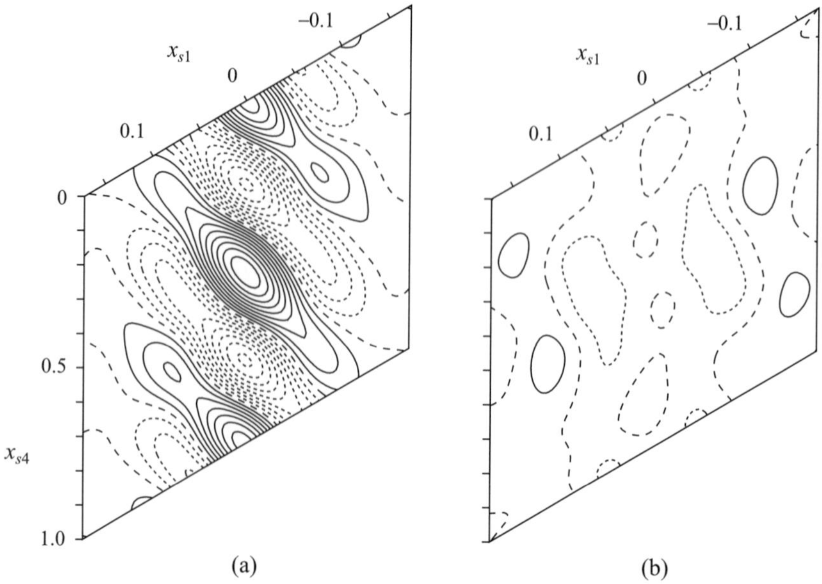

FIG. 8.4. (x_{s1}, x_{s4}) Sections of difference Fourier maps in superspace of $[LaS]_{1.14}[NbS_2]$. (a) Map corresponding to Fig. 8.1(a); $\rho^{min}/\rho^{\max} = -22.1/26.4$ electrons/Å^3; contours of constant density at intervals of 2.7 electrons/Å^3. (b) Map corresponding to Fig. 8.1(b); $\rho^{min}/\rho^{\max} = -2.4/2.0$ electrons/Å^3; contours at intervals of 1.35 electrons/Å^3. Negative densities are represented by dotted lines and zero density is represented by dashed lines. Coordinates from Jobst and van Smaalen (2002).

The principle of Maximum Entropy states that this most-probable 'image' is the 'image' that maximizes the entropy subject to constraints provided by the data.

The 'image' can be almost any object that is described by a strictly positive function of one or more coordinates. Examples include the signal obtained in Nuclear Magnetic Resonance (NMR) experiments, a celestial map of the intensity of the signal in radio astronomy and a photograph taken with visible light. Accordingly, coordinates are the frequency of the radiation, two angles defining the direction of observation and the position on the photograph. Image values are the strength of the signal or the intensity of the radiation at each particular set of values of the coordinates. Measured images are always noisy and they may be blurred, *e.g.* through the response function of the instrument like the instrumental contribution to the peak width in powder diffraction data. In one approach the MEM is used to obtain the most probable perfect or hidden image from the measured noisy and blurred image. The MEM thus becomes a method of resolution enhancement that is an alternative to other deconvolution and Fourier-filtering techniques.

In an alternative approach the data are indirectly related to the 'image'.

This situation is typical for crystallographic applications of the MEM, where the 'image' is the electron density and the data are provided by the integrated intensities of Bragg reflections. Applications of this kind include the use of the MEM in direct methods of phase determination and the determination of accurate electron densities from phased diffraction data (Gilmore, 1996; Gilmore *et al.*, 1999; Takata *et al.*, 1999; de Vries *et al.*, 1996). The latter method has been extended towards aperiodic crystals and will be discussed in detail below.

Entropy in the MEM refers to the informational entropy due to Shannon. Although the entropy is defined for continuous functions, numerical evaluation is facilitated by the use of electron densities defined on grids over the unit cells:

$$\rho_k = \rho(\mathbf{x}_k) \qquad \text{for} \qquad k = 1, \cdots, N_{\text{pix}}, \tag{8.5}$$

where ρ_k is the value of the electron density at point $\mathbf{x}_k$ in the unit cell. The grid has N_i divisions along unit cell axis $\mathbf{a}_i$, resulting in $N_1 \times N_2 \times N_3 = N_{\text{pix}}$ pixels for a periodic crystal. Useful pixel sizes are between $(0.05\,\text{Å})^3$ and $(0.15\,\text{Å})^3$. As discussed by Jaynes (2003) and others, the entropy of any trial density $\rho(\mathbf{x})$ is defined relative to a reference density $\tau(\mathbf{x})$ as

$$S = -\sum_{k=1}^{N_{\text{pix}}} \rho_k \ln(\rho_k/\tau_k) \tag{8.6}$$

with $\tau_k = \tau(\mathbf{x}_k)$. The reference density or PRIOR incorporates any information about the electron density that is known without reference to the data. In the most non-committal case it can be taken as a constant equal to the average electron density: $\tau_k = \langle\rho\rangle$. The principle of maximum entropy states that the most probable electron density is given by the values $\{\rho_k\}$ that maximize the entropy S. A straightforward computation shows that the maximum of S is obtained for $\rho_k = \tau_k$ for all N_{pix} values of k. This trivial solution demonstrates that in the absence of data the best choice for the electron density $\rho(\mathbf{x})$ is the PRIOR $\tau(\mathbf{x})$.

The problem of interest is to find the electron density that maximizes the entropy subject to constraints provided by the data. Employing the method of undetermined Lagrange multipliers, this amounts to finding the maximum of

$$Q = S - \lambda_0 C_0 - \lambda_1 C_1 \tag{8.7}$$

with respect to $\{\rho_k\}$ and the undetermined Lagrange multipliers λ_0 and λ_1. The maximum of Q with respect to λ_l $(l = 0, 1, \cdots)$ is defined by $\partial Q/\partial\lambda_l = 0$, and it results in the conditions

$$C_l = 0 \qquad \text{for} \qquad l = 0, 1, \cdots. \tag{8.8}$$

Accordingly, constraints are functions C_l of the electron density $\{\rho_k\}$ that need to fulfil eqn (8.8). The first constraint is normalization of the electron density:

$$C_0 = \frac{V}{N_{\text{pix}}} \sum_{k=1}^{N_{\text{pix}}} \rho_k - N_{\text{el}} = 0\,, \tag{8.9}$$

where N_{el} is the number of electrons in the unit cell and V is its volume. Diffraction data can be incorporated into the problem by the so-called F-constraint,

$$C_1 = C_{F2} = -1 + \frac{1}{N_F} \sum_{j=1}^{N_F} w_j \, |F_{\text{obs}}(\mathbf{H}_j) - F_{\text{MEM}}(\mathbf{H}_j)|^2 = 0\,. \tag{8.10}$$

N_F is the number of reflections for which an experimental structure factor, $F_{\text{obs}}(\mathbf{H}_j)$, is available. The MEM is an ill-posed problem for constraints based on amplitudes of structure factors. Therefore $F_{\text{obs}}(\mathbf{H}_j)$ includes the phase of reflection j with scattering vector $\mathbf{H}_j$, as it is usually obtained from the calculated structure factor of the same reflection of an approximate structure model. $F_{\text{MEM}}(\mathbf{H}_j)$ is obtained by Fourier transform of the trial electron density $\{\rho_k\}$. The weights w_j can be derived from the standard uncertainties of $|F_{\text{obs}}(\mathbf{H}_j)|$,

$$w_j = \frac{1}{[\sigma(\mathbf{H}_j)]^2}\,, \tag{8.11}$$

but alternative weighting schemes are possible too.

The maximum of Q with respect to ρ_k is defined by $\partial Q/\partial \rho_k = 0$. The constraint of normalization, $C_0 = 0$, can be used to eliminate the Lagrange multiplier λ_0 [eqn (8.9)], resulting in a formal solution to the Maximum Entropy problem given by

$$\rho_k = \frac{1}{Z(\lambda_1)} \, \tau_k \exp\left[-\lambda_1 \frac{\partial C_1}{\partial \rho_k}\right] \tag{8.12a}$$

for $k = 1, \cdots, N_{\text{pix}}$. The partition function is

$$Z(\lambda_1) = \sum_{k=1}^{N_{\text{pix}}} \tau_k \exp\left[-\lambda_1 \frac{\partial C_1}{\partial \rho_k}\right], \tag{8.12b}$$

and it ensures normalization of $\{\rho_k\}$. Equation (8.12) and the F-constraint [eqn (8.10)] together provide $N_{\text{pix}} + 1$ equations that uniquely define the N_{pix} density values $\{\rho_k\}$ and the Lagrange multiplier λ_1. We call this solution to the Maximum Entropy problem $\rho^{\text{MEM}}(\mathbf{x})$ with density values $\{\rho_k^{MEM}\}$. However an analytical solution for $\rho^{\text{MEM}}(\mathbf{x})$ cannot be obtained, because eqn (8.12) are non-linear equations in $\{\rho_k\}$, as is shown by the derivative of the F-constraint:

$$\frac{\partial C_1}{\partial \rho_k} = 2\,Re \sum_{j=1}^{N_F} w_j \, \Delta F_{\text{M}}(\mathbf{H}_j) \exp[-2\pi i \mathbf{H}_j \cdot \mathbf{x}_k]\,, \tag{8.13a}$$

where *Re* denotes real part, and

$$\Delta F_{\mathrm{M}}(\mathbf{H}_j) = F_{\mathrm{obs}}(\mathbf{H}_j) - F_{\mathrm{MEM}}(\mathbf{H}_j)\,. \tag{8.13b}$$

The derivative of the F-constraint depends on all values $\{\rho_k\}$ through its dependence on $F_{\mathrm{MEM}}(\mathbf{H}_j)$. $\rho^{\mathrm{MEM}}(\mathbf{x})$ cannot be found by systematic variation of the variables $\{\rho_k\}$ either, because the number of pixels is too large, with typical values of $N_{\mathrm{pix}} = 10^6$ for periodic crystals and of $N_{\mathrm{pix}} = 10^8$ for incommensurate crystals. Therefore, equations (8.10) and (8.12) form the basis for iterative procedures in which values $\{{\rho_k}^{(n)}\}$ and ${\lambda_1}^{(n)}$ are determined that provide better approximations to $\rho^{\mathrm{MEM}}(\mathbf{x})$ for increasing iteration step $n = 0, 1, 2, \cdots$. Depending on the procedure of iteration different variants of the MEM are obtained.

The Gull–Daniel algorithm is a double iteration procedure. Starting with a small value of λ_1 (${\lambda_1}^{(0)}$) and a density equal to the PRIOR, ${\rho_k}^{(0)} = \tau_k$, subsequent iterates of the density are computed according to

$$ {\rho_k}^{(n+1)} = \frac{1}{Z({\lambda_1}^{(0)})}\,\tau_k\,\exp\left[-{\lambda_1}^{(0)}\frac{\partial C_1}{\partial \rho_k}^{(n)}\right], \tag{8.14}$$

where the derivative of the constraint [eqn (8.13a)] is evaluated for density values $\{{\rho_k}^{(n)}\}$. Convergence of this iteration is defined when $\{{\rho_k}^{(n+1)}\}$ and $\{{\rho_k}^{(n)}\}$ are sufficiently close to each other. This iteration is then repeated with the next higher value of λ_1 (${\lambda_1}^{(1)}$), of course with $\{{\rho_k}^{(0)}\}$ of the new iteration equal to the final value $\{{\rho_k}^{(\infty)}\}$ of the cycle for the previous value of λ_1. The double iteration is considered to be converged, if the constraint equation $C_1 = 0$ is fulfilled. The values $\{{\rho_k}^{(n)}\}$ for which C_1 drops below zero provide the desired solution $\{\rho_k^{\mathrm{MEM}}\}$. However, convergence is not guaranteed for any of the available MEM algorithms. Particularly the Gull–Daniel algorithm is known to be unstable and it is not used in crystallographic applications anymore.

The Sakata–Sato algorithm can be understood as a variation of the Gull–Daniel algorithm (Kumazawa *et al.*, 1995). Instead of Equation (8.14) subsequent approximations $\{{\rho_k}^{(n)}\}$ are calculated according to

$$ {\rho_k}^{(n+1)} = \frac{1}{Z(\lambda_1)}\,{\rho_k}^{(n)}\,\exp\left[-\lambda_1\frac{\partial C_1}{\partial \rho_k}^{(n)}\right]. \tag{8.15}$$

Convergence is defined by the point where the value of the constraint C_1 drops below 0. Experience has shown that $\{\rho_k^{\mathrm{MEM}}\}$ obtained by the Sakata–Sato algorithm is independent of λ_1 as long as λ_1 is sufficiently small. Too large values of λ_1 are identified by the failure to converge. Too small a value of λ_1 results in too slow convergence. Originally the Sakata–Sato algorithm has used a single value of λ_1. Alternatively λ_1 can be increased in small steps as long as the value of the constraint decreases. An increasing value of the constraint indicates divergence, and it is opposed by decreasing the value of λ_1.

Replacing τ_k in eqn (8.12) by $\rho_k{}^{(n)}$ amounts to the use of the result of one iteration step as 'PRIOR' in the next step. This violates the principles of the MEM. However the Sakato–Sato algorithm has been shown to perform well in many applications (compare Section 8.4), and it thus is justified by the results rather than by a firm theoretical foundation.

In a different approach Skilling and co-workers have aimed at finding the most non-committal electron density that maximizes the probability that it is true (Gull and Skilling, 1999). Selecting appropriate probability distribution functions, and introducing several approximations, they finally arrive at the criterion that the most probable electron density is the density $\rho^{\mathrm{MEM}}(\mathbf{x})$ that maximizes

$$\alpha S - C_1 \tag{8.16}$$

under the additional constraint of normalization of the density. Identifying α with $1/\lambda_1$, this criterion is found to be equivalent to finding the maximum of Q [eqn (8.7)]. Starting with a large value, α is decreased in small steps, while at each step $\rho_k{}^{(\infty)}$ is determined in an iterative procedure that performs much better than the Gull–Daniel algorithm (Skilling and Bryan, 1984). The Cambridge algorithm (Gull and Skilling, 1999) is stable and performs well. It is firmly rooted in probability theory, and in this sense it is correct. Nevertheless, the final density $\rho^{\mathrm{MEM}}(\mathbf{x})$ need not be exact, because the condition eqn (8.16) has only been obtained after introducing approximations to the probability distributions. Artifacts do occur in $\rho^{\mathrm{MEM}}(\mathbf{x})$ as will be discussed in Section 8.4.

Up to this point the discussion has been oriented to electron densities of periodic crystals. The superspace description allows a straightforward generalization towards aperiodic crystals (Steurer, 1991; van Smaalen *et al.*, 2003), if $\rho(\mathbf{x})$ is replaced by the generalized electron density $\rho_s(\mathbf{x}_s)$ with

$$\rho_k = \rho_s(\mathbf{x}_{s,k}) \tag{8.17}$$

and $k = 1, \cdots, N_{\mathrm{pix}}$ now counts the points of a grid over the superspace unit cell. For (3+1)-dimensional superspace this is a $N_1 \times N_2 \times N_3 \times N_4$ grid with $N_{\mathrm{pix}} = N_1\, N_2\, N_3\, N_4$ grid points. $F_{\mathrm{MEM}}(\mathbf{H}_{s,j})$ can be obtained as the Fourier transform of one superspace unit cell of the generalized electron density (Section 2.5), and reflections with scattering vectors $\mathbf{H}_{s,j}$ $(j = 1, \cdots, N_F)$ now include both main reflections and satellite reflections, as far as structure factor amplitudes have been obtained from the experiment. The question remains, why maximizing the entropy of the generalized electron density should provide the most probable electron density of an aperiodic crystal. This can be rationalized as follows. The generalized electron density can be considered as the juxtaposition along $\mathbf{a}_{s4}$ of infinitely many copies of the non-periodic electron density in physical space (see Chapter 2). A rectangular grid over the superspace unit cell thus samples the physical-space electron density at N_4 different copies of its basic-structure unit cell [Fig. 8.5(a)]. Alternatively these points correspond to a regular grid over the three-dimensional unit cell of a N_4-fold supercell approximation to the

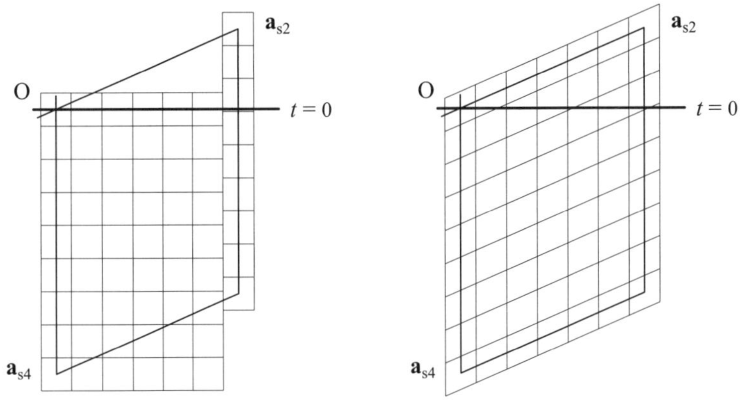

FIG. 8.5. (x_{s2}, x_{s4}) Sections of the superspace unit cell overlayed with a 8 × 6 grid. (a) Rectangular grid, and (b) regular grid. The pixels at $x_{s2} = 1$ represent the first column of pixels of the second unit cell.

incommensurate structure. For N_4 sufficiently large, this is a fair approximation that is not worse than the approximation introduced by the discretization of $\rho(\mathbf{x})$ in physical space. The transformation of a rectangular grid [Fig. 8.5(a)] towards a regular grid on the superspace unit cell [Fig. 8.5(b)] is a coordinate transformation with Jacobian 1. Values $\{\rho_k\}$ on a regular grid over the superspace unit cell can therefore be used to define the entropy of the electron density of an incommensurate crystal [eqn (8.6)].

The grid needs to be compatible with the space group or superspace group symmetry of the crystal. This implies that grid points are to be chosen either on the symmetry elements or between the symmetry elements (Fig. 8.6). The former choice is preferred, because the latter choice introduces artifacts into the density in the form of rows of pairs of pixels with identical density values [Fig. 8.6(b)]. Divisions N_i $(i = 1, \cdots, 3 + d)$ must be even integers in the presence of twofold axes or mirror planes, while higher-order symmetry elements may require N_i to be a fourfold or threefold integer in addition.

The unit cell can be divided into an asymmetric unit (au) and a remainder that depends on the asymmetric unit through the symmetry. The discrete electron density then is characterized by $N_{\mathrm{au}} \leqslant N_{\mathrm{pix}}$ unique pixels with independent density values ρ_k^{au}, while the remainder of the pixels have densities that depend on the unique density values. Arguments of continuity show that the entropy of the electron density with symmetry needs to be defined as [eqn (8.6)]

$$S = -\sum_{k=1}^{N_{\mathrm{au}}} m_k^{\rho}\, \rho_k^{\mathrm{au}} \ln\left(\rho_k^{\mathrm{au}} / \tau_k^{\mathrm{au}}\right), \tag{8.18}$$

where m_k^{ρ} is the multiplicity of pixel k. For most pixels m_k^{ρ} is equal to the number

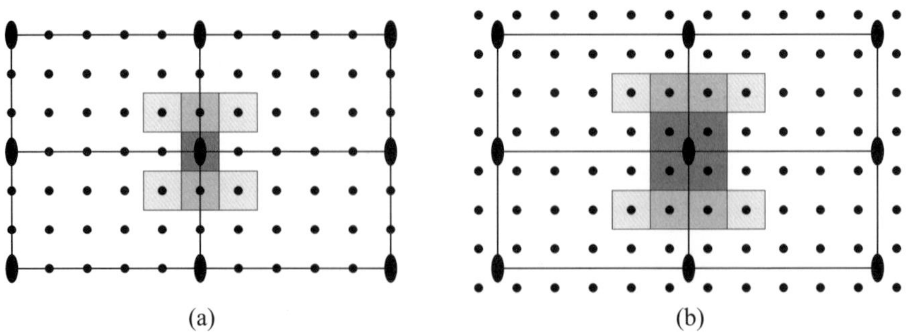

FIG. 8.6. One unit cell of a two-dimensional rectangular lattice with a symmetry-adapted grid with $N_1 = 6$ and $N_2 = 10$. (a) Preferred, and (b) less favourable choices of the positions of the grid points. Symmetry elements of $p2mm$ are indicated. Pixels that are equivalent by symmetry are indicated by the same shading. Adapted from van Smaalen *et al.* (2003).

of points of a general position of the space group or basic-structure space group, but m_k^ρ is smaller for pixels on symmetry elements. The formal solution to the maximum entropy problem is modified towards [eqn (8.12)]

$$\rho_k^{\mathrm{au}} = \frac{1}{Z(\lambda_1)}\,\tau_k^{\mathrm{au}}\,\exp\left[-\lambda_1\,\frac{1}{m_k^\rho}\,\frac{\partial C_1}{\partial \rho_k^{\mathrm{au}}}\right]. \tag{8.19}$$

m_k^ρ in eqn (8.19) cancels against the occurrence of m_k^ρ in the expression for the structure factor, such that

$$\frac{1}{m_k^\rho}\,\frac{\partial C_1}{\partial \rho_k^{\mathrm{au}}} = 2\,Re\sum_{j=1}^{N_F} w_j\,\Delta F_{\mathrm{M}}(\mathbf{H}_j)\,\exp[-2\pi i\mathbf{H}_j\cdot\mathbf{x}_k]\,. \tag{8.20}$$

Equal formal solutions are obtained for the cases with and without symmetry [compare eqn (8.13a)].

8.4 Maximum Entropy extensions

8.4.1 *Observed structure factors*

The maximum entropy method (MEM) requires data that are uniquely determined by the electron density. Computations are only feasible if this relation is linear, as it is the case for structure factors obtained by Fourier transformation of the electron density. Anomalous scattering cannot be incorporated into this scheme, and experimental data need to be corrected for this effect. The precise scale of the observed structure factors and any extinction correction need to be known too. The MEM thus requires the availability of a structure refinement of reasonable quality, from which the magnitudes of these effects as well as the phases of the structure factors can be obtained in good approximation. For the best-fitting structure model the Fourier transform of the electron density is obtained by a sum over all atoms in the unit cell with any anomalous contributions to the atom form factors set to zero:

$$F_{\mathrm{cal}}^{\mathrm{elec}}(\mathbf{H}_s) = A_{\mathrm{cal}}^{\mathrm{elec}}(\mathbf{H}_s) + i\,B_{\mathrm{cal}}^{\mathrm{elec}}(\mathbf{H}_s)\,. \tag{8.21}$$

The refinement provides $F_{\mathrm{cal}}(\mathbf{H}_s)$ that incorporates the effects of anomalous scattering as well as the scale factor (k_s) and a factor (k_E) representing the correction for extinction,

$$F_{\mathrm{cal}}(\mathbf{H}_s) = k_s\,k_E\,[A_{\mathrm{cal}}(\mathbf{H}_s) + i\,B_{\mathrm{cal}}(\mathbf{H}_s)]\,. \tag{8.22}$$

Corrected, scaled and phased experimental data can then be defined as

$$F_{\mathrm{obs}}^{\mathrm{elec}}(\mathbf{H}_s) = \frac{|F_{\mathrm{obs}}(\mathbf{H}_s)|}{|F_{\mathrm{cal}}(\mathbf{H}_s)|}\left[A_{\mathrm{cal}}^{\mathrm{elec}}(\mathbf{H}_s) + i\,B_{\mathrm{cal}}^{\mathrm{elec}}(\mathbf{H}_s)\right], \tag{8.23}$$

where $|F_{\mathrm{obs}}(\mathbf{H}_s)|$ are the observed data as they are obtained after data processing (Chapter 6).

One of the features of structure refinements is that phases of structure factors are better reproduced than their amplitudes, even if the structure model is not perfect. A structure model of reasonable quality will already provide correct phases. In this sense the MEM is a model-independent rather than a model-free method for reconstruction of the electron density.

Structure factors $F_{\mathrm{cal}}^{\mathrm{elec}}(\mathbf{H}_s)$ correspond to the known number of electrons in the unit cell of the approximate structure model. $F_{\mathrm{obs}}^{\mathrm{elec}}(\mathbf{H}_s)$ are the best estimates of the observed structure factors as they would correspond to the Fourier transform of a electron density with the same number of electrons. Nevertheless, the effects of anomalous scattering and extinction have been shown to deteriorate the quality of accurate electron densities, whether obtained by the MEM or multipole refinements. Better results can be expected if these effects are minimized, *e.g.* for compounds composed of light elements and for diffraction data obtained with high-energy X-rays.

8.4.2 *The F constraint*

Diffraction data are noisy. The supposed Gaussian distribution of the noise is represented by standard uncertainties ($\sigma[\mathbf{H}_j]$) of the structure factor amplitudes. In general, different values of $\sigma(\mathbf{H}_j)$ are obtained for different reflections $\mathbf{H}_j$. For a F constraint with weights equal to $1/[\sigma(\mathbf{H}_j)]^2$ [eqn (8.11)], one would expect $\Delta F_{\mathrm{M}}(\mathbf{H}_j)/\sigma(\mathbf{H}_j)$ [eqn (8.13b)] to obey a Gaussian distribution of unit width, once the MEM is converged and the constraint is fulfilled. It is a feature of the MEM that at convergence a few reflections have $\Delta F_{\mathrm{M}}(\mathbf{H}_j)/\sigma(\mathbf{H}_j) \gg 1$, while all other reflections are much better fitted than is allowed on the basis of $\sigma(\mathbf{H}_j)$ (Jauch, 1994). This suboptimal fit to the data leads to noisy features in $\rho^{\mathrm{MEM}}(\mathbf{x})$ that can hamper the extraction of the desired features from the density.

One way out of this problem is the use of alternative weights, that force the distribution of $\Delta F_{\mathrm{M}}(\mathbf{H}_j)/\sigma(\mathbf{H}_j)$ towards a Gaussian distribution. de Vries *et al.* (1994) have found that static weights,

$$w_j = \frac{1}{|\mathbf{H}_{s,j}|^n \, [\sigma(\mathbf{H}_j)]^2}, \tag{8.24}$$

lead to a dramatic improvement of the requirement of a Gaussian distribution of the final structure factor differences. The optimal value for the positive integer n needs to be determined on an *ad hoc* basis by analysing the distributions of $\Delta F_{\mathrm{M}}(\mathbf{H}_j)/\sigma(\mathbf{H}_j)$ at convergence for different n. Usually values within the range 2 to 6 are obtained for n.

In an alternative approach, generalized F constraints have been introduced as a constraint to higher-order moments of the Gaussian distribution (Palatinus and van Smaalen, 2002):

$$C_1 = C_{Fn} = -1 + \frac{1}{m_n}\frac{1}{N_F}\sum_{j=1}^{N_F} w_j \, |F_{\mathrm{obs}}(\mathbf{H}_j) - F_{\mathrm{MEM}}(\mathbf{H}_j)|^n = 0\,. \tag{8.25}$$

The classical F constraint corresponds to $n = 2$. The expectation values of the nth-order moments of the Gaussian distribution are m_n, with values $m_2 = 1$, $m_4 = 3$, $m_6 = 15$ and $m_8 = 105$. (Moments of odd order are zero for the Gaussian distribution.) Considerably improved distributions of $\Delta F_{\mathrm{M}}(\mathbf{H}_j)/\sigma(\mathbf{H}_j)$ have been obtained for MEM calculations with the C_{F4} and C_{F6} constraints (Palatinus and van Smaalen, 2002). Applications of the MEM have shown that it is advisable to use either static weights or F_n constraints, in order to find an optimum approximation to $\rho^{\mathrm{MEM}}(\mathbf{x})$.

Powder diffraction suffers from overlap of reflections (Section 7.7). Observed structure factor amplitudes can be extracted from the data for some reflections, but for other reflections only the sum of intensities is available of all reflections contributing to the group of overlapping reflections. These reflections can be taken into account by the so-called G-constraint that is defined as

$$\begin{aligned} C_G = -1 &+ \frac{1}{N_{\mathrm{all}}}\sum_{j=1}^{N_F} w_j \, |F_{\mathrm{obs}}(\mathbf{H}_j) - F_{\mathrm{MEM}}(\mathbf{H}_j)|^2 \\ &+ \frac{1}{N_{\mathrm{all}}}\sum_{j=N_F+1}^{N_{\mathrm{all}}} w_j \, |G_{\mathrm{obs}}(j) - G_{\mathrm{MEM}}(j)|^2 = 0 \end{aligned} \tag{8.26a}$$

with

$$G(j) = \left(\sum_{l=1}^{N_G(j)} |F(\mathbf{H}_l)|^2 \right)^{1/2}. \tag{8.26b}$$

The number of reflections contributing to the jth overlap group is $N_G(j)$, and $N_{\mathrm{all}} = N_F + N_G$. Convergence of the MEM can be achieved, if a sufficiently large fraction of the reflections is part of the first sum in C_G (Sakata *et al.*, 1990).

8.4.3 *The* PRIOR

The PRIOR and the constraint play independent but interrelated rôles in guiding the convergence of the MEM towards $\rho^{\mathrm{MEM}}(\mathbf{x})$ [compare eqn (8.12)]. The dynamic range of the maximum entropy problem is given by the difference between the PRIOR and the (unknown) true electron density. A better PRIOR implies a smaller dynamic range, and a better estimate for $\rho^{\mathrm{MEM}}(\mathbf{x})$ can be obtained. In case the quantity of interest is the redistribution of valence electrons due to chemical bonding, the appropriate PRIOR is the procrystal PRIOR. The latter is constructed from the electron density of the independent spherical atom model of the structure, as it is obtained from the best refinement.

Even in this case, series termination effects are present through all reflections for which an experimental structure factor is not available and thus are missing in the constraint. They become visible as artifacts in $\rho^{\mathrm{MEM}}(\mathbf{x})$. An improved convergence is obtained by the method of prior-derived F-constraints (PDC) (Palatinus *et al.*, 2005*b*). The data in the F constraint are augmented by high-angle reflections with structure factors obtained by Fourier transformation of the PRIOR $[F_{\mathrm{prior}}(\mathbf{H}_j)]$, resulting in the constraint:

$$
\begin{aligned}
C_1 = C_{F2}^{\mathrm{PDC}} = -1 + \frac{1}{N_{\mathrm{all}}} \sum_{j=1}^{N_F} w_j \left| F_{\mathrm{obs}}(\mathbf{H}_j) - F_{\mathrm{MEM}}(\mathbf{H}_j) \right|^2 \\
+ \frac{1}{N_{\mathrm{all}}} \sum_{j=N_F+1}^{N_{\mathrm{all}}} w_j \left| F_{\mathrm{prior}}(\mathbf{H}_j) - F_{\mathrm{MEM}}(\mathbf{H}_j) \right|^2 = 0 ,
\end{aligned}
\tag{8.27}
$$

with $N_{\mathrm{all}} = N_F + N_{\mathrm{prior}}$. Weights w_j of the N_{prior} PDC reflections are set equal to the value of the highest weight of observed data. The PDC directs the MEM towards a density that differs from the PRIOR only in their values of the low-angle structure factors, *i.e.* in those data that are affected by the valence electrons. With the procrystal PRIOR and the PDC, the MEM is an alternative to multipole refinements as method for accurate electron density studies.

Properties of chemical bonds can be derived from accurate electron densities with the aid of Bader's Atoms in Molecules (AIM) theory (Bader, 1994). The AIM theory considers topological properties of $\rho(\mathbf{x})$ and of its second derivative $\nabla^2\rho(\mathbf{x})$. It has been developed for electron densities obtained from quantum chemical calculations, but it can also be applied to densities obtained from multipole structure models and to $\rho^{\mathrm{MEM}}(\mathbf{x})$. For an account of the AIM theory I refer to the book by Bader (1994) and the extensive literature on multipole refinements (Koritsanszky and Coppens, 2001).

Because the AIM theory relies on derivatives of electron densities it cannot be applied to the generalized electron density in superspace. Instead, aperiodic crystals can be analysed within the framework of the AIM theory by consideration of t sections of generalized electron densities, *i.e.* by consideration of the electron density restricted to physical space. One calculation of this type has been made for aperiodic crystals, that is, for the incommensurately modulated structure of

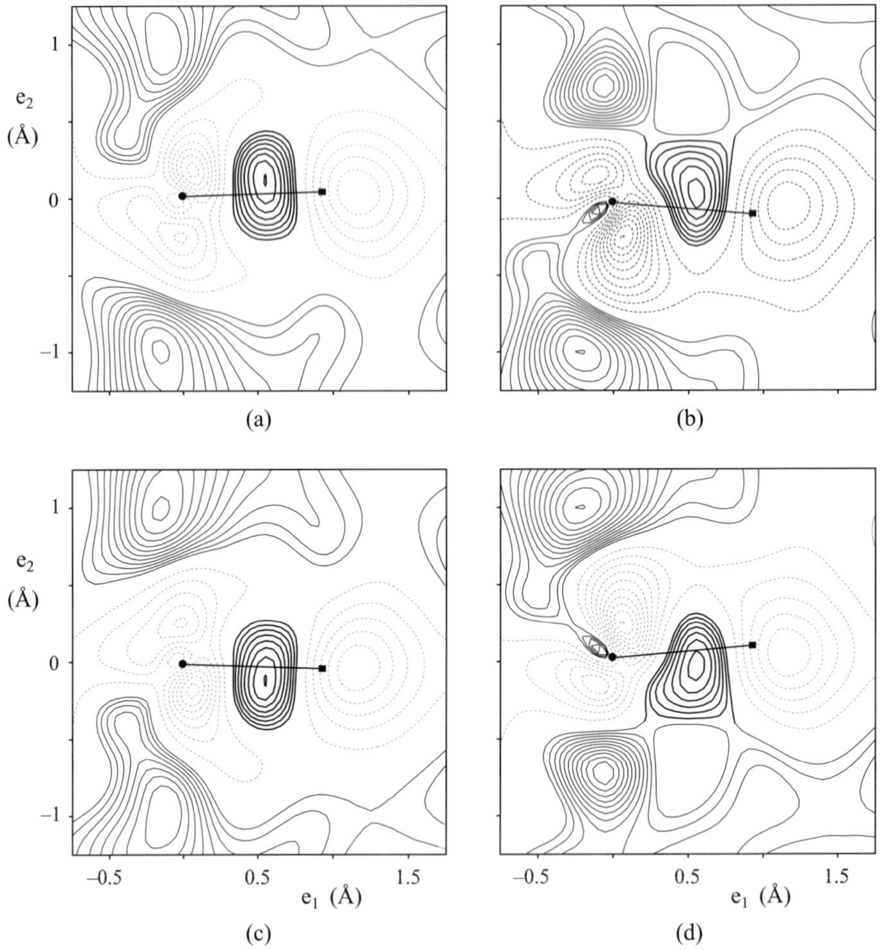

FIG. 8.7. Two-dimensional sections of sections t = constant of the generalized difference electron density $(\rho^{\text{MEM}} - \tau)$ of $(NH_4)_2BeF_4$. (a) $t = 0$, (b) $t = 0.25$, (c) $t = 0.5$ and (d) $t = 0.75$. The bond (straight line) between an N atom (dot) and an H atom (square) is indicated. Contours are drawn of positive (full line) and negative (dashed line) difference densities. The modulation of the difference electron density in the NH bond is highlighted. Adapted from Palatinus and van Smaalen (2004*a*).

$(NH_4)_2BeF_4$ (Palatinus and van Smaalen, 2004*b*). The electron density has been made visible in one of the N–H bonds of this compound. The modulation of the electron density in the N–H bond is approximately out of phase with the modulation of the positions of these two atoms (Fig. 8.7). However, the results are too preliminary in order to draw any conclusions from this observation.

The dynamic range of the MEM is large if a flat PRIOR is used instead of the procrystal PRIOR. Series termination produces noise and artifacts in $\rho^{\mathrm{MEM}}(\mathbf{x})$, that are larger than the effects of chemical bonding on the density, but that are still orders of magnitude smaller than the noise in Fourier maps calculated with the same data. The MEM with a flat PRIOR is the method of choice if the exact positions of the atoms are the quantity of interest. Examples of this type of problem include the description of disorder and anharmonic temperature vibrations as well as the shapes of modulation function of the atoms in incommensurate crystals (Section 8.5).

The flat PRIOR allows a MEM calculation with calculated structure factors as observed data. Only those calculated structure factors are included into the calculation, for which an observed structure factor is available in the data set. The resulting density will suffer from similar artifacts and noise as $\rho^{\mathrm{MEM}}(\mathbf{x})$ does. A comparison of the two densities will allow us to distinguish between artifacts produced by the MEM and true structural features in $\rho^{\mathrm{MEM}}(\mathbf{x})$.

8.4.4 *A comparison of MEM algorithms*

The quality of $\rho^{\mathrm{MEM}}(\mathbf{x})$ can be evaluated in several ways. The criterion $\partial Q/\partial \rho_k = 0$ for a maximum of Q implies that at convergence the relations

$$\frac{\partial S}{\partial \rho_k} = \lambda_1 \frac{\partial C_1}{\partial \rho_k} \tag{8.28}$$

must hold for all $k = 1, \cdots, N_{\mathrm{pix}}$. For simulated data of a known density (ρ^{true}) of crystalline oxalic acid dihydrate, these relations have been found to be almost fulfilled for a MEM calculation with the Cambridge algorithm, while the Sakata–Sato algorithm leads to a density $\rho^{\mathrm{MEM}}_{\mathrm{S\text{–}S}}$ that severely violates eqn (8.28) (Fig. 8.8) However, the comparison of $\Delta\rho(\mathrm{S\text{–}S}) = \rho^{\mathrm{MEM}}_{\mathrm{S\text{–}S}}$ - ρ^{true} with $\Delta\rho(\mathrm{Cambridge}) = \rho^{\mathrm{MEM}}_{\mathrm{Cambridge}}$ - ρ^{true} has shown that both algorithms produce MEM electron densities with remnant noise that is larger than the average difference between $\Delta\rho(\mathrm{S\text{–}S})$ and $\Delta\rho(\mathrm{Cambridge})$. In view of $\rho^{\mathrm{MEM}}(\mathbf{x})$ both algorithms have similar performances.

8.5 Modulation functions

The generalized electron density consists of strings of high density that are on the average parallel to $\mathbf{a}_{s4}$ of the superspace unit cell, and that are centred on coordinates (x_1^0, x_2^0, x_3^0) defining the basic-structure positions of the atoms of an incommensurately modulated structure (Chapter 2). The trace of maximum density along the string defines the modulation function for displacive modulation of this atom (compare Figs. 8.1 and 8.2). Points of maximum density must be determined as a function of t rather than as functions of x_{s4} or $\bar{x}_{s4}$, because electron densities of atoms in physical space follow from the generalized electron density as sections t = constant and not sections x_{s4} = constant. Accordingly a procedure has been developed that determines the electron density

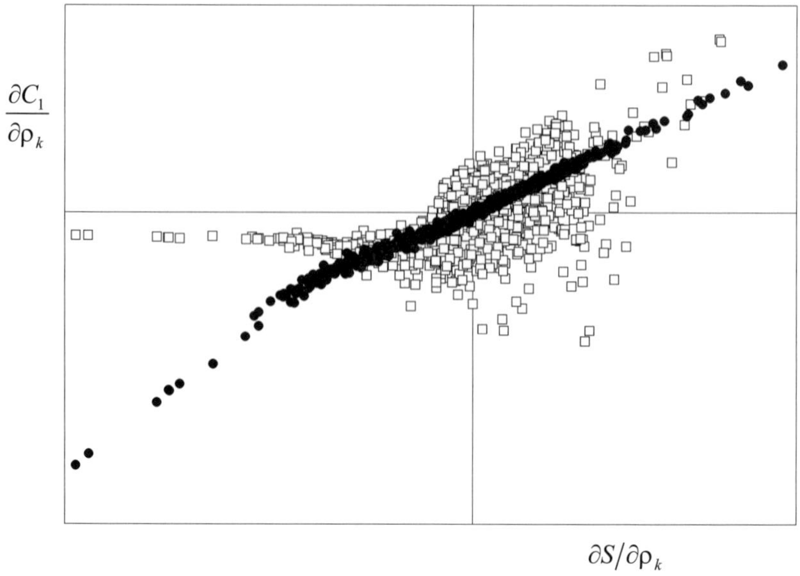

FIG. 8.8. Graphical representation of eqn (8.28) for a MEM calculation with the Cambridge algorithm (dots) and one with the Sakata–Sato algorithm (squares). Points that fulfil eqn (8.28) are on a straight line with slope λ_1. Adapted from van Smaalen *et al.* (2003), copyright (2003) by the IUCr.

on a three-dimensional grid in a t section by interpolation of the discrete generalized electron density (van Smaalen *et al.*, 2003). Each t section is analysed by standard methods according to Bader's AIM theory, providing—amongst other quantities—the position of maximum density in each string. The latter is identified with the atomic position. Repeating this procedure for sufficiently many values of t within one period along the x_{s4} axis ($0 \leqslant t < 1$) provides the positions of the atoms as functions of t, and also as functions of $\bar{x}_{s4}$, because $\bar{x}_{s4} = t + \mathbf{q}\cdot\mathbf{x}^0(\mu)$ and basic-structure coordinates $\mathbf{x}^0(\mu)$ are fixed quantities for each string.

This method of analysis has been applied to the modulated structure of the incommensurate composite crystal $[\mathrm{LaS}]_{1.14}[\mathrm{NbS}_2]$ (compare Section 8.1). The MEM in superspace has been employed to determine the generalized electron density $\rho_s^{\mathrm{MEM}}(\mathbf{x}_s)$ (van Smaalen *et al.*, 2003). The string of maximum density centred on $\mathbf{x}^0(\mathrm{La}) = (0, 0, 0.327)$ represents the La atom (Fig. 8.9(a)–(c); compare with the Fourier map in Fig. 8.1). The points of maximum density define the displacive modulation function of La [Fig. 8.9(d)–(f)]. The MEM electron density calculated with $F_{\mathrm{cal}}(\mathbf{H}_s)$ as data reproduces the modulation functions of La very well. The discrepancy between $\rho_s^{\mathrm{MEM}}(\mathbf{x}_s)$ and the model for $u_1^{\mathrm{La}}(\bar{x}_{s4})$ has been attributed to the relatively large pixel size of 0.18 Å along this direction in

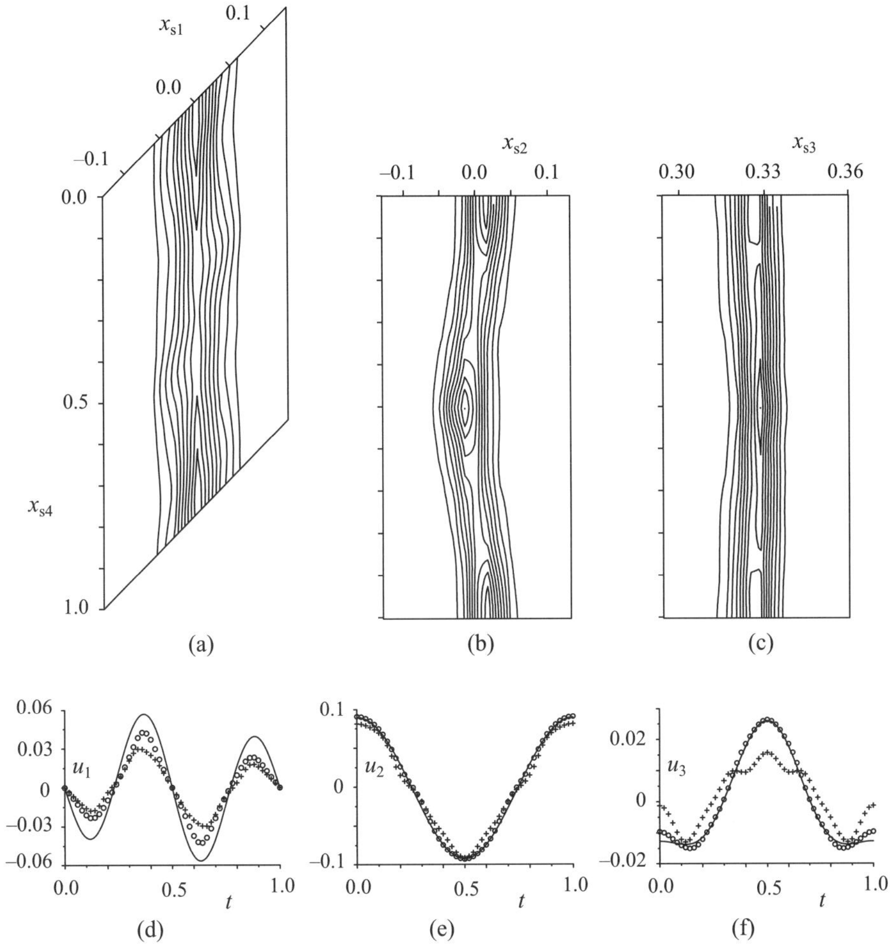

FIG. 8.9. Generalized electron density and modulation functions of $[LaS]_{1.14}[NbS_2]$. (a) (x_{s1}, x_{s4}) Section of $\rho_s^{\mathrm{MEM}}(\mathbf{x}_s)$ centred on La at (0, 0, 0.327) with $\rho^{\max} = 961.1$ electrons/Å^3. (b) (x_{s2}, x_{s4}) Section with $\rho^{\max} = 1409.9$ electrons/Å^3. (c) (x_{s3}, x_{s4}) Section with $\rho^{\max} = 1161.4$ electrons/Å^3. Contour lines are given at intervals of 10% of $\rho^{\max}$. (d) Component $u_1(t)$ of the displacive modulation function of La (in Å), (e) $u_2(t)$ and (f) $u_3(t)$. Modulation functions from the structure model with two harmonics (full line) are compared to modulation functions derived from $\rho_s^{\mathrm{MEM}}(\mathbf{x}_s)$ obtained with $F_{\mathrm{cal}}(\mathbf{H}_s)$ as data (circles) or $F_{\mathrm{obs}}(\mathbf{H}_s)$ as data (crosses). Reprinted with permission from van Smaalen *et al.* (2003), copyright (2003) by the IUCr.

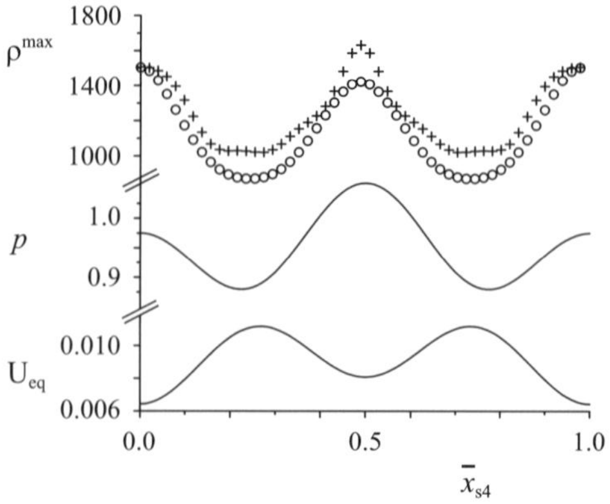

FIG. 8.10. Modulation of the La atom in $[LaS]_{1.14}[NbS_2]$. Top: value of the electron density along the trace of maximum density of the string in $\rho_s^{\mathrm{MEM}}(\mathbf{x}_s)$ representing the La atom ($\rho^{\max}$ in electrons/Å^3). $\rho_s^{\mathrm{MEM}}(\mathbf{x}_s)$ has been obtained with $F_{\mathrm{cal}}(\mathbf{H}_s)$ as data (circles) or $F_{\mathrm{obs}}(\mathbf{H}_s)$ as data (crosses). Middle: occupational modulation function (p). Bottom: modulation of the equivalent isotropic temperature parameter of La (U_{eq} in Å^2). Reprinted with permission from van Smaalen *et al.* (2003), copyright (2003) by the IUCr.

comparison with the small size of 0.06 Å of the modulation amplitude u_1 [Fig. 8.9(d)]. In general it has been observed that the accuracy of this method is limited by the interpolation procedure to about 10% of the pixel size of $\rho_s^{\mathrm{MEM}}(\mathbf{x}_s)$. The MEM electron density with observed data gives rise to a modulation function of which the component $u_2(\bar{x}_{s4})$ matches very well with the model and the modulation obtained from $\rho_s^{\mathrm{MEM}}(\mathbf{x}_s)$ with calculated data [Fig. 8.9(e)]. Components $u_3(\bar{x}_{s4})$ are different, when derived from $\rho_s^{\mathrm{MEM}}(\mathbf{x}_s)$ obtained with observed and calculated data, respectively. These discrepancies have been attributed to differences between the structure model with two harmonic functions and the true modulation (van Smaalen *et al.*, 2003).

The shape of the electron density about a local maximum as well as the value of the maximum, $\rho^{\max}(t)$, usually depend on t (compare Fig. 8.2). This may be an artifact due to series termination effects, as they especially will occur in Fourier maps, but to a much lesser extent in MEM electron densities. Alternatively, it may be a structural effect caused by modulations of temperature parameters, modulations of anharmonic temperature parameters or modulations of occupancies of different atomic species or vacancies on the site under consideration. For $[LaS]_{1.14}[NbS_2]$ the function $\rho^{\max}(t)$ has been found to match very well with the combined effect of vacancy ordering and modulations of the temperature parameters (Fig. 8.10).

Incommensurate crystals are studied by the MEM with the purpose of obtaining answers to the following questions:

1. What are the true modulation functions?
2. What information about the modulation can be extracted from a given set of diffraction data?
3. What resolution of data is required to determine the desired features of the modulation?

The real interest is point 1, but in practice we always deal with the second and third questions. Series termination effects in Fourier maps and parameterized functions in structure models may introduce artifacts into the structure model, while the MEM is supposed to provide the least biased answer to these questions.

The very few applications of the MEM to incommensurate crystals already suggest the following answers to these questions. Diffraction data with satellite reflections up to orders $m = 1$ or $m = 2$ can often completely be described by structure models with modulation functions containing one or two harmonics. A proof of the absence of higher-order harmonics in modulation functions requires an experimental proof of the absence of significant scattered intensity at the positions of higher-order Bragg reflections. It is thus suggested to include higher-order satellite reflections in data collections, even if their intensities appear as insignificant. Standard uncertainties on these zero values then provide an upper bound on possible higher-order harmonic contributions to the modulation functions.

The MEM will provide better estimates for modulation functions than structure refinements do, if the true modulation cannot be described by simple parameterized functions in the form of a few harmonics or a saw-tooth or block wave. This feature has been demonstrated for simulated data of an artificial structure with one atom in the basic-structure unit cell and with a longitudinal modulation of modified block-wave shape and with $\mathbf{q} = 0.345\ \mathbf{a}_1^*$ [Fig. 8.11(a)]. The MEM is found to reproduce the modulation function much better than a refinement of up to third-order harmonic functions does, when simulated data are employed with satellite reflections up to order $m = 3$. In this way the MEM has been used to determine the modulation functions of the crystallographically independent Bi atoms in the self-hosting incommensurate composite crystal of the high-pressure phase Bi-III [Fig. 8.11(b)] (van Smaalen and Palatinus, 2004; McMahon and Nelmes, 2004).

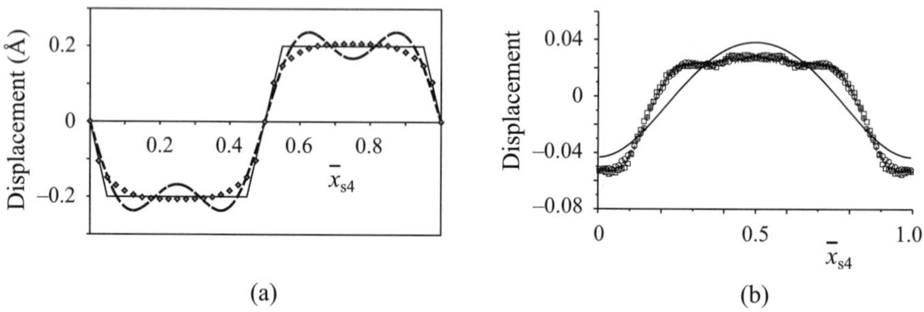

FIG. 8.11. (a) True modulation function (full line) of an artificial incommensurately modulated structure (see text) compared to modulation functions recovered by the MEM (diamonds) and a structure refinement (dashed line) applied to simulated diffraction data with satellite reflections up to order m = 3. Reprinted with permission from Palatinus and van Smaalen (2004*b*), copyright (2004) by Oldenburg Verlag. (b) Modulation $u_1(\bar{x}_{s4})$ of the single crystallographically independent atom in the first subsystem ('host') of the incommensurate composite crystal Bi-III. The full line represents the two-harmonic model of the best refinement. Circles, squares and crosses denote values obtained from $\rho_s^{\mathrm{MEM}}(\mathbf{x}_s)$ calculated with the Cambridge algorithm or the Sakata–Sato algorithm with $F2$ or $F4$ constraints. Reprinted with permission from van Smaalen and Palatinus (2004), copyright (2004) by Taylor and Francis.

9

DETERMINATION OF THE SUPERSPACE GROUP

9.1 Indexing of the diffraction pattern

The first step in analysing the diffraction of crystals is finding an integer indexing of the Bragg reflections. Experimentally each Bragg reflection is characterized by a specific orientation of the crystal and the direction of the diffracted beam. Four-circle diffractometers with point detectors produce a list of Bragg reflections in a random search procedure. Each reflection is described by its values for three setting angles describing the crystal orientation, and by its value for the diffraction angle 2θ corresponding to the detector position. Diffractometers with area detectors produce two-dimensional diffraction images. Intensity maxima define Bragg reflections at positions given by a pair of coordinates (x, y) that define the direction of the diffracted beam. To each image belongs a range of crystal orientations, characterized by the values of one to three setting angles and by a range parameter, $\Delta\omega$ or $\Delta\phi$, defining the range of rotation about a selected axis, during which the scattering has been collected onto the detector. The orientation of a crystal is defined by an orientation matrix A. This 3×3 matrix gives the coordinates of the three basis vectors of the reciprocal lattice of the crystal with respect to a Cartesian coordinate system fixed to the diffractometer, for an orientation of the crystal with all setting angles equal to zero. For periodic crystals A refers to the reciprocal lattice, for modulated crystals it refers to the reciprocal lattice of the basic structure, while for composite crystals the basic-structure reciprocal lattice of any of the subsystems is a suitable choice for the definition of the orientation matrix.

Robust methods exist for periodic crystals, that determine the orientation matrix along with the integer reflection indices (Duisenberg, 1992; Steller *et al.*, 1997). Setting angles of Bragg reflections can be accurately measured with a point detector. For each orientation matrix A they can be used to compute accurate values of the reflection indices, $(h_1^{\rm exp}\ h_2^{\rm exp}\ h_3^{\rm exp})$. Reflections belonging to the reciprocal lattice defined by A will have values $h_i^{\rm exp}$ close to integers, typically differing from integers by less than 0.01, and this property can be used to distinguish them from non-matching reflections. Subsequent inspection or automated analysis of the real-valued indices $(h_1^{\rm exp}\ h_2^{\rm exp}\ h_3^{\rm exp})$ of non-matching reflections will reveal possible modulation wave vectors or the second lattice of a composite crystal [eqn (2.5)], or it might reveal secondary lattices for twinned and multi-phase crystals (Duisenberg, 1992).

Accurate values are not available for the setting angles of Bragg reflections measured by area detectors, because scattered radiation is collected onto a single image for a range of values of one of the angles describing rotation of the crystal, *e.g.* with $\Delta\omega$ between $0.1°$ and $1.0°$, while steps in centring procedures with point detectors are chosen as small as $0.001°$ to $0.01°$. For images with $\Delta\omega \sim 0.5°$, the centre of the scan can be assigned to each reflection on this image as its setting angle ω. The relatively large errors in setting angles causes experimental reflection indices to deviate from integers by considerable amounts, *e.g.* by 0.1, even if these reflections belong to the reciprocal lattice. Successful indexing procedures rely on the assumption that all reflections belong to a single reciprocal lattice (Steller *et al.*, 1997). Integer indices $(h_1\, h_2\, h_3)$ are assigned to the reflections, where h_i is equal to the integer nearest to h_i^{exp}. This procedure fails for incommensurate crystals, because it will assign integer indices to satellite reflections, and it thus will fail to distinguish between main reflections and satellite reflections. The fine slicing technique ($\Delta\omega = 0.05°$–$0.1°$) allows a more accurate determination of the setting angles of reflections. Extended automated procedures then can distinguish between main reflections and satellite reflections, if indexing procedures are combined with refinements of the orientation matrix and refinements of the components of the modulation wave vectors (Schönleber *et al.*, 2001; Pilz *et al.*, 2002). In an alternative approach, the orientation matrix for the strong, presumably main, reflections can be used to compute an undistorted diffraction image in reciprocal space (Estermann and Steurer, 1998). Visual inspection of these images may reveal satellite reflections or secondary reciprocal lattices.

The result of a successful indexing is an orientation matrix and a minimal set of reciprocal vectors $M = \{\mathbf{a}_1^*, \cdots, \mathbf{a}_{3+d}^*\}$ that is necessary to obtain an integer indexing of all observed Bragg reflections (Section 2.2).

9.2 Point symmetry

Point symmetry of the diffraction pattern of aperiodic crystals is given by a three-dimensional point group (Chapter 3). For incommensurately modulated crystals this is a crystallographic point group that is compatible with the lattice of the basic structure. All possible symmetry elements can be tested by comparing intensities of related reflections, whereby groups of equivalent reflections exclusively contain main reflections or exclusively contain satellite reflections of a single satellite order m. The result is the appropriate three-dimensional point group describing the diffraction symmetry.

Tests for symmetry are often performed through consideration of the internal R index (compare Section 7.1):

$$R_{\mathrm{int}} = \frac{\sum_{j=1}^{N_{\mathrm{ref}}} |I(\mathbf{H}_j) - \langle I(j)\rangle|}{\sum_{j=1}^{N_{\mathrm{ref}}} \langle I(j)\rangle}, \tag{9.1}$$

where $\langle I(j)\rangle$ is the average intensity computed as the average over all reflections equivalent to reflection $\mathbf{H}_j$. For modulated crystals it may occur that deviations from certain symmetries appear only in the modulation and not in the basic-structure parameters. The partial R_{int} index for main reflections $[R_{\text{int}}(0)]$ will have a low value in accordance with this pseudo symmetry, but R_{int} for all reflections might be small too, if modulation amplitudes are small and satellite reflections are weak. It is thus of utmost importance to consider the partial index $R_{\text{int}}(m)$ for satellite reflections, in order to reveal the true point symmetry. As discussed in Section 6.3, twinning might prevent the correct point symmetry to be determined. Several examples of incommensurate crystals are known, for which pseudo-merohedral twinning has resulted in diffraction patterns with symmetries higher than the symmetries of the crystal structures.

Once the three-dimensional point group has been obtained, point symmetry operators in superspace can be determined from the consideration of the action of each operator R on the additional reciprocal vectors $\mathbf{a}_{3+1}^*$ through $\mathbf{a}_{3+d}^*$.

9.3 Reflection conditions

Intrinsic translations of screw and glide operators in superspace groups can be found from the analysis of possible reflection conditions on the subgroup of the reflections with scattering vectors that are left invariant by the point symmetry element under consideration. This analysis is feasible only if the basic-structure unit cell is transformed towards a supercell, $\{\mathbf{A}_1, \mathbf{A}_2, \mathbf{A}_3\}$, such that transformed modulation wave vectors, $\mathbf{q}_i$, have their rational component equal to zero (Section 3.9.1). Of course, a transformation is not required if $\mathbf{q}_r = 0$ from the outset [eqn (3.4)]. The set of invariant reciprocal points is most easily defined in physical space: invariant points are reciprocal points that lie in the mirror plane or on the rotation axis of the physical-space part of the symmetry operator. They always include a reciprocal lattice plane or reciprocal lattice line of main reflections. Depending on the components of the modulation wave vectors, satellite reflections may or may not be part of the set of invariant points. For example, satellite reflections are never invariant points for operators with $\epsilon = -1$, while they may fail to be invariant points for operators with $\epsilon = 1$, if the modulation wave vector contains rational components (Section 3.4).

The choice of a supercell for the basic structure introduces a centring of the superspace lattice. Furthermore, superspace lattices may already be centred for the standard choice of basic-structure unit cell. Any lattice centring can be analysed within the framework of reflection conditions, if they are considered as intrinsic translations of the unit operator E. All reflections are invariant points for this operator. For example, a reflection condition

$$(H_1\, H_2\, H_3\, H_4)\ :\ H_2 + H_3 + H_4 = 2\,n$$

might have been observed. This condition indicates the presence of an A' centring $(0, \frac{1}{2}, \frac{1}{2}, \frac{1}{2})$ (Table 3.9), as it might be found for a non-standard setting of the superspace group $Ammm(\sigma_1\, 1\, 0)\bar{1}00$ (No. 14 in Table 3.11).

In general, reflection conditions for reflections $(H_1\, H_2\, H_3\, H_4)$ are of the form [eqn (3.57)]

$$p_1 H_1 + p_2 H_2 + p_3 H_3 + p_4 H_4 = p_0 n, \tag{9.2}$$

with p_i and n integers. The quantity p_i is set to zero for all $i = 1, \cdots, 4$ for which $H_i = 0$ in the set of invariant points. Reflections may have non-zero intensities, if they fulfil the condition eqn (9.2) for some integer n, while reflections that violate eqn (9.2) must have zero intensity. The condition eqn (9.2) then corresponds to the intrinsic translation [eqn (3.58)]

$$\mathbf{v}_s^{\mathrm{I}} = \left(\frac{p_1}{p_0}, \frac{p_2}{p_0}, \frac{p_3}{p_0}, \frac{p_4}{p_0} \right). \tag{9.3}$$

For example consider a sixfold rotation on the hexagonal lattice with modulation wave vector $\mathbf{q} = (0,\, 0,\, \sigma_3)$. Invariant reciprocal lattice points are $(0\,0\,h_3\,h_4)$, and a possible reflection condition is

$$(0\,0\,h_3\,h_4) \;:\; h_3 + 3\,h_4 = 6\,n\,,$$

i.e. $p_1 = p_2 = 0$, $p_3 = 1$, $p_4 = 3$ and $p_0 = 6$. This condition corresponds to the intrinsic translation $(0,\, 0,\, \frac{1}{6},\, \frac{1}{2})$. The symmetry operator is $(6_1, s)$, but a different choice for the modulation wave vector would lead to the standard setting $(6_1, 0)$ of this sixfold screw operator, with the alternate reflection condition (Table 3.12)

$$(0\,0\,h_3\,h_4) \;:\; h_3 = 6\,n\,.$$

Intrinsic translations obtained from indexed diffraction patterns may correspond to non-standard settings of superspace groups, and they do not necessarily appear in the list of superspace groups in the *International Tables for Crystallography Vol. C* (Janssen *et al.*, 1995). A non-standard setting is often preferred, and one should not hesitate to keep it in the further structural analysis.

10

METHODS OF STRUCTURE SOLUTION

10.1 Introduction

The phase problem is the fundamental problem in structural analysis of crystals. Structure refinements require the input of a structure model with atomic positions that are accurate to approximately 0.1 Å, while initial values of temperature parameters can be set equal to standard values. A structure model of this quality can sometimes be constructed on the basis of the Patterson function (Section 10.3) and other considerations, like general crystal-chemical rules and analogies to similar compounds with known crystal structures. An approximate structure model of sufficient quality solves the phase problem, because the phases of the calculated structure factors of this model can be used in Fourier analysis (Chapter 8).

Phases of structure factors can be obtained from the measured intensities of Bragg reflections without an intervening structure model by direct methods (Section 10.4) or by one of several more modern methods, including charge flipping as it is presented in Section 10.5. Fourier maps with these phases contain local maxima, whose positions provide the atomic positions for an approximate structure model that can be used as start model in structure refinements. A structure model derived from a Fourier map may sometimes lead to improved estimates of the reflection phases. The latter can be used to compute an improved Fourier map, from which an improved structure model can be derived. This process can be continued until convergence, thus resulting in the best possible approximate structure model (Fourier cycling).

Basic structures of modulated compounds can be identified with crystal structures of periodic high-temperature phases, if the modulated structure is the result of a periodic-to-incommensurate phase transition. In this not uncommon case, the basic structure is thus accessible through standard methods of structural analysis for periodic crystals applied to diffraction data of the periodic phase. If a non-modulated phase is not available, basic structures can often be found by application of standard methods of structural analysis to the intensities of main reflections. Alternatively, a structure solution can sometimes be obtained, if main reflections are employed with modified intensities, where the latter are defined as the sum of intensities of main reflections and nearby satellite reflections. The idea behind this 'method' is that intensities of satellite reflections are 'borrowed' from intensities of main reflections (compare the discussion in Section 2.5.2). If all attempts to determine the basic structure fail, direct methods in superspace or charge flipping in superspace can be applied to all reflections at once (Sections 10.4 and 10.5).

Diffraction data by twinned crystals can be used in structure refinements, if the twin law is known (Section 6.3): the measured intensities of reflections are compared to the sums of intensities of Bragg reflections of the different domains, that contribute to the diffraction maxima. However, twinning almost invariably leads to a failure of systematic methods of structural analysis, because these methods depend on the knowledge of the intensities of Bragg reflections, and the sum of intensities of several Bragg reflections represents insufficient information for them. This problem is aggravated by the fact that ordinary diffraction experiments might not reveal the twinning (if present), and advanced experimental methods are necessary to determine the true symmetry of the modulated structure (Section 6.3). Twinning occurs frequently in incommensurate crystals that have been obtained through a phase transition from a phase with a periodic crystal structure. One of the first steps in a structural analysis therefore must be the careful consideration of the possibility that the measured data have originated in a twinned crystal.

An alternative to the superspace methods of structure determination (Sections 10.3–10.5) is the application of standard methods of structure determination to a supercell approximation of the modulated structure. On the other hand, superstructures that represent a small distortion from some basic structure can often be solved easier and refined better with superspace techniques, even if the modulation is commensurate.

10.2 Modulation functions by trial and error

With a known basic structure, the modulation can often be obtained by structure refinement against main and satellite reflections of a model with arbitrary but small starting values for the Fourier coefficients of the modulation functions. Preferred starting values for displacive modulation amplitudes are below 0.001 Å. More structures can be solved by this method of trial and error, if modulation amplitudes are gradually introduced into the model. Initially, first-order harmonic amplitudes are refined against main reflections and first-order satellite reflections. This is followed by the introduction of higher-order harmonics and refinements against all data. Furthermore, modulation amplitudes can be refined for some atoms first (*e.g.* the heavy atoms), while modulation parameters of other atoms are introduced at a later stage. Multiple start models can be tried in this way, while the measure of success of this method is a low R index towards the end of the refinement.

Estimates of the relative importance of modulation amplitudes of different atoms can be obtained from the anisotropic temperature parameters retrieved from the basic-structure refinement against main reflections. The effects of modulations on main reflections can be mimicked by additional contributions (U_{ij}^{mod}) to the temperature parameters [compare the smearing of density in the average structure due to the modulation—Fig. 8.2(c)]:

$$U_{ij} = U_{ij}^{0} + U_{ij}^{\mathrm{mod}} , \tag{10.1}$$

with U_{ij}^0 denoting the true temperature parameters, and $i, j = 1, 2, 3$. A comparison of the first terms in the series expansions of the Debye–Waller factor [eqn (6.11)] and the zeroth-order Bessel function, shows that

$$U_{ij}^{\text{mod}} = \frac{1}{4\pi a_i^* a_j^*} A_i A_j \,. \tag{10.2}$$

Anomalously large components $U_{ii}(\mu)$ of the temperature tensor of atom μ in the basic-structure refinement indicate that the corresponding modulation amplitudes A_i will be large. Parameters of this kind are good candidates to start the refinement of the modulation structure with. Analysis of the temperature parameters provides magnitudes of modulations at best, and it does not provide any phase relations between different modulation parameters. This method is highly inaccurate, because the true temperature parameters are not known *a priori*. Only relatively large modulation amplitudes ($A_i > 0.1$ Å) can be detected in this way.

10.3 Patterson function methods

The Patterson function is defined as the inverse Fourier transform of the diffracted intensity. For periodic crystals only Bragg reflections of the reciprocal lattice contribute to the inverse Fourier transform,

$$P(\mathbf{z}) = \frac{1}{V} \sum_{\mathbf{H}} |F(\mathbf{H})|^2 \exp[2\pi i \, \mathbf{H} \cdot \mathbf{z}] \,, \tag{10.3}$$

The resulting function is periodic according to the crystal lattice, and it comprises of local maxima at positions $\mathbf{z}'$ that correspond to interatomic vectors of the crystal structure. The Patterson function can be obtained from an X-ray diffraction experiment by the substitution $|F(\mathbf{H})| = |F_{\text{obs}}(\mathbf{H})|$. It is of fundamental importance in structural analysis, because it provides a direct-space image of the information contained in the scattered intensities, without the need of additional information or assumptions, like phases of reflections.

The Patterson function has its maximum value at the origin ($\mathbf{z} = 0$): the origin peak. With N atoms in the unit cell, the electron density contains N local maxima centred on the positions of the atoms. The density about each local maximum approximately represents the electron density of an atom, with a maximum proportional to the atomic number and with a width, σ, of the order of 1 Å (compare Fig. 8.2). The Patterson function would contain $N(N-1)$ local maxima of widths $\sim\sqrt{2}\sigma$ each, but due to the large number and the finite widths, overlap of local maxima is intrinsic to Patterson functions, such that the actual number of resolved maxima is smaller than $N(N-1)$. This property has prevented the Patterson function being used as a general-purpose tool of *ab initio* structure solution. Nevertheless it has proven to be helpful as part of structure solution procedures in many applications, including rotation and translation functions in protein crystallography and Harker lines and Harker sections in connection with an analysis of space group symmetry.

The Patterson function has been generalized towards non-periodic and partially ordered systems, and then is called a pair distribution function (pdf). Isotropic substances and microcrystalline powders have diffraction patterns for which the intensity is a function of the scattering angle (2θ) only. The inverse Fourier transform reduces to an integral over 2θ of the scattered intensity, and it includes both Bragg scattering and diffuse scattering. The pair distribution function reduces to the radial distribution function that depends on $z = |\mathbf{z}|$ only. Local maxima indicate interatomic distances that occur frequently in the material.

In a similar way, the Patterson function of an aperiodic crystal is defined as the inverse Fourier transform in physical space of all diffracted intensity, *i.e.* of both main reflections and satellite reflections [compare eqn (8.3)]:

$$P(\mathbf{z}) = \frac{1}{V} \sum_{\mathbf{G}} \sum_{h_4=-\infty}^{\infty} |F(\mathbf{G}, h_4)|^2 \exp[2\pi i\,(\mathbf{G} + h_4\mathbf{q})\cdot\mathbf{z}] \,. \tag{10.4}$$

The summation over $\mathbf{G}$ extends over all reciprocal lattice vectors of the basic structure, and $\mathbf{H} = \mathbf{G} + h_4\mathbf{q}$ (Section 1.3). The Patterson function is not periodic, but it does show local maxima at positions corresponding to interatomic vectors in the aperiodic crystal. A special function is the Patterson function of the average structure of a modulated crystal, that is defined as

$$P_{\mathrm{ave}}(\mathbf{z}) = \frac{1}{V} \sum_{\mathbf{G}} |F(\mathbf{G})|^2 \exp[2\pi i\mathbf{G}\cdot\mathbf{z}] \,. \tag{10.5}$$

The summation extends over main reflections only. $P_{\mathrm{ave}}(\mathbf{z})$ is periodic. Local maxima are broader than atomic maxima for the same reasons that local maxima in partial Fourier maps are broader than atomic maxima in Fourier maps (Fig. 8.2).

The generalized Patterson function is defined as the inverse Fourier transform in superspace of the diffracted intensity,

$$P_s(\mathbf{z}_s) = \frac{1}{V} \sum_{\mathbf{H}_s} |F(\mathbf{H}_s)|^2 \exp[2\pi i\,\mathbf{H}_s\cdot\mathbf{z}_s] \,. \tag{10.6}$$

The summation extends over all reciprocal lattice vectors in superspace, and thus includes both main reflections and satellite reflections. This function is periodic according to the superspace unit cell. Calculated structure factors can be obtained as the Fourier transform in superspace of one unit cell of the generalized electron density [eqn (2.30)]. Employing this property one can show that the generalized Patterson function is equal to the autocorrelation function of the generalized electron density [eqn (10.6)],

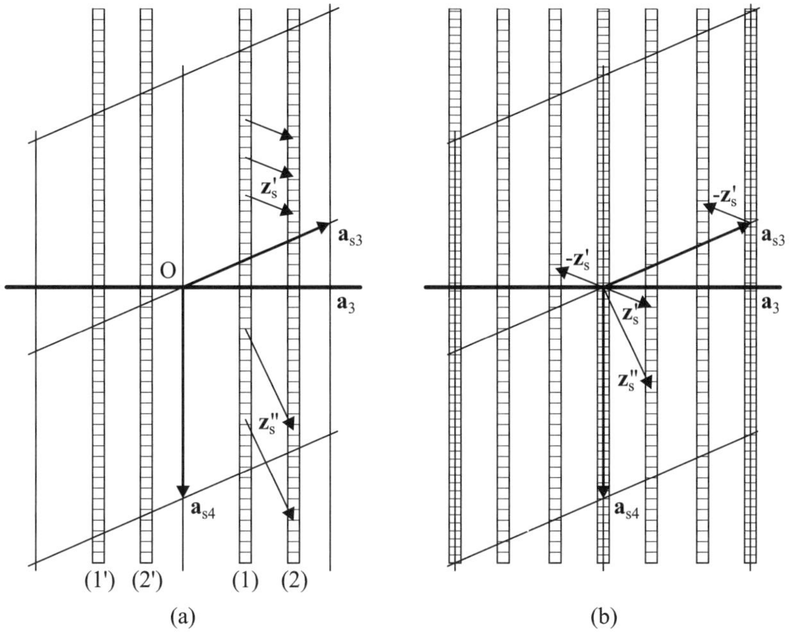

FIG. 10.1. Incommensurately modulated structure with two atoms in the basic-structure unit cell ($\mu = 1, 2$), and with zero modulation amplitudes. (a) (x_{s3}, x_{s4}) Section of the generalized electron density, $\rho_s(\mathbf{x}_s)$. (b) (z_{s3}, z_{s4}) Section of the generalized Patterson function, $P_s(\mathbf{z}_s)$. The magnitude of the electron density or the magnitude of the Patterson function is indicated by the spacing of the hatching along each string. Two different vectors, $\mathbf{z}_s'$ and $\mathbf{z}_s''$, are indicated [eqn (10.7)].

$$P_s(\mathbf{z}_s) = \int_{\text{unit cell}} \rho_s(\mathbf{x}_s)\, \rho_s(\mathbf{x}_s + \mathbf{z}_s)\, \mathrm{d}\mathbf{x}_s \, . \tag{10.7}$$

A formal analogy exists to Patterson functions of periodic crystals, that are autocorrelation functions of periodic density functions in physical space. However, constituent elements of generalized electron densities are atomic strings instead of point-like atoms in electron densities in physical space (Section 2.3). Consequently, features of $P_s(\mathbf{z}_s)$ are more complicated than features of $P(\mathbf{z})$.

The generalized Patterson function of an incommensurately modulated crystal can be used to determine basic-structure coordinates and modulation amplitudes (Steurer, 1987). The structure of $P_s(\mathbf{z}_s)$ is most easily understood through the initial consideration of an incommensurately modulated structure in the limit of zero modulation amplitudes. Atomic strings appear as straight lines parallel to $\mathbf{a}_{s4}$ (Fig. 10.1; compare to Fig. 3.3). Non-zero contributions to the integral in eqn (10.7) are obtained for all $\mathbf{x}_s$ on the string of atom 1, while $\mathbf{x}_s + \mathbf{z}_s$ is on

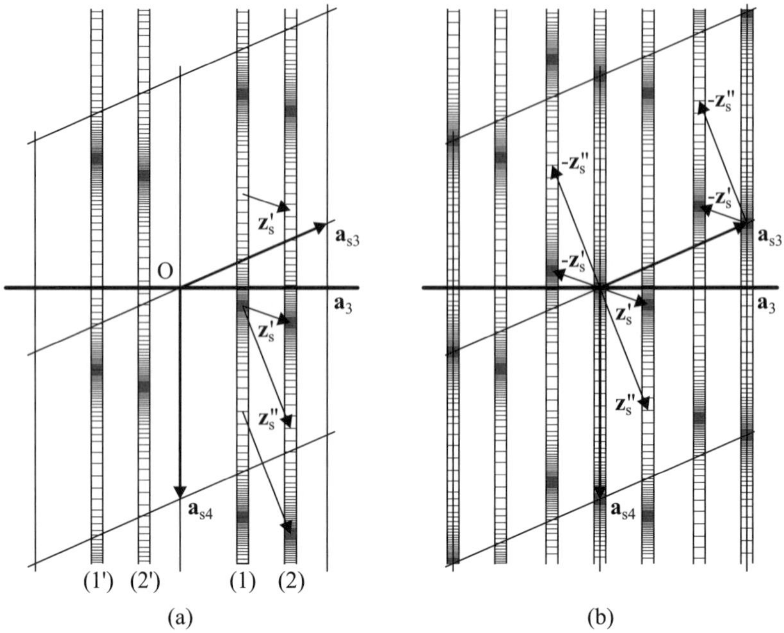

FIG. 10.2. Incommensurately modulated structure with two atoms in the basic-structure unit cell ($\mu = 1, 2$), and with occupational modulation. (a) (x_{s3}, x_{s4}) Section of the generalized electron density. (b) (z_{s3}, z_{s4}) Section of the generalized Patterson function. The magnitude of the electron density or Patterson function is indicated by the spacing of the hatching along each string. Two different vectors, $\mathbf{z}'_s$ and $\mathbf{z}''_s$, are indicated [eqn (10.7)].

the string of atom 2. Inspection of Fig. 10.1(a) shows that the resulting value of $P_s(\mathbf{z}_s)$ is independent of the component z_{s4} of the vector $\mathbf{z}_s$. The maximum value of $P_s(\mathbf{z}_s)$ is obtained, if $\mathbf{z}_s$ connects the maximum of string 1 with the maximum of string 2 in $\rho_s(\mathbf{x}_s)$. An incommensurately modulated structure with two atoms in the basic-structure unit cell and with zero modulation amplitudes thus gives rise to a generalized Patterson function containing two local maxima in the form of lines parallel to $\mathbf{a}_{s4}$ and centred on $\pm\mathbf{z}^0(1,2) = \pm[\mathbf{x}^0(2) - \mathbf{x}^0(1)]$ [Fig. 10.1(b)]. The value of $P_s(\mathbf{z}_s)$ is constant along each string, but it may be different for different strings. The absolute maximum is the origin peak: a string parallel to $\mathbf{a}_{s4}$ centred on $\mathbf{z} = 0$. Each section t = constant gives the Patterson function in physical space—which is periodic in this example, because intensities of satellite reflections are zero for zero modulation amplitudes. The widths of the strings are equal to $\sqrt{2}\sigma$, where σ is the width of the atomic strings in $\rho_s(\mathbf{x}_s)$.

An occupational modulation implies that electron-density values vary along the atomic strings in dependence on x_{s4}, but that the strings of maximum density remain straight lines, if displacement modulations are zero [Fig. 10.2(a)]. For example, consider two atoms (μ = 1,2) with different basic-structure coordinates $\mathbf{x}^0(\mu)$, and with occupational modulation functions

$$p^{\mu}(\bar{x}_{s4}) = \tfrac{1}{2} + P^1(\mu)\cos\left[2\pi(\bar{x}_{s4} - \varphi^1[\mu])\right], \tag{10.8}$$

where $P^1(\mu)$ is the amplitude and $\varphi^1(\mu)$ is the phase of the harmonic modulation function of atom μ. These modulation functions describe partial vacancy ordering, while complete order would require block-wave functions (Section 7.6). Maximum occupational probability is found for $\bar{x}_{s4}(1) = \varphi^1(1)$ and $\bar{x}_{s4}(2) = \varphi^1(2)$. Elementary properties of integrals show that a maximum value of $P_s(\mathbf{z}_s)$ is obtained for vectors $\mathbf{z}_s$ that connect points of maximum density as well as points of minimum density to each other. This condition is fulfilled for $\mathbf{z}_s = \mathbf{0}$, and $P_s(\mathbf{z}_s)$ has its absolute maximum at the origin. Smaller values of $P_s(\mathbf{z}_s)$ are obtained for vectors $\mathbf{z}_s$ that connect points of different densities. The vector $\mathbf{z}_s$ = (0, 0, 0, 0.5) connects a point of maximum density with a point of minimum density on each atomic string, thus leading to the minimum of $P_s(\mathbf{z}_s)$ on the line (0, 0, 0, z_{s4}). The origin peak is a string along $\mathbf{a}_{s4}$ with a maximum density that depends on z_{s4}. The largest value is found at the origin of superspace and the smallest value is at (0, 0, 0, 0.5) [Fig. 10.2(b)]. In a similar way, local maxima of $P_s(\mathbf{z}_s)$ are obtained as strings parallel to $\mathbf{a}_{s4}$ and centred on $\pm\mathbf{z}^0(1,2) = \pm[\mathbf{x}^0(2) - \mathbf{x}^0(1)]$. The maximum value of $P_s(\mathbf{z}_s)$ along these strings varies with z_{s4} and it is largest for $z'_{s4} = \varphi^1(2) - \varphi^1(1)$, while it is smallest for $z''_{s4} = 0.5 + \varphi^1(2) - \varphi^1(1)$.

The position of the maximum of $P_s(\mathbf{z}_s)$ on the string centred at $[\mathbf{x}^0(2) - \mathbf{x}^0(1)]$ in the generalized Patterson function directly provides the phase difference, $\varphi^1(2) - \varphi^1(1)$, between the modulation functions of atoms 1 and 2. For the single-harmonic function of eqn (10.8), the values of the generalized Patterson function can be evaluated for all points along the strings,

$$\begin{aligned} P_s(0,\,0,\,0,\,z_{s4}) &\propto \tfrac{1}{2} + \tfrac{1}{2}\left(\left[P^1(1)\right]^2 + \left[P^1(2)\right]^2\right)\cos[2\pi\, z_{s4}] \\ P_s(\mathbf{z}^0(1,2),\,z_{s4}) &\propto \tfrac{1}{4} + \tfrac{1}{2}P^1(1)P^1(2)\cos\left[2\pi(z_{s4} - [\varphi^1(2) - \varphi^1(1)])\right]. \end{aligned} \tag{10.9}$$

The product of occupational modulation amplitudes, $P^1(1)P^1(2)$, of atoms 1 and 2 can be obtained from the ratio of maximum to minimum values of $P_s(\mathbf{z}^0(1,2),\,z_{s4})$. If strings of all pairs of atoms can be evaluated in this way (including the origin peak), then complete information on the modulation amplitudes of all atoms can be obtained from the generalized Patterson function. However—as mentioned before—overlap of peaks is intrinsic to Patterson functions, and only a few strings can be resolved in most cases.

Features in generalized Patterson functions are more complicated for crystals with displacive modulations than they are in the case of occupational modulations. Consider two atoms in the basic-structure unit cell with displacive modulation functions

$$u_i^{\mu}(\bar{x}_{s4}) = A_i(\mu)\cos[2\pi(\bar{x}_{s4} - \varphi_i[\mu])], \tag{10.10}$$

for $i = 1,2,3$. $\mathbf{A}(\mu) = [A_1(\mu), A_2(\mu), A_3(\mu)]$ are the amplitudes and $[\varphi_1(\mu), \varphi_2(\mu), \varphi_3(\mu)]$ are the corresponding phases of this single-harmonic modulation function. The vector $\mathbf{z}'_s$ connects points of equal phases on the strings of atoms 1 and 2 [Fig. 10.3(a)]. If the amplitudes and phases are equal along all directions $i = 1,2,3$, a single vector $\mathbf{z}'_s$ connects any point of string 1 with a point of string 2. Accordingly, $P_s(\mathbf{z}_s)$ has a local maximum in $\mathbf{z}'_s$, that has a width equal to the normal atomic width of $\sqrt{2}\sigma$. The position along z_{s4} of the maximum at $\mathbf{z}'_s = [\mathbf{x}^0(2) - \mathbf{x}^0(1), \varphi_i(2) - \varphi_i(1)]$ can be used to determine the relative phases of the modulation functions of atoms 1 and 2. If phases $\varphi_i(\mu)$ are different for different directions i, local maxima are found for three different vectors $\mathbf{z}'_s$. The direction i of narrowest width of $P_s(\mathbf{z}'_s)$ indicates that $z'_{s4} = \varphi_i(2) - \varphi_i(1)$, providing the relative phase of the ith components of the modulation functions.

The vector $\mathbf{z}''_s$ connects the point with $\bar{x}_{s4} = \varphi_i(1)$ on string 1 with the point $\bar{x}_{s4} = \frac{1}{2} + \varphi_i(2)$ on string 2. Inspection of Fig. 10.3(b) shows that all vectors $\mathbf{z}_s$ with fourth components $z_{s4} = \frac{1}{2} + \varphi_i(2) - \varphi_i(1)$ have different components z_{si} ($i = 1,2,3$), if they should connect points of strings 1 and 2. Extremal values of $\mathbf{z} = (z_{s1}, z_{s2}, z_{s3})$ are found for $\mathbf{z}''_s$ with $\mathbf{z}'' = [\mathbf{x}^0(2) - \mathbf{x}^0(1)] - [\mathbf{A}(1) + \mathbf{A}(2)]$ and for $\mathbf{z}'''_s$ with $\mathbf{z}''' = [\mathbf{x}^0(2) - \mathbf{x}^0(1)] + [\mathbf{A}(1) + \mathbf{A}(2)]$. Values of $P_s(\mathbf{z}_s)$ at these points are smaller than the value of $P_s(\mathbf{z}'_s)$, because only parts of the atomic strings give non-zero contributions to the integral in eqn (10.7). Taking into account the 'density of states' of the cosine function then shows that $P_s(\mathbf{z}_s)$ comprises of double strings of maximum value in dependence on z_{s4} [Fig. 10.3(d)]. The largest value is obtained in $\mathbf{z}'_s$ and the smallest value of the maximum of $P_s(\mathbf{z}_s)$ is found at $\mathbf{z}''_s$ and $\mathbf{z}'''_s$. Unequal modulation amplitudes of atoms 1 and 2 make the narrowest peak at $\mathbf{z}'_s$ a double peak centred at $\mathbf{z}' = [\mathbf{x}^0(2) - \mathbf{x}^0(1)] \pm [\mathbf{A}(2) - \mathbf{A}(1)]$. Different shapes of the displacive modulation functions will lead to different shapes of the double string in $P_s(\mathbf{z}_s)$. Non-zero contributions to $P_s(\mathbf{z}_s)$ are found for all points with $\mathbf{z}_s$ between the extremal values. Furthermore, the atomic width of $\sqrt{2}\sigma$ is larger than the modulation amplitudes in most crystals, such that the double string may appear as a single string with a width and a maximum value that depends on z_{s4}. In all cases the phase difference of modulation functions can be estimated as the coordinate z'_{s4} of the point $\mathbf{z}'_s$ of narrowest width of $P_s(\mathbf{z}_s)$. By a poor-man's deconvolution procedure, the sum of amplitudes can be estimated from the difference between the widths of $P_s(\mathbf{z}_s)$ in the points of narrowest and widest width, as measured along each of the three directions i for the respective amplitudes $[A_i(1) + A_i(2)]$ ($i = 1,2,3$).

Structure determination with the aid of the generalized Patterson function is particularly fruitful for crystals with non-trivial symmetries. A pair of atoms ($\mu = 1, 1'$) related by symmetry possesses related modulated functions, with $\mathbf{A}(1') = \mathbf{A}(1)$ and with a definite phase relation that depends on the particular symmetry operation. Next to the values of $[\mathbf{A}(1') + \mathbf{A}(1)]$ and $[\varphi_i(1') - \varphi_i(1)]$ as obtained from $P_s(\mathbf{z}_s)$, symmetry provides a second relation between these parameters, such that the modulation function of atom 1 can be determined

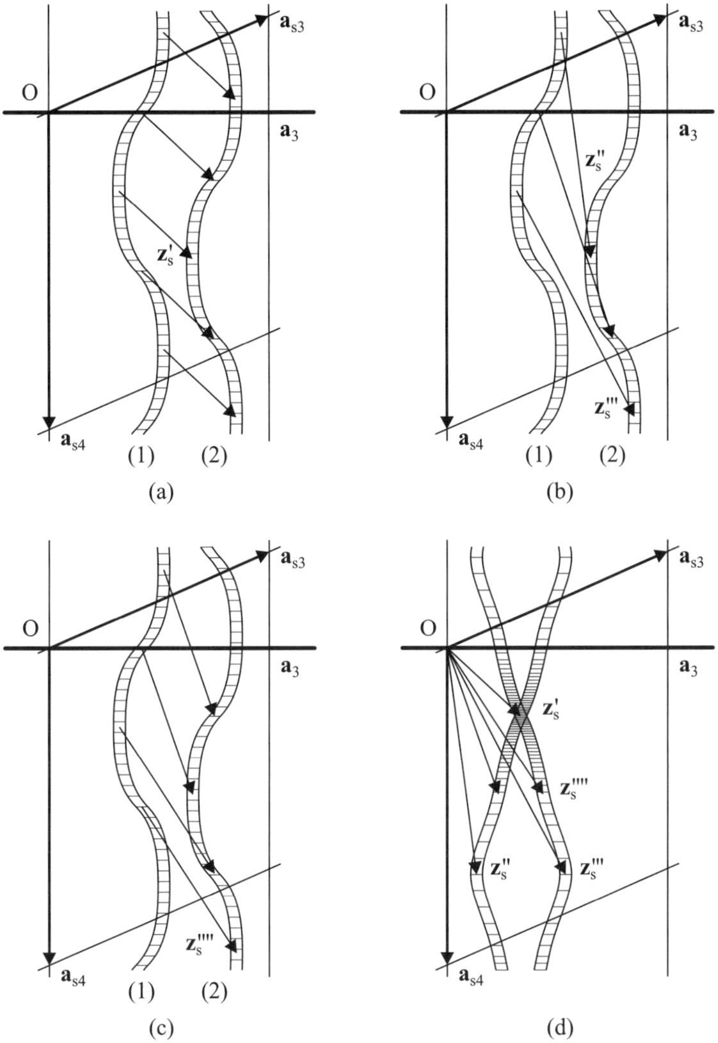

FIG. 10.3. Incommensurately modulated structure with two atoms in the basic-structure unit cell ($\mu = 1, 2$), and with displacive modulation. (a)–(c) (x_{s3}, x_{s4}) Section of the generalized electron density. (a) The vector $\mathbf{z}_s'$ connects points of equal phases between the two atomic strings. (b) Vectors $\mathbf{z}_s$ with $z_{s4} = z_{s4}' + 0.5$ connect points of phase difference 0.5. (c) Vectors $\mathbf{z}_s$ with $z_{s4} = z_{4s}' + 0.25$. (d) (z_{s3}, z_{s4}) Section of the generalized Patterson function. The origin peak and the double string centred on $-\mathbf{z}_s'$ are not shown. The magnitude of the electron density or the Patterson function value is indicated by the spacing of the hatching along each string. Phases of the modulation functions are $\varphi_3(1) = -0.123$ and $\varphi_3(2) = 0.192$ [eqn (10.10)].

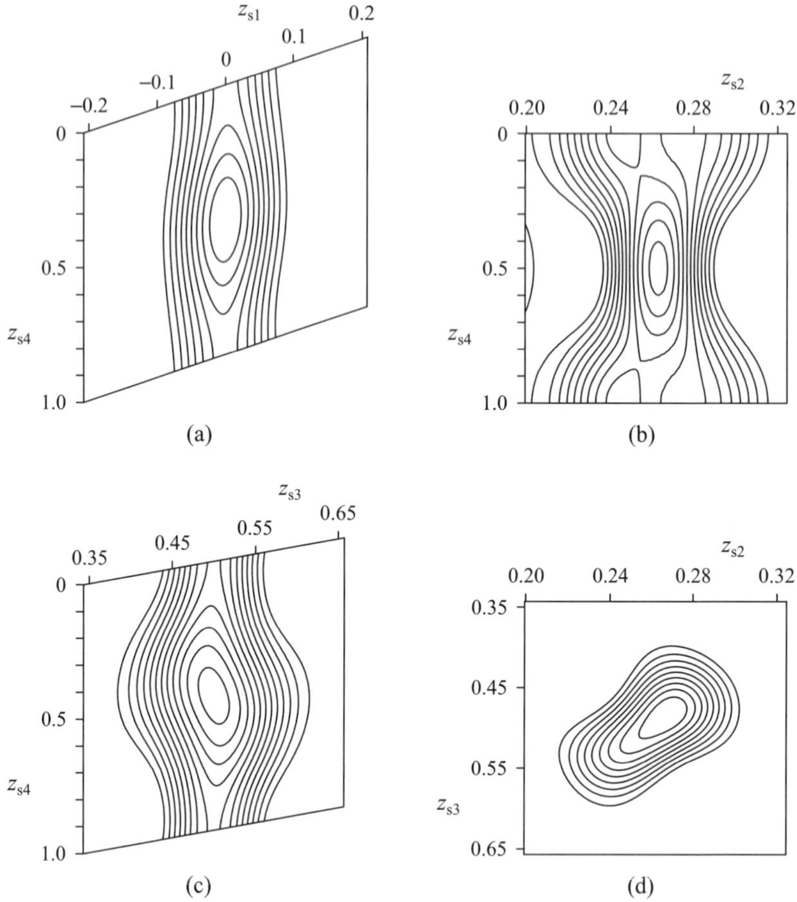

FIG. 10.4. Generalized Patterson function of incommensurately modulated $KAsF_4(OH)_2$, centred on $(0,\, 2x_2^0 - 1,\, 0.5)$ with $x_2^0 \approx 0.63$. (a) (z_{s1}, z_{s4})-section, (b) (z_{s2}, z_{s4})-section, (c) (z_{s3}, z_{s4})-section, and (d) (z_{s2}, z_{s3})-section at $z_{s4} = 0.25$. Basic-structure coordinates of As are $(0,\, x_2^0,\, 0.25)$. Diffraction data from Peterkova *et al.* (1998).

from the analysis of a single double string in the generalized Patterson function.

This type of analysis has been applied to incommensurately modulated $KAsF_4(OH)_2$ with the (3+1)-dimensional superspace group $C2/c(\sigma_1\, 0\, \sigma_3)\bar{1}0$ and $\sigma_1 = 0.8607$ and $\sigma_3 = 0.5585$ (Peterkova *et al.*, 1998). Atom As is located on the twofold axis at $(0,\, x_2^0,\, 0.25)$. Application of the c-glide $(x_{s1},\, -x_{s2},\, \frac{1}{2} + x_{s3},\, x_{s4})$ shows that a double string of local maximum of $P_s(\mathbf{z}_s)$ is centred on $\mathbf{z}^0 = (0,\, 2x_2^0,\, 0.5)$ for a pair of symmetry-related As atoms (Fig. 10.4). Two-dimensional sections (z_{si}, z_{s4}) of $P_s(\mathbf{z}_s)$ then provide estimates for the modulation amplitudes $A_1(\text{As}) = 0.007$, $A_2(\text{As}) = 0.014$ and $A_3(\text{As}) = 0.026$, as derived from the differences between maximum and minimum widths along the strings (Fig. 10.4). The positions of the maxima of $P_s(\mathbf{z}_s)$ at $z_{s4} = 0$ for $i = 1, 3$ and at $z_{s4} = \frac{1}{2}$ for $i = 2$ indicate the relative phases of the modulation functions of As and As′. Information about the phases of the modulation functions of As cannot be obtained from these values, because the phase differences are entirely determined by the c-glide operation. Symmetry restrictions due to the twofold axis lead to $\varphi_1(\text{As}) = \pm\frac{1}{4}$, $\varphi_2(\text{As}) = 0$ or $\frac{1}{2}$ and $\varphi_3(\text{As}) = \pm\frac{1}{4}$, thus leaving the signs of the amplitudes undetermined. The problem of the relative phases of displacement modulation functions along the three directions can be solved through the consideration of further sections of the generalized Patterson function. The phase of the first function can arbitrarily be chosen at $\varphi_2(\text{As}) = 0$. The section (z_{s2}, z_{s3}) of $P_s(\mathbf{z}_s)$ at $z_{s4} = \frac{1}{4}$ is elongated along the direction $z_{s2} - z_{s3}$ [Fig. 10.4(d)]. This indicates that a positive value of the displacement along $\mathbf{a}_2$ at $\bar{x}_{s4}(\text{As}) = 0$ coincides with a negative displacement along $\mathbf{a}_3$ at $\bar{x}_{s4}(\text{As}') = 0.25$, thus showing that $\varphi_3(\text{As}) = -\frac{1}{4}$ [eqn (10.10)]. A similar analysis of the section (z_{s1}, z_{s2}) may provide $\varphi_1(\text{As})$.

The generalized Patterson function of incommensurate composite crystals exhibits the same features as that of modulated crystals. Strings or double strings of maximum value are found to be centred on $\mathbf{x}_\nu^0(2) - \mathbf{x}_\nu^0(1)$, and they are on the average parallel to the incommensurate direction of subsystem ν. The generalized Patterson function thus provides information about the common reciprocal lattice plane and the incommensurate directions of the two subsystems, and it shows that this information is contained in the diffracted intensities without the application of special methods of structure solution. The incommensurate axes of the subsystems can most easily be obtained from the origin peak, that exhibits string-like features precisely along the incommensurate directions of each subsystem (Fig. 10.5). Both subsystems contribute to the value of $P_s(\mathbf{z}_s)$ at the origin, and the generalized Patterson function has its absolute maximum in the point $\mathbf{z}_s = \mathbf{0}$.

10.4 Direct methods in superspace

Direct methods aim at the determination of the phases of the structure factors from knowledge of their amplitudes. Without specific information on the crystal structure, the atomic character of matter implies that different sets of phases have different probabilities, given the set of amplitudes of the structure factors. The phases of maximum probability are usually sufficiently close to the true phases, in order to reveal the atomic positions as local maxima in a Fourier map (Giacovazzo *et al.*, 2002).

The phase of the structure factor of a single Bragg reflection depends on the position of the origin of the coordinate system, and thus is an arbitrary number. Direct methods are concerned with structure invariants: products of structure factors that possess phases that are independent of the origin of the coordinate system. Most extensively used is the triplet invariant. The product of three structure factors,

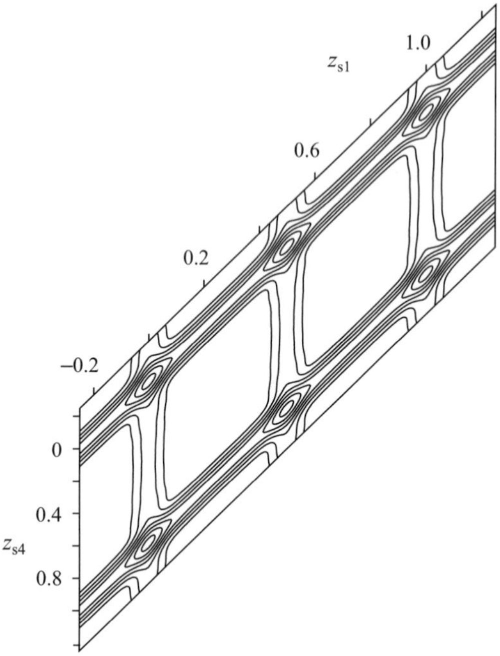

FIG. 10.5. Generalized Patterson function of the incommensurate composite crystal $[LaS]_{1.14}[NbS_2]$. The (z_{s1}, z_{s4}) section shows string-like features along the incommensurate directions of both the LaS and NbS_2 subsystems. The string at $z_{s1} = 0.5$ is the result of the particular superspace group symmetry. Diffraction data from Jobst and van Smaalen (2002).

$$F(\mathbf{H}_1)F(\mathbf{H}_2)F(\mathbf{H}_3) = |F(\mathbf{H}_1)||F(\mathbf{H}_2)||F(\mathbf{H}_3)| \exp[i\,(\varphi(\mathbf{H}_1) + \varphi(\mathbf{H}_2) + \varphi(\mathbf{H}_3))]\,, \quad (10.11)$$

has a phase

$$\Phi = \varphi(\mathbf{H}_1) + \varphi(\mathbf{H}_2) + \varphi(\mathbf{H}_3) \quad (10.12)$$

that is independent of the origin of the coordinate system if (Giacovazzo, 1998)

$$\mathbf{H}_1 + \mathbf{H}_2 + \mathbf{H}_3 = \mathbf{0}\,. \quad (10.13)$$

For acentric crystals the probability distribution of Φ is given by the Cochran distribution,

$$P_\Phi(\Phi) = [2\pi\, I_0(G_{1,2})]^{-1} \exp[G_{1,2}\cos(\Phi)]\,, \quad (10.14)$$

where $I_0(G_{1,2})$ is a modified Bessel function of the first kind. $P_\Phi(\Phi)$ depends on the normalized structure factor amplitudes as well as on the size of the unit cell through the parameter

$$G_{1,2} = \frac{2\mathcal{S}_3}{\mathcal{S}_2^{\frac{3}{2}}} |E(\mathbf{H}_1)||E(\mathbf{H}_2)||E(\mathbf{H}_3)| \approx \frac{2}{\sqrt{N}} |E(\mathbf{H}_1)||E(\mathbf{H}_2)||E(\mathbf{H}_3)| , \quad (10.15)$$

where N is the number of atoms in the unit cell of a periodic structure, and $\mathcal{S}_3$ is defined in analogy with $\mathcal{S}_2$ [eqn (6.22)]. Reflection phases are restricted to 0 and π for centrosymmetric crystals. Possible values for Φ are 0 and π too [eqn (10.12)]. The probability of Φ being 0 is given by

$$P_+ = \tfrac{1}{2} + \tfrac{1}{2} \tanh\left[\tfrac{1}{2} G_{1,2}\right] . \quad (10.16)$$

The most probable value of Φ is zero, but a consistent set of structure factor phases does not exist that would give $\Phi = 0$ for all triplet invariants. Instead, the set of structure factor phases is searched for, that maximizes the joint probability of all triplet invariants. Direct methods involve a series of, computationally demanding, procedures for the determination of the most probable set of structure factor phases, possibly including the consideration of probabilities of quartet and higher-order structure invariants.

The concept of structure invariant applies to incommensurate crystals in unaltered form (Hao *et al.*, 1987). A triplet invariant is defined by eqn (10.13), where $\mathbf{H}_i$ $(i = 1, 2, 3)$ can be scattering vectors of main reflections as well as of satellite reflections. Incommensurability ensures that eqn (10.13) can be replaced by the corresponding relation between reciprocal lattice vectors in superspace (Section 2.2),

$$\mathbf{H}_{s1} + \mathbf{H}_{s2} + \mathbf{H}_{s3} = \mathbf{0}_s . \quad (10.17)$$

It is then obvious that a pair of scattering vectors $(\mathbf{H}_{s1}, \mathbf{H}_{s2})$ completely determines the integer indices $(h_1^{(3)}\ h_2^{(3)}\ h_3^{(3)}\ h_4^{(3)})$ of the third vector $\mathbf{H}_{s3}$. Consequently, triplet invariants may involve three main reflections (mmm type invariant) or one main reflection and two satellites (sms) or three satellite reflections (sss). The latter type of triplet invariant occurs only if satellite reflections of order $m \geqslant 2$ are included. Two main reflections with one satellite reflection never is a structure invariant.

With normalized structure factors of modulated crystals obeying the same distribution functions as those for periodic crystals (Fig. 6.10), one may expect that direct methods should work for incommensurate structures too. However, the Cochran distribution has been derived for periodic structures only. For the same reasons that the normalized average diffracted intensities ('Wilson plot') follow different functional forms for main reflections and satellite reflections (Section 6.5.2), the probability distribution of triplet invariants will be different from the Cochran distribution, and it might be different for the various types of triplets mmm, sms and sss. At least the factor $G_{1,2}$ will have to incorporate the effects of the modulations. Indeed, de Gelder *et al.* (1996) have found empirically that triplet invariants follow the Cochran distribution [eqns (10.14) or (10.16)], if $G_{1,2}$ is replaced by

$$G'_{1,2} = \alpha\, G_{1,2} , \quad (10.18)$$

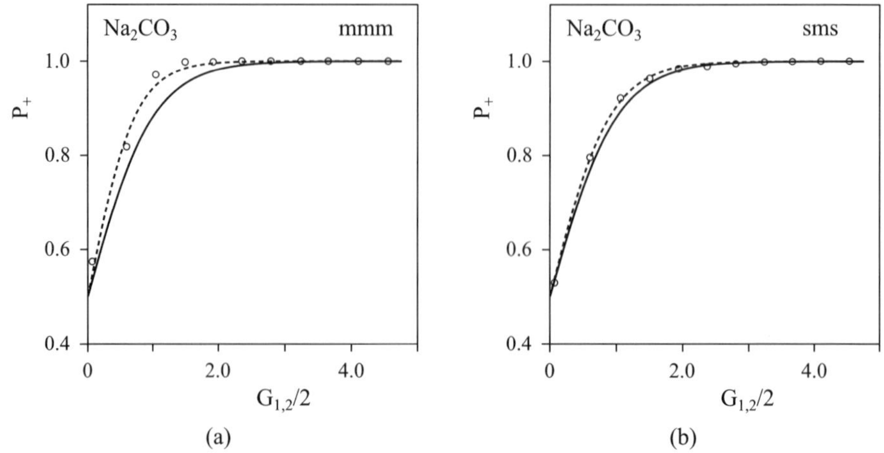

FIG. 10.6. Probability distribution P_+ for diffraction data of Na_2CO_3. (a) Triplets of main reflections (mmm). (b) Triplets of two satellite reflections with one main reflection (sms). Circles denote experimental values; solid curves represent the Cochran distribution of eqn (10.16). Dashed curves are fits to the data of $P_+(\frac{1}{2}\alpha\, G_{1,2})$, resulting in $\alpha = 1.4$ for mmm triplets and $\alpha = 1.11$ for sms triplets. Reprinted with permission from de Gelder *et al.* (1996), copyright (1996) by the IUCr.

where α is a parameter that depends on the compound and on the type of triplet invariant, mmm or sms. α is obtained by a fit of the modified Cochran distribution to the data. Values have been found within the range $0.2 < \alpha < 3.5$, while $\alpha(\text{sms}) < \alpha(\text{mmm})$ for all cases studied [Fig. 10.6; see de Gelder *et al.* (1996)].

Probability relations between structure factor phases of incommensurate crystals are unexplored territory. Only Peschar *et al.* (2001) have tested probability functions for doublet invariants of modulated structures, that may be employed to determine phases of satellite reflections, if phases of main reflections are known.

Disregarding modifications to probability functions, Fan and coworkers have applied direct methods to modulated and composite crystals, by solving the Sayre equation extended to superspace (Hao *et al.*, 1987; Fu and Fan, 1994),

$$F(\mathbf{H}_{s1}) = \frac{\Theta}{V} \sum_{\mathbf{H}_{s2}} F(\mathbf{H}_{s2})\, F(\mathbf{H}_{s1} - \mathbf{H}_{s2}), \tag{10.19}$$

where Θ has the same meaning as for periodic structures and V is the volume of the unit cell of the basic structure. They employed structure factor amplitudes instead of normalized structure factors in probability relations, thus implicitly assuming that larger amplitudes $|F(\mathbf{H}_1)||F(\mathbf{H}_{s2})||F(\mathbf{H}_{s3})|$ define a larger probability for $\Phi = 0$. This method has been successfully applied to diffraction data of several incommensurately modulated structures and incommensurate composite crystals (Fu *et al.*, 1995; Lam *et al.*, 1995; Mo *et al.*, 1996; Fan *et al.*, 1998).

10.5 Charge flipping

Direct methods have been the most important method of structure determination in small-molecule crystallography for many years. Despite their successes direct methods sometimes fail to find the solution, while in other cases direct methods fail by principle. Most prominently this is in structure solution of proteins, where the large number of atoms in the unit cell leads to nearly flat probability distributions of the phases of structure invariants, as it follows from the factor $1/\sqrt{N}$ in eqn (10.15). Direct methods often fail for powder diffraction data, because they require complete data sets of structure-factor amplitudes, and powder diffraction cannot provide such data sets due to the intrinsic problem of overlap of reflections.

Alternative methods of structure solution have been proposed, that employ atomicity and positivity of electron densities in ways different than direct methods do. Many of these methods can be considered as extensions of direct methods procedures, that should lead to solutions where the latter fail. They include entropy maximization, the shake-and-bake algorithm and Patterson search techniques based on partial structure information (Bricogne, 1988; Miller *et al.*, 1994; Giacovazzo *et al.*, 2002). Popular methods for structure determination from powder diffraction data include global search procedures in direct space and simulated annealing (David *et al.*, 2002). A particular class of methods have become known as density-modification methods (Terwilliger, 2000; Elser, 2003). These methods employ information on the electron density beyond mere positivity or atomicity. For example, protein crystals are known to be composed of protein and solvent regions that each occupy about 30% to 70% of the volume of the unit cell. Solvent molecules are disordered and the electron density in the solvent region is expected to be constant at a value much lower than the densities at the atomic maxima in the protein region. This observation has led to the methods of solvent flattening and solvent flipping (Wang, 1985; Abrahams, 1997). In each cycle of an iterative procedure the electron density in the solvent region is modified, with the effect of improving the phases of structure factors and improving the structure model of the protein molecule. An *ab initio* method of structure solution based on charge flipping has been proposed by Oszlányi and Sütő (2004, 2005). This method has been successfully applied to incommensurate crystals (Palatinus, 2004). Therefore it will be presented in detail below.

Charge flipping is an iterative procedure that eventually converges towards a model for the electron density, $\rho^{\mathrm{FLIP}}(\mathbf{x})$, and a set of phases for the structure factors. The electron density is described through its values ρ_k $(k = 1, \cdots, N_{\mathrm{pix}})$ on a grid over the unit cell, as it has been defined in Section 8.3 in connection with the maximum entropy method. For the same reasons as discussed in Section 8.3, this feature allows charge flipping to be applied to periodic crystals as well as aperiodic crystals, without modifications to the algorithm. The $N_{\rm pix}$ pixels k refer to a grid over the unit cell in physical space or to a grid over the unit cell in $(3+d)$-dimensional superspace, while calculated structure factors, $F_{CF}(\mathbf{H})$, can be obtained by discrete Fourier transform of one (superspace) unit cell for any trial electron density $\{\rho_k\}$. The effects of anomalous scattering are neglected in this procedure.

The charge flipping procedure starts with the observed structure-factor amplitudes, $|F_{\rm obs}(\mathbf{H})|$, and a set of randomly chosen reflection phases that satisfies Friedel's law, $\varphi^{(0)}(-\mathbf{H}) = \varphi^{(0)}(\mathbf{H})$:

$$F^{(0)}_{\rm obs}(\mathbf{H}) = \begin{cases} |F_{\rm obs}(\mathbf{H})| \exp[i\,\varphi^{(0)}(\mathbf{H})], & |F_{\rm obs}(\mathbf{H})| \text{ is known} \\ 0, & |F_{\rm obs}(\mathbf{H})| \text{ not measured.} \end{cases} \qquad (10.20)$$

The initial estimate of the electron density, $\{\rho^{(0)}_k\}$ is defined as the inverse Fourier transform of the structure factors $F^{(0)}_{\rm obs}(\mathbf{H})$. The nth iteration cycle starts with the modification of $\{\rho^{(n-1)}_k\}$ towards a density $\{\rho^{(n)}_{CF,k}\}$, according to the charge-flipping principle,

$$\rho^{(n)}_{CF,k} = \begin{cases} \rho^{(n-1)}_k, & \text{if } \rho^{(n-1)}_k > \delta \\ -\rho^{(n-1)}_k, & \text{if } \rho^{(n-1)}_k < \delta, \end{cases} \qquad (10.21)$$

for all pixels $k = 1, \cdots, N_{\rm pix}$ and with $\delta > 0$. That is, any negative density value is turned into its opposite, positive value, while only small positive densities are made negative. The parameter δ is discussed below. Calculated structure factors, $F^{(n)}_{CF}(\mathbf{H})$, are obtained by Fourier transform of $\{\rho^{(n)}_{CF,k}\}$. Structure factors of the nth cycle are then constructed according to

$$F^{(n)}_{\rm obs}(\mathbf{H}) = \begin{cases} |F_{\rm obs}(\mathbf{H})| \exp[i\,\varphi^{(n)}_{CF}(\mathbf{H})], & |F_{\rm obs}(\mathbf{H})| > F_{\rm threshold} \\ F^{(n)}_{CF}(\mathbf{H}) \exp[i\,\frac{1}{2}\pi], & |F_{\rm obs}(\mathbf{H})| < F_{\rm threshold} \\ 0, & |F_{\rm obs}(\mathbf{H})| \text{ not measured.} \end{cases} \qquad (10.22)$$

The threshold is chosen such that the structure factors keep their observed amplitudes for a reasonable fraction of the data, *e.g.* 80%. Finally, the electron density of the nth cycle of the iteration, $\{\rho^{(n)}_k\}$, is computed as the inverse Fourier transform of the structure factors $F^{(n)}_{\rm obs}(\mathbf{H})$. In alternative versions of the algorithm, $F^{(n)}_{\rm obs}(\mathbf{0})$ can be set equal to zero or to $F^{(n)}_{CF}(\mathbf{0})$.

The parameter δ can be chosen as a fraction of the value of the density in the local maxima of light atoms, but it should be above the noise level as follows from the inverse Fourier transform of the observed data. Alternatively, δ can be chosen such as to always flip the charge of a fraction of, for example, 50% of the pixels, or it can be chosen as to always flip a fraction of $\sim 50\%$ of the integrated charge. An optimal value of δ can be found by trial and error, while the value of

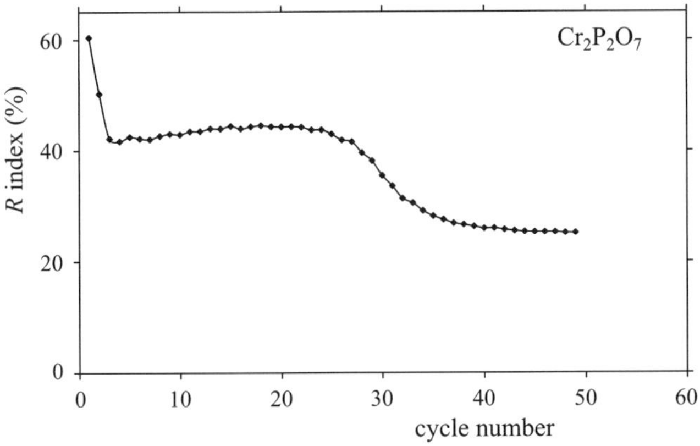

FIG. 10.7. R index as a function of the iteration cycle in one run of charge flipping in superspace, leading to the solution of the incommensurately modulated crystal structure of $Cr_2P_2O_7$ [eqn (10.23)]. Courtesy L. Palatinus.

δ can be reduced once convergence has been reached. Progress of the iterations can be monitored by a conventional R index,

$$R = \frac{\sum_{j=1}^{N_{\text{ref}}} \left| |F_{\text{obs}}(\mathbf{H}_j)| - \left| F_{CF}^{(n)}(\mathbf{H}_j) \right| \right|}{\sum_{j=1}^{N_{\text{ref}}} |F_{\text{obs}}(\mathbf{H}_j)|}, \tag{10.23}$$

where the summation extends over all observed data. After a sharp decrease in the first few cycles R remains almost constant at a high value for many cycles. Convergence is indicated by a rapid drop of R towards a lower value (Fig. 10.7). This 'low' value typically is between 20% and 30%. A peculiar property of the charge-flipping algorithm is that true convergence is never obtained. Densities below δ change sign between consecutive cycles, but the absolute values $|\rho_k^{(n)}|$ do converge. Density values $\rho_k^{(n)} > \delta$ do reach convergence at values ρ_k^{FLIP}. Local maxima of $\{\rho_k^{\text{FLIP}}\}$ then provide atomic positions and modulation function, employing procedures discussed in Section 8.5.

As for many other methods of structure solution, rigorous mathematical conditions of convergence have not been formulated for charge flipping. Indeed, charge flipping might fail to converge or variations of the method might perform better. However, an analysis of several mathematical aspects has illustrated the importance of key features of the method (Oszlányi and Sütő, 2004, 2005). The charge-flipping algorithm relies on the property that electron densities are close to zero over a major fraction of the volume of the unit cell. This property follows from the atomic character of matter, and atomicity thus is indirectly at the basis of the method. The inverse Fourier transform of any finite data set will produce both positive and negative values in the low-density region, because of series termination effects. This suggests that a change of sign of ρ_k might indeed result in the exploration of a larger region of phase space than, for example, replacing negative ρ_k by zero. The explored region of phase space is enhanced by the phase shift of $\frac{1}{2}\pi$ as applied to the weak reflections in the second half of each iteration cycle [eqn (10.22)]. An important feature of the charge-flipping method is that symmetry is not used. All structures are solved in space group $P1$ or superspace group $P1(\sigma_1\,\sigma_2\,\sigma_3\,)0$. The reason for this choice becomes apparent, if we consider a centrosymmetric structure. Enforcing inversion symmetry on the density requires the charge-flipping method to find a singular solution amongst all possible phase sets. On the other hand, with an unrestricted density, charge flipping needs to find just one out of an infinite number of good solutions that differ from each other in the position of the origin with respect to the inversion centre only. Arriving at one good solution amongst many good solutions is more probable than arriving at the single good solution.

Charge flipping has been successfully applied to X-ray diffraction data of a series of aperiodic crystals, including incommensurately modulated crystals, incommensurate composite crystals and one quasicrystal (Palatinus, 2004; Katrych *et al.*, 2007). The method does not require knowledge of basic structures. It automatically finds a model for the generalized electron density together with the complementary set of phases of structure factors of both main and satellite reflections. The generalized electron density, $\rho_s^{\mathrm{FLIP}}(\mathbf{x}_s)$, is close to the Fourier map computed with the phases of the refined structure (Fig. 10.8). Analysis of $\rho_s^{\mathrm{FLIP}}(\mathbf{x}_s)$ by standard methods will produce the structure model, including basic-structure coordinates and modulation-function amplitudes of the atoms as well as the locations of possible superspace symmetry elements.

Charge flipping in superspace has the promise of becoming the method of choice for structure solution of aperiodic crystals. However, it fails for data from twinned crystals, because the required $|F_{\mathrm{obs}}(\mathbf{H})|$ data are not available for reflections originating in a twinned crystal. Yet different methods are needed to solve the phase problem for the—not untypical—case of a twinned, incommensurately modulated crystal.

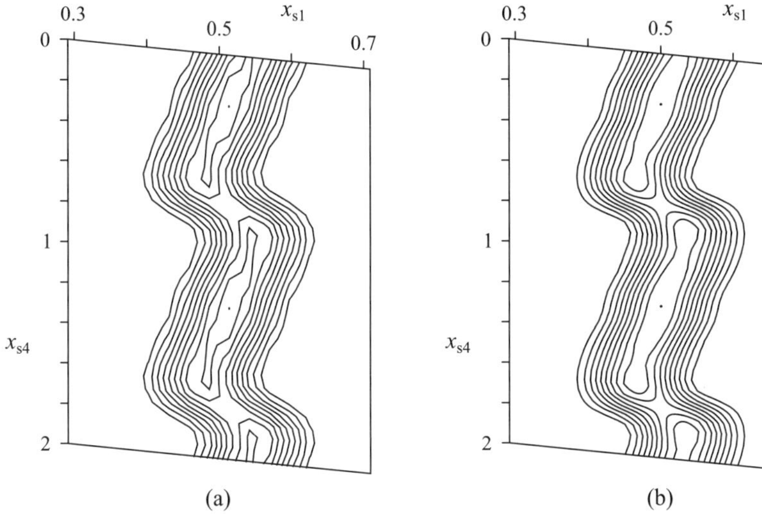

FIG. 10.8. (x_{s1}, x_{s4}) Sections of generalized electron densities of incommensurately modulated $Cr_2P_2O_7$, centred on the position of the Cr atom. (a) $\rho_s^{\mathrm{FLIP}}(\mathbf{x}_s)$ obtained by charge flipping in superspace. (b) Fourier map with phases from the refined structure model. Notice the acute angle between the $\mathbf{a}_{s1}$ and $\mathbf{a}_{s4}$, that indicates a negative value of the first component of the modulation wave vector $\mathbf{q} = (-0.361, 0, 0.471)$. Adapted from Palatinus *et al.* (2006), copyright (2006) by the IUCr.

11

SYSTEMATIC CRYSTAL CHEMISTRY

11.1 Introduction

Incommensurately modulated structures form the basis for qualitative and quantitative analyses towards understanding the chemical stability of incommensurate compounds. Other subjects of interest are the question about the origin of incommensurability as opposed to periodicity, and physical properties that either do or don't show behaviour related to the aperiodic character of the material.

The superspace description provides the tools for a quantitative analysis of incommensurate structures: the so-called t-plots. t-Plots can be made for any quantity that can be computed from the structural parameters, including the values of the modulation functions themselves, interatomic distances, bond angles and orientations of rigid groups. t-Plots have been introduced in Section 2.4. They give the value of the quantity under consideration as a function of the global phase of the modulation wave, as it is represented by the parameter t in the definition of $\bar{x}_{s4}$ in eqn (2.20). The periodicity of the modulation implies that any quantity is periodic in t, with all values occurring anywhere in the structure summarized in the interval $0 \leqslant t < 1$. The important property is the correlation between t-plots of different quantities. Consider a selection of atoms in the first basic-structure unit cell of a modulated structure. These atoms are located in the physical-space section $t = 0$ of the first unit cell of the superspace model for this structure. For a different value of t—say t_0—these atoms are found in the section $t = t_0$ of the first superspace unit cell, that is equivalent to a basic-structure unit cell somewhere in physical space defined by the section $t = 0$ (Section 2.3). For values of t close to t_0, different basic-structure unit cells are obtained in the physical-space section $t = 0$ that are far from the unit cell implied by t_0. However, the original set of atoms always will be found in a single unit cell of the section $t = 0$ for any value of t. Thus distances from a central atom towards all its neighbours in the first basic-structure unit cell represent distances between the central atom and neighbouring atoms in other unit cells for different values of t. Consideration of these distances as a function of t for $0 \leqslant t < 1$ then provides a summary of the variation of the first coordination shell of the central atom as it varies throughout physical space. For example, t-plots of interatomic distances show that the coordination of Au by Te in $AuTe_2$ varies between twofold coordination at $t = 0$ and 0.5 and fourfold coordination at $t = 0.25$ and 0.75 (Fig. 11.1). The interpretation of t-plots of composite crystals is slightly more complicated (Section 4.8). Applications of t-plots to crystal-chemical considerations have been given in Section 2.4.2 for incommensurately modulated $Sr_2Nb_2O_7$ and

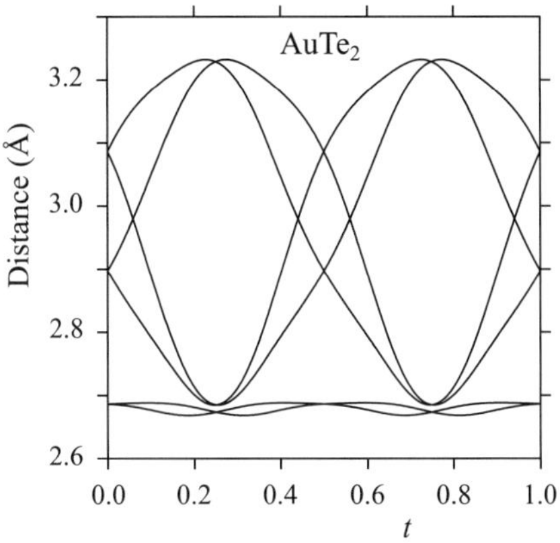

FIG. 11.1. t-Plot of distances between Au and neigbouring Te atoms in the incommensurately modulated structure of $AuTe_2$. Coordinates from Schutte and de Boer (1988).

in Sections 4.8 and 8.5 for the incommensurate composite crystal $[LaS]_{1.14}[NbS_2]$.

Applications of t-plots can be taken further than mere considerations of correlations between structural parameters. The distances about a central atom can be used to apply the bond-valence method to incommensurate crystals, thus providing the atomic valence of the central atom as a function of t (Section 11.5). Molecular mechanics is a semi-empirical method that defines the energy of a molecule or crystal through additive contributions depending on the bond lengths, bond angles, torsion angles and non-bonded distances of the structure (Boeyens and Comba, 2001). Each pair of chemical elements is assigned a few transferable parameters that define a function describing the dependence on the distance of the contribution to the energy of any pair of atoms of this kind. Together with parameters for bond angles and torsion angles they define a force field of transferable parameters. t-Plots of the individual contributions to the energy of an incommensurate crystal can thus be constructed on the basis of t-plots of geometrical parameters of the structure. A straightforward generalization of molecular mechanics then follows through the definition of the energy of an incommensurate crystal as the integral over one period of t of the sum of all individual contributions to the energy, where individual contributions are to be considered for the atoms in one basic-structure unit cell (Gao and Coppens, 1989). In this way, the total binding energy of incommensurate structures can be compared to energies of basic structures and to energies of any commensurate superstructure. Molecular dynamics studies structures at finite temperatures by solving Newton's equations of motion with a force field defining the potential energy. It involves derivatives with respect to the atomic coordinates, for which the proper generalization towards superspace is not available.

11.2 Compounds with incommensurate crystal structures

Incommensurate compounds are found across entire chemistry, encompassing elements, inorganic and organic compounds and minerals. Incidental occurrences of incommensurability have been reported, resulting in the recognition that Na_2CO_3, biphenyl and many other compounds possess incommensurately modulated structures (van Aalst *et al.*, 1976; Baudour and Sanquer, 1983). While incommensurability is a widespread phenomenon, it is also concentrated in a few classes of compounds that are responsible for the majority of incommensurate compounds. They include the A_2BX_4 type crystals (Section 11.4), inorganic misfit layer compounds (Section 11.5), charge-density waves in transition-metal di- and trichalcogenides (van Smaalen, 2005), phosphate bronzes (Ottolenghi and Pouget, 1996) and the chemical elements (Section 11.3). Although incommensurability is found for almost all compounds with the β-K_2SO_4 structure type, the correct structure type is not a guarantee for an incommensurate crystal structure. For example, many binary and ternary transition metal trichalcogenides follow from similar building principles, but only a few of these compounds develop a charge-density wave. The trichalcogenides with incommensurate structures do not belong to a single structure type, but they seem to have been picked at random from the different structure types amongst this class of materials.

Many so-called incommensurate crystals develop their incommensurately modulated structure at low temperatures or at high pressures. They undergo a phase transition at a critical temperature or a critical pressure from a state with a periodic structure towards a state with an incommensurate structure. At still lower temperatures or still higher pressures a second phase transition often occurs, at which the modulation becomes commensurate. Superspace is the perfect tool to study relations between structures in the different phases, as it is illustrated in Section 11.4.

A special application of superspace is in a unified approach to the crystal structures of the various members of homologous series, including commensurate and incommensurate compounds (Perez-Mato *et al.*, 1999). This approach is discussed in Section 11.6 along with a detailed presentation of the superspace description of $[Sr]_{1+x}[TiS_3]$ $(1.05 < x < 1.22)$ compounds.

Compilations of incommensurate compounds can be found in several review articles. An early account of insulating incommensurately modulated compounds has been given by Cummins (1990). A_2BX_4 compounds have been discussed by Hogervorst (1986) and Fabry and Perez-Mato (1994). More recent accounts are by Yamamoto (1993, 1996) and van Smaalen (1995). Inorganic misfit layer sulfides and selenides have been reviewed by Wiegers (1996). An oxidic counterpart ('misfit cobaltates') has been discovered more recently (Grebille *et al.*, 2004). Inorganic charge-density-wave compounds are presented by van Smaalen (2005) and Tsinde and Doert (2005). The incommensurate structures of the elements have been reviewed by McMahon and Nelmes (2004). Databases of incommensurate materials have been made available on the internet at the Bilbao Crystallography Server (`http://www.cryst.ehu.es/icsdb/index.html`) and by Caracas

(2002) at `http://www.mapr.ucl.ac.be/~crystal/index.html`.

11.3 The elements

The α form of uranium (space group $Cmcm$) is the stable modification of this element at ambient conditions. α-U transforms into an incommensurately modulated crystal at $T_{c1} = 43$ K (Lander *et al.*, 1994). Four, symmetry-related modulation wave vectors have been identified. Each has three incommensurate components, but information is not available on the symmetry or crystal structure of this phase. A second transition, at $T_{c2} = 38$ K, results in an incommensurately modulated structure with superspace group $P2(\frac{1}{2}\sigma_2\sigma_3)\bar{1}$. The components of the modulation wave vector continuously vary with temperature within the ranges $\sigma_2 = 0.15 \pm 0.02$ and $\sigma_3 = 0.20 \pm 0.02$. A third transition, at $T_{c3} = 23$ K, results in a commensurate modulation with $\mathbf{q} = (\frac{1}{2}, \frac{1}{6}, \frac{5}{27})$. This lock-in transition is accompanied by the appearance of high-order satellite reflections up to $|h_4| = 9$, representing a squaring of the modulation wave towards low temperatures (Marmeggi *et al.*, 1990; van Smaalen and George, 1987).

A charge-density-wave instability is believed to be responsible for the formation of the incommensurate modulation in α-U. Evidence for this interpretation has come from various experiments and from electronic band-structure calculations. It is in agreement with the small magnitude of the modulation (largest atomic displacement is 0.05 Å along $\mathbf{a}_1$) and with the observation of a soft phonon above T_{c1}. α-U is unique among the elements, because it is the only element that develops a charge-density wave, and because it is the only element with an incommensurate phase at ambient pressure.

Many elements undergo a series of phase transitions on increasing pressure. Consecutive phases are enumerated by roman numbers, starting with phase I at ambient conditions. High-pressure phases Sr-V, Rb-IV, Ba-IV, As-III, Sb-II and Bi-III have a self-hosting incommensurate composite crystal structure of tetragonal symmetry (see McMahon and Nelmes (2004) and references therein). The host or first subsystem forms a framework structure with channels parallel to the tetragonal axis. These channels accommodate chains of atoms comprising the second subsystem or guest (Fig. 11.2). Basic-structure periodicities of the host and the guest are mutually incommensurate along the tetragonal axis, with, for Bi-III, $a_{13}/a_{23} = 1.309$. This value is close to 4/3, but incommensurability has definitely been established from accurately measured positions of host and guest main reflections in X-ray powder diffraction in dependence on pressure.

Incommensurately modulated structures have been found at high pressures for the group-6 elements sulfur, selenium and tellurium. Required pressures are generally higher than for the metallic elements, with a record-high for sulfur. The incommensurately modulated phase S-IV is stable between 83 GPa and 153 GPa (Degtyareva *et al.*, 2005). Iodine is incommensurately modulated between pressures of 23.2 and 25.5 GPa (Takemura *et al.*, 2004).

Explanations for the ubiquitous occurrence of incommensurate phases among the elements follow two types of reasoning. First the observation is made that

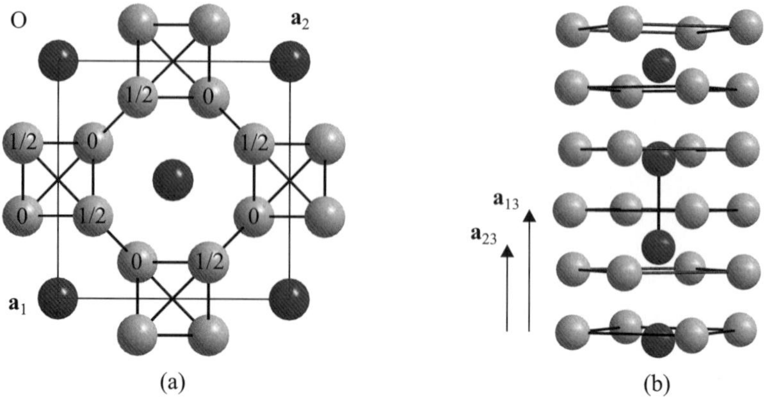

FIG. 11.2. The incommensurate composite crystal structure of Bi-III. (a) Projection of the basic structure along the mutually incommensurate axes $\mathbf{a}_{13}$ and $\mathbf{a}_{23}$. Atoms of the host (light grey) are located at $x_{13}^0 = 0$ and $1/2$ as indicated. Atoms of the guest are dark grey. (b) Perspective view of one channel, showing one supercell period of the supercell approximation $a_3^s \approx 3\,a_{13}$ $\approx 4\,a_{23}$ Coordinates from McMahon *et al.* (2000).

many elements exhibit complex crystal structures with large unit cells at high pressures, before they turn into simpler sphere packings at still higher pressures (Schwarz *et al.*, 1998; McMahon and Nelmes, 2004). Complex structures are found for the elements discussed above as well as for elements for which an incommensurate phase has not been reported. Then the observation is made that incommensurability is not a special state of matter, but is one form of a complex crystal structure. The same kind of competing interactions that are responsible for a stable periodic state with a large unit cell in one case, will lead to an incommensurate structure in other cases. Ample evidence has been compiled for the competition between periodic and aperiodic structures from numerical studies of simplified models, like the ANNNI and DIFFOUR models (Janssen, 1986). The problem reduces to a search for the explanation of the occurrence of complex structures at high pressures as opposed to the densest sphere packings that naively might have been expected. The source of competing interactions will be found in the electronic structure of the elements, but a detailed discussion of them is beyond the scope of the present exposure.

Incommensurability instead of periodicity can sometimes be rationalized on the basis of a crystal-chemical analysis. An example is provided by the composite crystal structure of Bi-III (Fig. 11.2). The incommensurate ratio of $a_{13}/a_{23} = 1.309$ is well approximated by $4/3$. Accordingly, an approximate crystal structure can be given on the basis of a threefold supercell of the host, alternatively described as a fourfold supercell of the guest. Following the procedures described in Section 5.3, and noticing that only the section $t = 1/6$ of superspace describes the

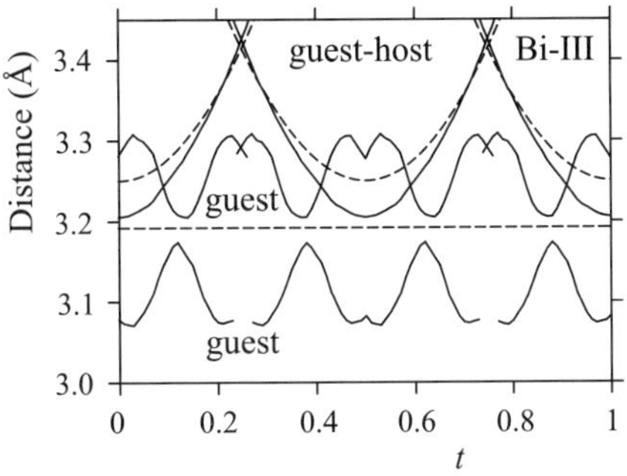

FIG. 11.3. t-Plot of interatomic distances in the incommensurate composite crystal structure of Bi-III. The central atom is the only crystallographically independent atom of the guest subsystem. Dashed lines represent distances in the basic structure (McMahon *et al.*, 2007).

diffraction data of the commensurate structure well, the supercell space group $P4/ncc$ is obtained from the superspace group $I'4/mcm(0\,0\,\sigma_3)0\bar{1}00$. t-Plots of interatomic distances indicate a quasi-pairing of Bi within the guest (Fig. 11.3). The intra-dimer distance of 3.12 Å is the shortest Bi–Bi distance in the structure, while quasi-dimers are only obtained if the inter-dimer distance of 3.26 Å is sufficiently long. The formation of quasidimers probably is the driving force for incommensurability. The guest subsystem is compressed along the tetragonal axis in the putative commensurate structure $P4/ncc$ as compared to the incommensurate structure. The average Bi–Bi distance along the chains is equal to a_{23} = 3.195 Å in the incommensurate structure, but it is only 3.151 Å in the approximate superstructure. Apparently, this difference of 0.04 Å would make too short intra-dimer distances—then destabilising the structure by Born repulsion—or it would make too short inter-dimer distances—then leading to a more regular chain and cancelation of the dimerisation, thus raising the energy of the crystal.

11.4 Phase transitions in A_2BX_4-type ferroelectrica

A_2BX_4 compounds with the β-K_2SO_4 type structure crystallize in space group $Pnma$ with four formula units in the unit cell. Two crystallographically independent A atoms, the single crystallographically independent B atom and two of the three crystallographically independent X atoms are located in the mirror plane, while each BX_4 tetrahedral group is completed by a pair of symmetry-related X atoms across the mirror plane (Fig. 5.4). In standard models of the structure, the shortest A–X distance appears shorter than the length of a stable A–X bond. More realistic descriptions of the atomic arrangements involve disorder of BX_4 tetrahedra over two orientations, or anharmonic temperature parameters describing large-amplitude librations of these groups, resulting in true A–X distances that are longer than the average distance observed in standard struc-

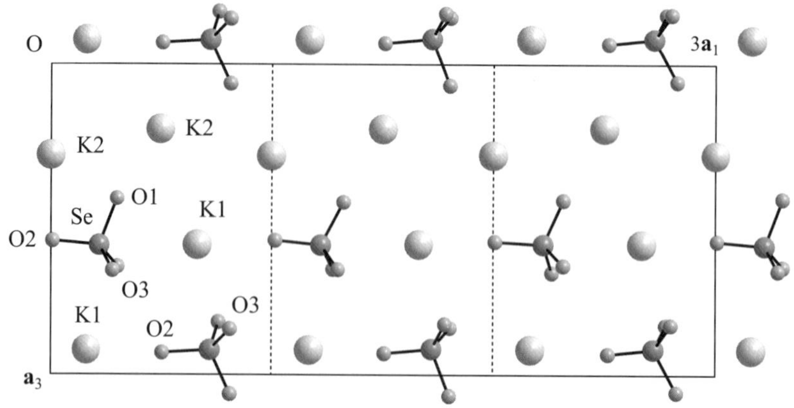

FIG. 11.4. Projection along $\mathbf{a}_2$ of the $3a_1 \times a_2 \times a_3$ superstructure of K_2SeO_4, stable below $T_{\mathrm{lock-in}} = 93$ K. Atoms occurring about the plane $x_3 = 0.25$ are shown. Three unit cells of the basic structure are separated by dashed lines. Compare to Fig. 5.4. Coordinates from Yamada *et al.* (1984).

ture models. Many A_2BX_4 compounds undergo phase transitions on cooling, leading to incommensurate and commensurate superstructures based on the development of a static modulation wave of differently oriented BX_4 groups (Fig. 11.4). Alleviation of the internal strain about the shortest A–X contact is generally believed to be the driving force for this transition (Fabry and Perez-Mato, 1994).

The typical series of phase transitions of A_2BX_4 compounds is, as a function of temperature,

$$\begin{array}{ccccc} \text{threefold} & \xleftrightarrow{T_{\mathrm{lock-in}}} & \text{incommensurately} & \xleftrightarrow{T_i} & Pnma \\ \text{superstructure} & & \text{modulated} & & \text{structure,} \end{array}$$

with $T_i > T_{\mathrm{lock-in}}$. On approaching T_i from above, a soft mode develops in the Σ_2 phonon branch at $\mathbf{q} \approx (\frac{2}{3}, 0, 0)$ (Iizumi *et al.*, 1977). Atomic displacements of Σ_2 symmetry are displacements of A into the direction perpendicular to the mirror plane, and rotations of the BX_4 group about an axis parallel to the mirror plane and going through the location of the B atom. Below T_i the soft mode condenses into a static wave with an incommensurate wave vector $\mathbf{q} = (\frac{2}{3}+\delta, 0, 0)$ and with atomic displacements that still follow the Σ_2 symmetry. $\delta \approx 0.02$ and it jumps to zero at $T_{\mathrm{lock-in}}$, but in first approximation the Σ_2 mode again defines the pattern of atomic displacements in the superstructure. Phonons are sinusoidal displacement waves. On approaching $T_{\mathrm{lock-in}}$, higher harmonic displacements develop within the incommensurate phase, as it has been established for Rb_2ZnCl_4 by temperature-dependent X-ray diffraction (Aramburu *et al.*, 2006). Higher harmonic waves may represent normal modes of different symmetries than the Σ_2 symmetry of the fundamental wave. The superspace approach summarizes all possible distortions of the structure as those harmonic displacement waves that are compatible with the single superspace group of the structure. The same superspace group defines the displacements in the incommensurate and commensurate phases. The supercell space groups of the latter are uniquely defined by the superspace group, as it is given in Table 5.2 for A_2BX_4 compounds. Relations between incommensurate and commensurate superstructures, and between sinusoidal and anharmonic displacement waves are better described by superspace groups than by consideration of the various normal modes contributing to the distortion.

11.5 Atomic valences in aperiodic crystals

The concept of valence is central to chemistry. In a simplistic picture, the valence of an atom is equal to the absolute value of the ionic charge in ionic compounds, and it is equal to the number of chemical bonds formed by this atom in covalent compounds. Valence is an integral number, but many elements may appear with different valences (different oxidation states), depending on the chemical compound they have formed.

Most chemical bonds are of mixed ionic/covalent character. Valence can then be identified with the total amount of bonding in which an atom participates. This definition is utilized in the bond-valence method, in which the valence of a central atom is related to the distances between this atom and the surrounding atoms, irrespective of the kind of bonding between them (Brown, 2002). The bond-valence method is a semi-empirical method that assigns an amount of bonding, ν_{ij}, to each pair of atoms i and j according to

$$\nu_{ij} = \exp[(R_0 - d_{ij})/b] \, . \tag{11.1}$$

The distance between atoms i and j is d_{ij}. R_0 is a parameter whose value depends on the chemical nature of i and j only, and b is a universal constant equal to 0.37 Å. Consideration of all atoms $j = 1, \cdots, N_i^{\mathrm{shell}}$ in the first coordination shell of atom i gives the valence of i as

$$V_i = \sum_{j=1}^{N_i^{\mathrm{shell}}} \nu_{ij} \, . \tag{11.2}$$

Values of R_0 have been determined for all pairs of elements by a fit of eqn (11.2) to structural data of compounds with atoms in well defined oxidation states. For example, R_0(Na–Cl) = 2.15 Å is directly obtained from the crystal structure of NaCl, assuming valence 1 for both Na and Cl, and employing six Cl atoms in the first coordination shell of Na. A complete and consistent set of bond-valence parameters has been determined by Brese and O'Keeffe (1991). They can be used to compute atomic valences from crystal structures for any compound.

One of the paradigms of chemistry is that atoms of each element have valences equal to one of the integral values typical for this element in all its compounds.

This value is recovered by the bond-valence method. Employing this property, the bond-valence method is used to distinguish between correct and incorrect crystal structures and to determine chemical order in cases where diffraction data might not be able to distinguish between elements differing by a few electrons only (Brown, 2002). Mixed-valence compounds are compounds in which atoms of a single element occur in two different oxidation states. Anharmonic temperature parameters of large magnitude often describe temporal variations of the environment of a particular atomic site, that correspond to environments accommodating the two possible valences of the atom at this site. As these variations are not captured by the average atomic positions in the structure model, the bond-valence method will lead to apparent fractional valence values in these cases, that represent the average of the two integral valences of this element.

Many aperiodic crystals are thermodynamically stable compounds. The constituent atoms should therefore have the same integral valence values as atoms of these elements in periodic crystals. For incommensurately modulated crystals the first coordination shell is well defined for all atoms, but distances of a central atoms towards its neighbours vary from one unit cell to the next. This variation has been summarized in t-plots (Fig. 11.1). Distance variations can be interpreted as variations of bond strengths. Accordingly, a t-plot of the individual bond-valences ν_{ij} can be defined as [eqn (11.1)],

$$\nu_{ij}(t) = \exp[(R_0 - d_{ij}(t))/b] . \tag{11.3}$$

Employing the property that t-plots are correlated to each other for all N_i^{shell} atoms in the first coordination shell of atom i, a t-plot of the valence of atom i is defined as [eqn (11.2)],

$$V_i(t) = \sum_{j=1}^{N_i^{\text{shell}}} \nu_{ij}(t) . \tag{11.4}$$

$V_i(t)$ summarizes the various valences of atom i, as they are found in the incommensurately modulated structure.

For $AuTe_2$ t-plots indicate large variations of the bond-valence of Au–Te bonds [(Fig. 11.5(a)], supporting the idea of a variation of the coordination of Au. Around $t = 0.25$ and 0.75 they indicate a fourfold coordination of Au by Te. However, the interpretation is less clear for values of t about 0.0 and 0.5. A linear coordination of Au—as originally proposed on the basis of distance plots—is not supported by the computed valences. Although strongest bonding is found for two Te atoms, the four other Te atoms in the first coordination shell together are responsible for about half of the valence of Au. A better description of the coordination thus is as a highly distorted octahedral coordination of Au by Te. The computed valence of Au varies between 1.81 and 2.16 [(Fig. 11.5(b)]. This variation is too small to be interpreted as a mixed-valence compound of Au^+ and Au^{3+}. Instead it indicates a striving towards constant valence of Au throughout the structure. Loss of bonding towards one Te (elongated bond) is compensated

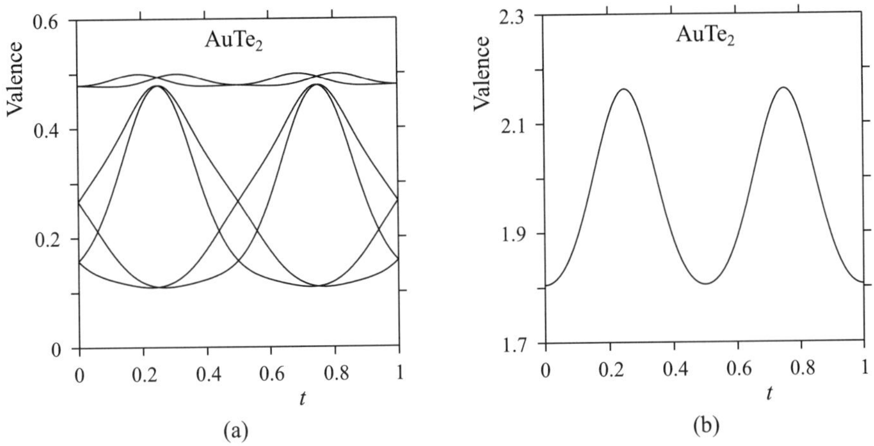

FIG. 11.5. Application of the bond-valence method to the incommensurately modulated structure of $AuTe_2$. (a) t-Plot of bond-valences of bonds between a central Au atom and six Te atoms in the first coordination shell. (b) t-Plot of the valence of Au. Coordinates from Schutte and de Boer (1988).

by shortening of another Au–Te bond, as it is apparent from the valence calculation: The variation of computed valence of Au is equal to the variations of bond-valences of each of the four longest Au–Te bonds. The remaining variation of the valence of Au can be due to inaccuracies in the modulation functions, and especially to missing higher-order harmonics in these functions.

Applications of the bond-valence method to a series of incommensurate compounds have confirmed that each atom has its required valence throughout the crystal structure. Variations with t of the distance of one pair of atoms are always compensated by complementary variations of the distance towards another neighbouring atom (van Smaalen, 1999). In all calculations on incommensurate crystals, values of R_0 have been employed as they have been derived from periodic crystal structures (Brese and O'Keeffe, 1991). This again affirms the equal basis of chemical bonding in incommensurate and periodic crystals.

t-Plots of interatomic distances in incommensurate composite crystals have been introduced in Section 4.8. Characteristic of composite crystals is that distances between atoms of different subsystems assume any value larger than a distance of minimum approach [Fig. 4.11(f)]. This prevents a meaningful definition of a first coordination shell of atoms, because for any threshold there will be atoms in the structure with coordinating atoms at distances just below and other atoms with coordinating atoms at distances just above the threshold value. Bond-valence parameters have been derived from periodic crystal structures for summations over the first coordination shell only [eqn (11.2)]. However, the exponential dependence of bond-valences on interatomic distances [eqn (11.1)] determines that nearly equal valence values are obtained if all atoms are included into the sums of eqns (11.2) and (11.4). A renormalization of bond-valence parameters would be required, but the corrections are small, and the available values of R_0 can be employed in good approximation for computation of the valence according to the alternate definition of valence (van Smaalen, 1992*a*):

$$V_i(t) = \sum_{j=1}^{\infty} \nu_{ij}(t) \,. \tag{11.5}$$

The sum extends over all atoms $j \neq i$ in the structure. Contributions to the valence are negligible for neighbours beyond $\sim$5 Å, and this distance can be used as cut-off in the numerical evaluation of the valence.

Bond-valences have been computed for the inorganic misfit layer compound $[LaS]_{1.14}[NbS_2]$ (Jobst, 2003). The t-plot of bond-valences between La and the five nearest neighbour S atoms within the LaS subsystem (S2 atoms) is typical for a t-plot of an incommensurately modulated structure [Fig. 11.6(a)]. Increased bonding (higher valence value) towards one S atom is compensated by reduced bonding (lower valence value) towards another S atom. Bond-valences between La as central atom and all neighbouring S atoms of the NbS_2 subsystem produce a t-plot with an inverted appearance as compared to the t-plot of interatomic distances [Fig. 11.6(b); compare to Fig. 4.11(f)]. It indicates significant non-zero contributions to the valence of La from four or five S1 atoms. Instead of a first coordination shell of 5 + 1 atoms, the composite character implies up to five partial bonds between La and S atoms of the NbS_2 subsystem.

The computed valence of La is obtained as 3.13, with a remaining variation in dependence on t of ± 0.034 [Fig. 11.6(e)]. This value is in accordance with the expected oxidation state of rare earth elements. Contributions to the valence of La can separately be computed for S1 and S2 type atoms [Fig. 11.6(c),(d)]. Compensating modulations leading to a nearly constant valence are found to be valid within each subsystem, with a partial valence of 2.15 within the LaS subsystem and a partial valence of 0.98 for bonding of La towards S atoms of the NbS_2 subsystem. Similar calculations for a series of inorganic misfit layer compounds have been used to characterize different types of these compounds (Petricek *et al.*, 1993). Compounds $[ReS]_x[NbS_2]$ with Re = rare earth have strong intersubsystem bonding, with a contribution of intersubstem Re–S contacts of $\sim$1 towards the valence of 3 of Re. Compounds with Re = Pb, Sn have weak intersubsystem bonding with a contribution of intersubsystem Re–S contacts of $\sim$0.3 to the valence of 2 of Re.

11.6 Homologous series of compounds

The homologous series Mo_nO_{3n-1} (n = integer) is defined as a collection of compounds with chemical compositions and crystal structures that depend on the parameter n in a systematic way (Magnéli, 1953). Crystal structures are derived from the ReO_3 structure type. The latter is characterized by an infinite array of corner-sharing MoO_6 octahedral groups with cubic symmetry, thus defining

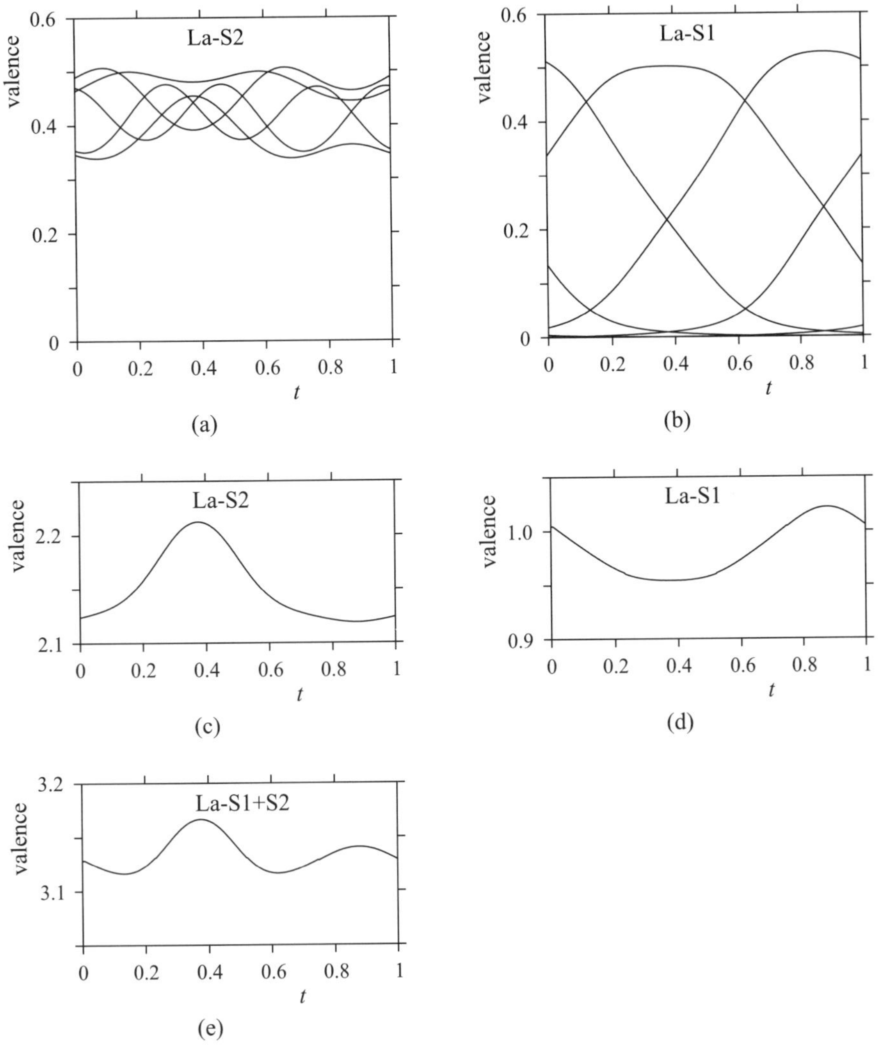

FIG. 11.6. The bond-valence method applied to the inorganic misfit layer compound $[LaS]_{1.14}[NbS_2]$. (a) Bond valences $\nu_{i,j}(t)$ between i = La and five nearest-neighbour S atoms of the LaS subsystem [eqn (11.3)]. (b) Bond valences between La and S atoms of the NbS_2 subsystem. (c) Contribution to the valence of La from S atoms of the LaS subsystem (average value 2.15). (d) Contribution to the valence of La from S atoms of the NbS_2 subsystem (average value 0.98). (e) Valence of La (average value 3.13). Coordinates from Jobst (2003).

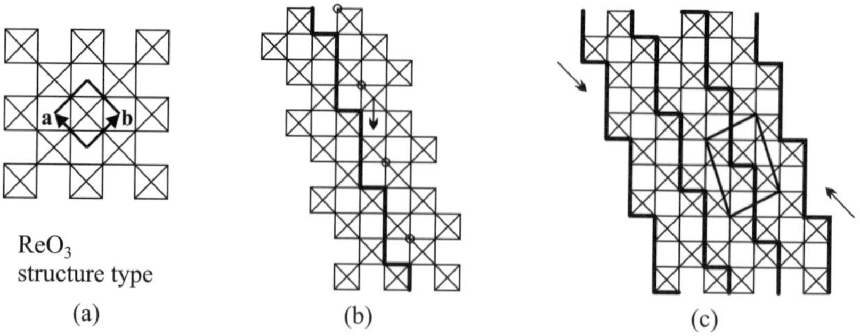

FIG. 11.7. Generation of the homologous series Mn_nO_{3n-1} by shear. (a) One layer of the ReO_3 structure type. The primitive cubic unit cell is indicated. (b) Patch of ReO_3 type structure with the shear plane $(\bar{1}\,2\,0)$ and the shear vector $[\frac{1}{2}, \frac{1}{2}, 0]$ superimposed. Circles denote oxygen atoms that need to be removed after the shear operation is completed. (c) Sheared structure for $n = 4$. Each block of ReO_3 structure is n octahedra wide, as measured in the direction of $\mathbf{a}_{\text{cubic}}$ (direction of the arrows). The supercell is indicated. A square with a cross represents the top view of an octahedral group MnO_6.

the $n = \infty$ end member of the homologous series [Fig. 11.7(a)]. A stacking fault can be introduced on the $(\bar{1}\,2\,0)$ plane of the ReO_3 structure type [heavy line in Fig. 11.7(b)] by displacement of the material to the right of this plane over a distance $[-\frac{1}{2}, -\frac{1}{2}, 0]$ with respect to material to the left of this plane [arrow in Fig. 11.7(b)]. This operation brings forth a condensation of MoO_6 octahedra into groups of four octahedra that share edges instead of vertices on the fault plane [Fig. 11.7(c)]. Edge-sharing octahedra require less oxygen atoms than vertex-sharing octahedra do, and some of the oxygen atoms of the ReO_3 structure type are superfluous after the shear operation is completed [circles in Fig. 11.7(b)], because they would occupy positions already occupied by other oxygen atoms. Introduction of this type of stacking fault at regular intervals produces a periodic structure composed of slabs with the ReO_3 structure type, that are separated from each other by the 'faulty' stacking described above. In particular, each slab can be made n octahedra wide, then resulting in a superstructure with n octahedra in the supercell, and with a composition Mn_nO_{3n-1}.

The volume of the supercell is slightly less than n times the volume of the cubic ReO_3 type unit cell, because of the increased condensation of octahedra. Additional complications arise due to the strain on the shear planes, that is relieved by further structural distortions, leading to a doubling of the supercell. Compounds $n = 8$, 9 and 10 have been synthesized for the homologous series Mo_nO_{3n-1}, while different n have been realized if Mn is substituted by W (Magnéli, 1953). A property of homologous series is that different unit cells are required to describe the crystal structures of different members of the series, despite the obvious relations between their structures, and despite the nearly

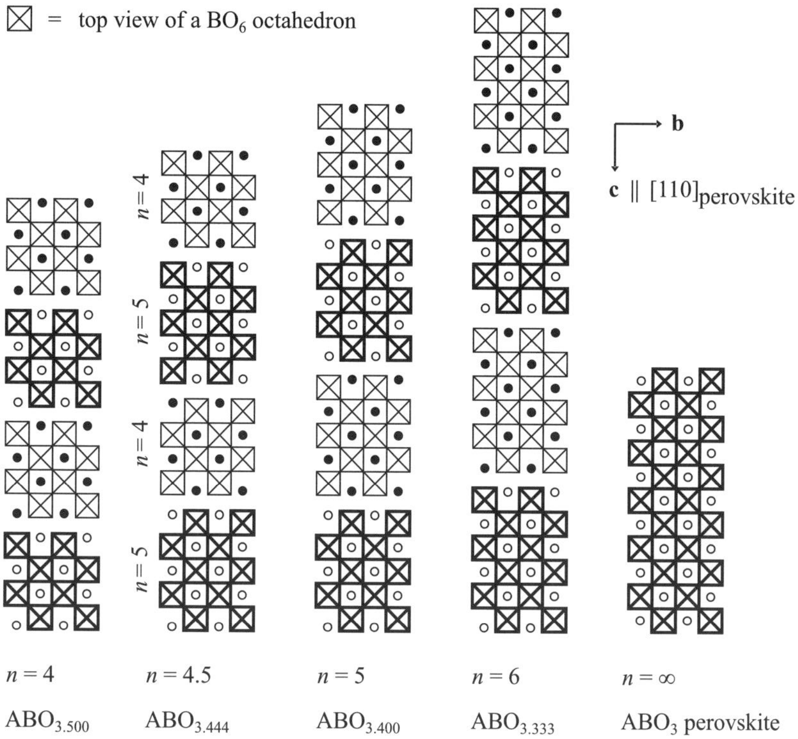

FIG. 11.8. Schematic representations of crystal structures of the homologous series $A_nB_nO_{3n+2}$ for n = 4, 4.5, 5, 6 and ∞. Squares with crosses represent top views of BO_6 octahedra, with thick lines and thin lines indicating octahedra with different values of the projected coordinate. Filled and open circles denote A atoms with different values of the projected coordinate. Courtesy L. Lichtenberg. Based on Lichtenberg *et al.* (2001).

equal compositions. For Mn_nO_{3n-1} the chemical composition can be expressed as $MnO_{3-(1/n)}$; for $n = 8$, 9 the oxygen contents differs by 0.5 % only.

The perovskite structure type can be obtained from the ReO_3 structure type by the introduction of cations A into the cavities bordered by eight octahedra BO_6, resulting in the composition ABO_3. The lattice is cubic with one unit ABO_3 in the unit cell. The presence of A cations allows oxygen-rich compounds to be formed, with additional oxygen located on shear planes, as it is, for example, realized in the homologous series $A_nB_nO_{3n+2}$ with chemical compositions ABO_{3+x} ($x = 2/n$; Lichtenberg *et al.*, 2001). Slabs of n octahedra wide of perovskite-type structure are separated by shear planes (1 1 0). Across these planes the structure is displaced by the shear vector $[\frac{1}{2}, \frac{1}{2}, \frac{1}{2}]$ (Fig. 11.8). Strain is present due to the irregular environments of the A type cations located at the shear planes. It is resolved by the formation of superstructures or by an incommensurate modulation of the positions of the A type cations and of the orientations of the BO_6 octahedra. An incommensurate superstructure is, for example, realized in the $n = 4$ member $Sr_2Nb_2O_7$ (Section 2.4.2). The $A_nB_nO_{3n+2}$ homologous series has a rich chemistry, with A equal to a mixture of alkaline, alkaline earth and rare earth elements and with B equal to a mixture of Nb, Ta and Ti (Lichtenberg *et al.*, 2001). Interest in this class of materials derives from properties like high-temperature ferroelectricity in insulating $Sr_2Nb_2O_7$ ($n = 4$) and low-dimensional electronic properties in $Ca_5Nb_5O_{17}$ ($n = 5$). Compounds have been synthesized for $n = 2$, 4, 5 and 6, but also for the mixed-layer sequences $\cdots 454545 \cdots$ and $\cdots 445445445 \cdots$, corresponding to compositions $n = 4.5$ and $n = 4.33$, respectively (Fig. 11.8). Like Magnéli phases, different compounds (different n) possess different unit cells despite related compositions and related structures. Furthermore, different compounds may have different space groups, with $n =$ odd being centrosymmetric and $n =$ even being acentric.

Interrelations between compounds within a homologous series are revealed by superspace models for their crystal structures, as it has been proposed by Perez-Mato *et al.* (1999) for the homologous series $A_{3n+3m}A'_nB_{3m+n}O_{9m+6n}$ (n, m integers). These compounds are based on the $2H$ hexagonal perovskite structure ABO_3, the $m = 1$, $n = 0$ end member of this series. The hexagonal perovskite structure is characterized by columns of face-sharing BO_6 octahedra that are separated by collinear rows of A atoms (Fig. 11.9). The $m = 0$, $n = 1$ end member of composition $A_3A'BO_6$ is obtained from the hexagonal perovskite structure by replacing every second octahedron by a $A'O_6$ trigonal prismatic group that shares its trigonal faces with the faces of neighbouring BO_6 octahedra. Different members of this homologous series correspond to different sequences of octahedral and trigonal prismatic groups. Many compounds have been synthesized that belong to either of the five structure types displayed in Fig. 11.9 (Stitzer *et al.*, 2001). It is easily conceived, however, that an irrational ratio of the numbers of prisms and octahedra should be possible too. These compounds are more easily visualized when the the chemical composition is rewritten as $[A]_{1+x}[A'_xB_{1-x}O_3]$ with $x = \frac{n}{3m+2n} \leqslant \frac{1}{2}$. The ratio of numbers of prisms and octahedra is defined by x as $x : (1 - x)$. The variable x also defines the number of A cations that is present in the structure for each $(A', B)O_6$ polyhedron. The value of $(1 + x)$ reflects the property that prisms are longer than octahedra, as measured along the direction of the columns.

Almost equal values of x require widely different combinations of n and m. In a conventional crystallographic description they would be assigned unit cells of quite different sizes, despite similar compositions. The superspace approach describes these compounds as composite crystals, with $(A'_xB_{1-x})O_3$ as first subsystem and A as second subsystem. The superspace model depends continuously on the composition x, and thus provides the required relationships between compounds of similar compositions. The superspace group is $R\bar{3}m(0\,0\,\sigma_3)\bar{1}s$, while space groups of commensurate members of the homologous series can be obtained as special sections of superspace, employing the methods presented in Chapter 5.

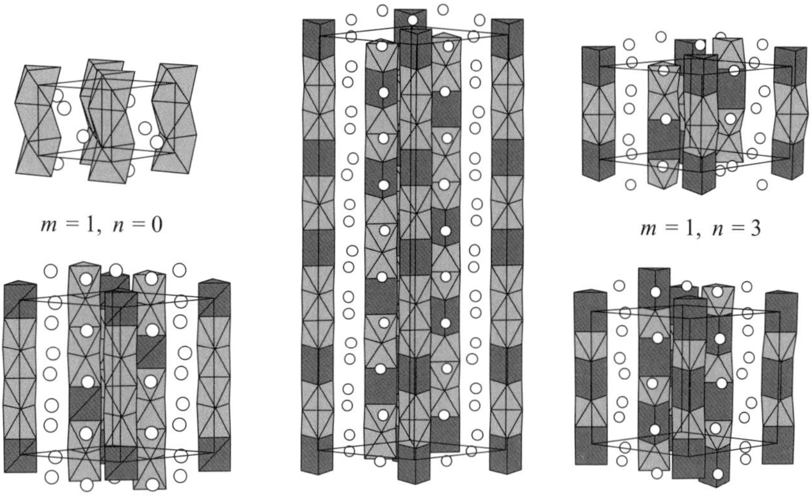

FIG. 11.9. Schematic representations of crystal structures of the homologous series $A_{3n+3m}A'_nB_{3m+n}O_{9m+6n}$ for $m = 1$ with $n = 0, 1, 2, 3$ and for $m = 0$ with $n = 1$. Octahedra light grey, trigonal prisms dark grey, and A atoms as circles. Reprinted with permission from Stitzer *et al.* (2002), copyright (2002) by Elsevier.

The pecularities of the superspace descriptions of homologous series are introduced here for a sulfide counterpart of hexagonal perovskites: $[\mathrm{Sr}]_{1+x}[\mathrm{TiS}_3]$. A superspace model based on harmonic modulation functions was first proposed by Onoda *et al.* (1993), who also reported the synthesis of a series of compounds $[\mathrm{Sr}]_{1+x}[\mathrm{TiS}_3]$ with compositions $0.05 < x < 0.22$. Here the analysis by Gourdon *et al.* (2000) is followed. Compounds $[\mathrm{Sr}]_{1+x}[\mathrm{TiS}_3]$ fit into the homologous series of hexagonal perovskites with A = Sr and $A' = B$ = Ti. The two subsystems share the basal plane of a hexagonal lattice, with $\mathbf{a}_1 = \mathbf{a}_{11} = \mathbf{a}_{21}$ and $\mathbf{a}_2 = \mathbf{a}_{12} = \mathbf{a}_{22}$ [Fig. 1.4(d)]. The first subsystem consists of columns of composition TiS_3. It has a rhombohedral basic structure (R centring in Table 3.9), with a basic-structure periodicity $\mathbf{a}_3 = \mathbf{a}_{13}$ along the threefold axis. Accounting for the different chemistry of sulfides and oxides, Ti is found in octahedral coordination by S (Oh coordination) and in a coordination intermediate between octahedral and trigonal prismatic coordinations [PO coordination; compare Fig. 1.4(c)]. The latter type of coordination replaces the trigonal prismatic coordination as it is found in the oxides. Nevertheless, PO groups are longer than Oh groups, and the basic-structure period $a_3 = a_{13}$ is determined by the relative concentrations of PO and Oh coordinations, *i.e.* by the ratio $x : (1 - x)$. The second subsystem consists of chains of Sr atoms, with a basic-structure periodicity $\mathbf{a}_{23}$ along the threefold axis. The lattice is trigonal, but the basic-structure unit cell $\{\mathbf{a}_1, \mathbf{a}_2,$

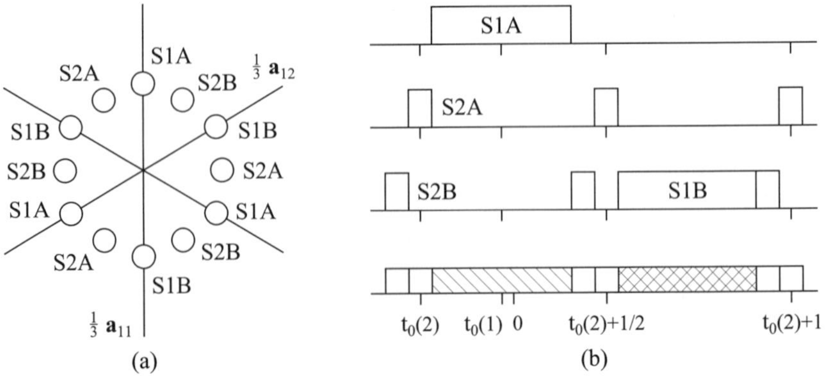

FIG. 11.10. Crystal structure of $[Sr]_{1+x}[TiS_3]$. (a) Twelve sulfur sites in the basic-structure unit cell at $x_{13}^0 \approx 0.5$ (compare Fig. 1.4). (b) Block wave functions representing the occupational modulation functions of the atoms S1A, S2A, S1B and S2B, and all atoms merged together (from top to bottom). In each case the region of value 1 is indicated. Coordinates from Gourdon *et al.* (2000).

$\mathbf{a}_{23}\}$ is H centred. With a matrix W^2 corresponding to the interchange of axes $\mathbf{a}_3^* = \mathbf{a}_{13}^*$ and $\mathbf{a}_4^* = \mathbf{a}_{23}^*$ [eqn (4.13)], the superspace lattice of the second subsystem appears H' centred (Table 3.9). The relation between the subsystems is characterized by the fourth reciprocal vector $\mathbf{a}_4^* = (0, 0, \sigma_3)$ with $\sigma_3 = a_{13}/a_{23} = \frac{1}{2}(1+x)$.

The basic-structure unit cell of TiS_3 contains 12 different sulfur positions for each column, that have almost the same coordinate $x_{13}^0 \approx 0.5$ [Fig. 11.10(a)]. All known commensurate superstructures possess threefold rotational symmetry, so that these 12 positions split into four groups of three positions. The sites within a group are simultaneously occupied or vacant, while only one of the four groups of sites—denoted by S1A, S1B, S2A and S2B, respectively—can be occupied in each basic-structure unit cell. This feature is encoded within the superspace approach by complementary modulation functions for the four sulfur sites.

The occupational modulation function of a single site is given by a block wave that is defined by value 1 for an interval $\Delta(\mu)$ centred on $x^{bw}(\mu)$ on the $\bar{x}_{s4}$ axis [eqn (7.27)]. Figure 11.10(b) gives $p^\mu(\bar{x}_{s4})$ as a function of the physical space section $t = \bar{x}_{x4}(\mu) - \sigma_3\, x_{13}^0(\mu)$, with the centre of the block wave at $t_0(\mu) = x^{bw}(\mu) - \sigma_3\, x_{13}^0(\mu)$. Site S1A is located on a twofold axis of the superspace group $R\bar{3}m(0\,0\,\sigma_3)\bar{1}s$ at $(x_{11}^0, x_{11}^0, \frac{1}{2})$ in the basic-structure unit cell ($x_{11}^0 \approx 0.169$). The centre of the block wave is restricted by symmetry to $x^{bw}(\text{S1A}) = 1/4$, while its width $\Delta(\text{S1A})$ may be chosen to match the desired concentration of S1A type sites in the structure. Experimentally observed compositions of $x \approx 1/8$ correspond to modulation wave vectors $\sigma_3 = \frac{1}{2}(1+x) \approx 9/16$, and a fraction of $2(2\sigma_3 - 1) = 2x$ of S1A sites is replaced by S2A sites, as it is expressed by the occupational modulation function [top row of Fig. 11.10(b)],

$$\Delta(\mathrm{S1A}) = \tfrac{1}{2} - (2\sigma_3 - 1) = \tfrac{1}{2} - x \tag{11.6a}$$
$$x^{bw}(\mathrm{S1A}) = \tfrac{1}{4}\,. \tag{11.6b}$$

Sites S1B are obtained by application of the mirror operator $\{m_x\, 1\,|\, 0, 0, 0, \frac{1}{2}\}$ to site S1A. Basic-structure positions are transformed towards $(0,\, x_{11}^0,\, \frac{1}{2})$ and the phase of the modulation is shifted by 1/2, resulting in a modulation function of S1B that is shifted by exactly 1/2 on the t axis with respect to the modulation function of S1A [third row of Fig. 11.10(b)]. As a consequence, sites S1A and S1B are never simultaneously occupied within the same copy of the basic-structure unit cell, as long as $\Delta(\mathrm{S1A}) = \Delta(\mathrm{S1B}) \leqslant \frac{1}{2}$. The limiting case corresponds to $x = 0$, and it defines a structure with octahedral coordinations of Ti only.

Site S2A is located on the mirror $\{m_x\, 1\,|\, 0, 0, 0, \frac{1}{2}\}$ at basic-structure position $(x_{11}^0,\, 2x_{11}^0,\, x_{13}^0)$ with $x_{11}^0 \approx 0.095$ and $x_{13}^0 \approx 0.52$. The glide component along $\bar{x}_{s4}$ determines that the occupational modulation function of S2A consists of two disjunct parts that are shifted with respect to each other by $\frac{1}{2}$. The interval of occupancy 1 for the first part follows from the parameters [eqn (7.27)],

$$\Delta(\mathrm{S2A}) = \tfrac{1}{2}[\tfrac{1}{2} - \Delta(\mathrm{S1A})] = \tfrac{1}{2}x \tag{11.7a}$$
$$x^{bw}(\mathrm{S2A}) = \sigma_3(x_{13}^0(\mathrm{S2A}) - \tfrac{1}{2}) + x^{bw}(\mathrm{S1A}) - \tfrac{1}{2}\Delta(\mathrm{S1A}) - \tfrac{1}{2}\Delta(\mathrm{S2A})\,, \tag{11.7b}$$

while the second interval has a centre shifted by 1/2 [second row of Fig. 11.10(b)]. Sites S2B and their modulation functions are obtained from those of S2A by the application of a twofold axis.

The condition that at each value of t exactly one of the sites S1A, S1B, S2A and S2B is occupied determines that the widths of the occupational block waves must add up to 1. It is easily checked that $2\Delta(\mathrm{S1A}) + 4\Delta(\mathrm{S2A})$ is equal to 1 indeed. Secondly, this condition implies a distance between the centres of the block waves of $t_0(\mathrm{S1A}) - t_0(\mathrm{S2A}) = \frac{1}{2}\Delta(\mathrm{S1A}) + \frac{1}{2}\Delta(\mathrm{S2A})$ on the t axis [bottom row of Fig. 11.10(b)]. Translation of this condition towards a relation between centres on the $\bar{x}_{s4}$ coordinate axis requires the different basic-structure coordinates of S1A and S2A to be taken into account. With σ_3 as the only non-zero component of the modulation wave vector this amounts to an additional phase shift equal to the first term in eqn (11.7b). Notice that the phase shift is zero if $x_{13}^0(\mathrm{S2A}) = 0.5$.

A concise representation of occupational modulation functions of complementary sites is obtained by juxtaposition of differently coloured blocks along the $\bar{x}_{s4}$ axis [bottom row of Fig. 11.10(b)]. Each block represents the interval on the $\bar{x}_{s4}$-axis where the corresponding site is occupied. This representation allows the construction of two-dimensional projections of the superspace structure model of $[\mathrm{Sr}]_{1+x}[\mathrm{TiS_3}]$. Figure 11.11(a) gives the projection of the occupational modulation functions onto the (x_{s3}, x_{s4})-plane. It shows that Ti (a continuous block centred on the origin) is coordinated by a group of three sulfur atoms at

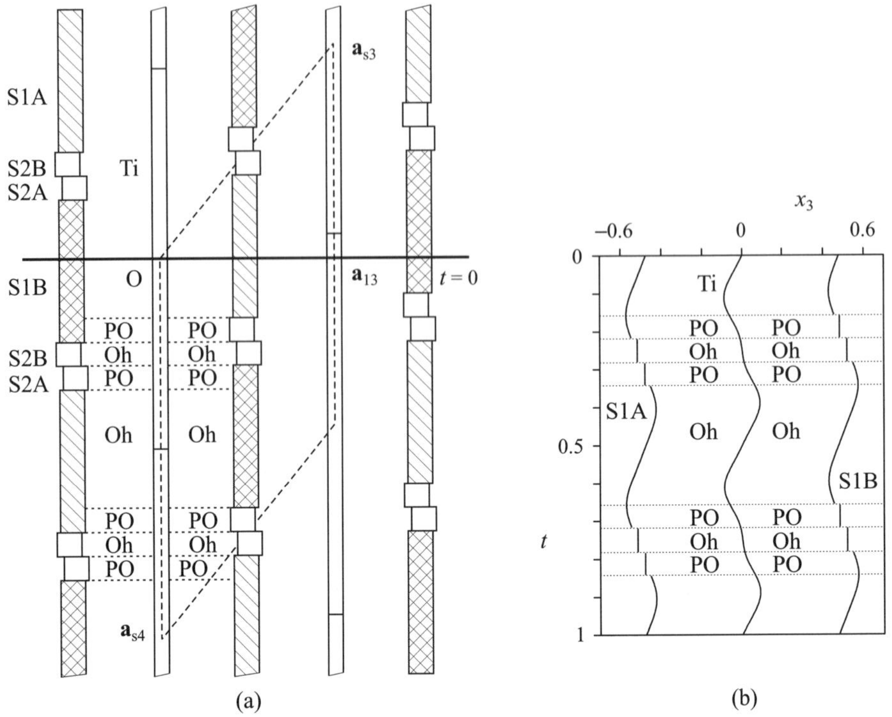

FIG. 11.11. Projection onto the (x_{s3}, x_{s4})-plane of the crystal structure of the TiS_3 subsystem of $[Sr]_{1+x}[TiS_3]$. (a) Occupational modulation functions of atoms Ti, S1 and S2. (b) t-Plot of the modulated positions of atoms Ti, S1 and S2, showing the modulation along $\mathbf{a}_3$. Coordinates from Gourdon *et al.* (2000).

smaller x_3^0 and by a group of three sulfur atoms at higher x_3^0, resulting in two types of coordinations: Oh coordination for S1A–Ti–S1B and S2A–Ti–S2B sequences, and PO coordination for S1A–Ti–S2A, S1A–Ti–S2B, S1B–Ti–S2A, and S1B–Ti–S2B sequences. Other sequences, like S1A–Ti–S1A, do not occur. Figure 11.11 demonstrates the relations between crystal structures of compounds of different compositions. A different composition (different x) results in a different length of the modulation wave vector with $\sigma_3 = \frac{1}{2} + \frac{1}{2}x$, but it also implies different widths of the occupational domains [eqns (11.6a) and (11.7a)], such that physical-space sections of superspace contain Oh and PO coordinations of Ti only, while sequences like S1A–Ti–S1A remain forbidden.

In this way, a unified superspace approach is obtained for all members of a homologous series. All compounds are described by a single superspace group and almost equal superspace and basic-structure unit cells. Structure models of different compounds depend on the composition in a systematic way, through the continuous dependence of the widths of the occupational domains on x. Matching widths of occupational domains of different sulfur sites—as expressed by eqns (11.6) and (11.7) and as displayed in Fig. 11.11(a)—define the closeness condition for occupational domains, similar to the condition introduced for quasicrystals (Cornier-Quiquandon *et al.*, 1991).

Chemistry requires metal-to-sulfur distances to lie within narrow ranges about the optimal values of metal–sulfur bond lengths. A basic structure combined with occupational modulations is but a poor approximation to the real structure, and displacive modulations are intrinsic to superspace models for crystal structures of modular compounds. Major part of the modulation is a displacement of the atoms along $\mathbf{a}_3$. For the specific structure type of hexagonal perovskites (Fig. 11.9), Perez-Mato *et al.* (1999) have shown that the condition of equal metal–sulfur distances implies a saw-tooth-shaped modulation function. This shape is well reproduced by the crystal structure of $[\mathrm{Sr}]_{1.2872}[\mathrm{NiO}_3]$. For the slightly more complicated crystal structure of $[\mathrm{Sr}]_{1+x}[\mathrm{TiS}_3]$, modulation functions have a zig-zag shape, that is again well reproduced by the refined structure model [Fig. 11.11(b)].

11.7 The layer model for homologous series

A second approach to the crystal structures of modular compounds is based on the stacking of atomic layers, reminiscent of the description of crystal structures of closest packings of spheres in terms of various stackings of three types of layers. This approach has been formulated for the hexagonal-pervoskite homologous series by Schönleber *et al.* (2006), and it has been applied to the oxygen-rich perovskite family $A_nB_n\mathrm{O}_{3n+2}$ (Fig. 11.8).

Elcoro *et al.* (2001) have shown that the perovskite structure type can be described as the alternate stacking along $[1,\, 1,\, 0]_{\mathrm{perovskite}}$ of layers of composition ABO and layers of pure oxygen. The latter are denoted by O, while ABO layers occur in two positions, displaced with respect to each other by $\frac{1}{2}\mathbf{a}_2 = [-\frac{1}{2},\, \frac{1}{2},\, 0]_{\mathrm{perovskite}}$, and denoted by the symbols M and N, respectively. The perovskite structure is represented by the periodic stacking sequence $\cdots MONO \cdots$. Stacking faults leading to the $A_nB_n\mathrm{O}_{3n+2}$ homologous series can be described by a missing ABO layer, that is indicated by the symbol $\emptyset$. Layers are displaced by $\frac{1}{2}\mathbf{a}'_1 = [0,\, 0,\, \frac{1}{2}]_{\mathrm{perovskite}}$ across a stacking fault, and the displaced layers are given the symbols M', N' and O'. The ideal crystal structures of $A_nB_n\mathrm{O}_{3n+2}$ compounds can then be given as periodic sequences of layers, *e.g.* as $\cdots OMONOMONO\emptyset O'N'O'M'O'N'O'M'O'\emptyset \cdots$ for $n = 4$ [Fig. 11.12(a)]. The supercell is orthorhombic with basis vectors $\mathbf{a}'_1$ and $\mathbf{a}_2$ parallel to the layers and with $\mathbf{a}_3 \approx (n+1)\; [1,\, 1,\, 0]_{\mathrm{perovskite}}$ perpendicular to the layers. The superspace description is based on an orthorhombic basic-structure with basis vectors $\mathbf{a}'_1$, $\mathbf{a}_2$ and $\mathbf{a}_3^0 \approx [1,\, 1,\, 0]_{\mathrm{perovskite}}$. The same occupational modulation functions apply to all atoms within a single layer, while different layers M, M', N, N', O and O' obtain complementary or matching modulations. The perovskite structure is not modulated ($\mathbf{q} = \mathbf{0}$), but a superspace model can be constructed of

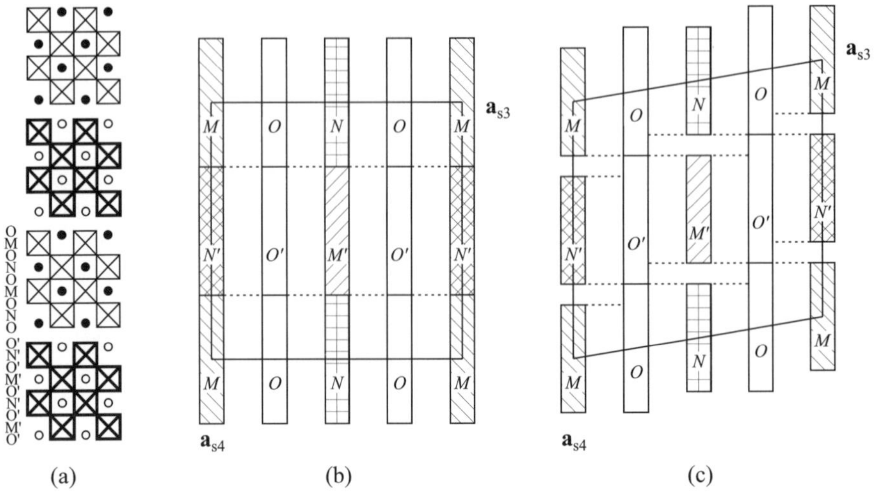

FIG. 11.12. Occupational modulation functions of the superspace model for the idealized crystal structures of compounds within the $A_nB_nO_{3n+2}$ homologous series. (a) Layer symbols superimposed onto the physical-space structure of $n = 4$ (compare Fig. 11.8). (b) Perovskite structure ($n = \infty$) with zero modulation. (c) Finite n with $\mathbf{q} = (0,\, 0,\, \frac{1}{n+1})$ and with the closeness condition indicated by dotted lines. Domains of occupancy 1 are indicated for each of the six layer types, M, M', N, N', O and O'. Based on Elcoro *et al.* (2001).

occupational modulation functions of widths $\Delta = 0.5$ and centres $x^{bw} = 0$ for M, N and O, and $x^{bw} = 0.5$ for M', N' and O' [eqn (7.27)]. A physical-space section either gives a periodic structure $MONO$ or a periodic structure $M'O'N'O'$, depending on the value of t [Fig. 11.12(b)]. Both sequences are equivalent representations of the perovskite structure. One out of $n+1$ layers ABO is missing in the compound $A_nB_nO_{3n+2}$. In the superspace model this is described by reducing the widths of the occupational domains of M, M', N and N' layers towards,

$$\Delta^{(n)} = \frac{1}{2}\left(1 - \frac{1}{n+1}\right).$$

Concomitantly the modulation wave vector is $\mathbf{q}^1 = (0,\, 0,\, \frac{1}{n+1})$, with the result that primed and unprimed layers never stack upon each other, but are always separated by an 'empty' layer [Fig. 11.12(c)]. The superspace group of this idealized structure is $F''mmm(0\,0\,\sigma_3)00\bar{1}$ (Table 3.9).

The idealized structure incorporates unfavourable coordinations of the atoms on the shear planes. Displacive modulations thus are intrinsic to the real structures of these compounds. The situation is more complicated than for the hexagonal perovskites, because displacive modulations include a second modulation with modulation wave vector $\mathbf{q}^2 = (\sigma_1,\, 0,\, 0)$ that has been found to be either

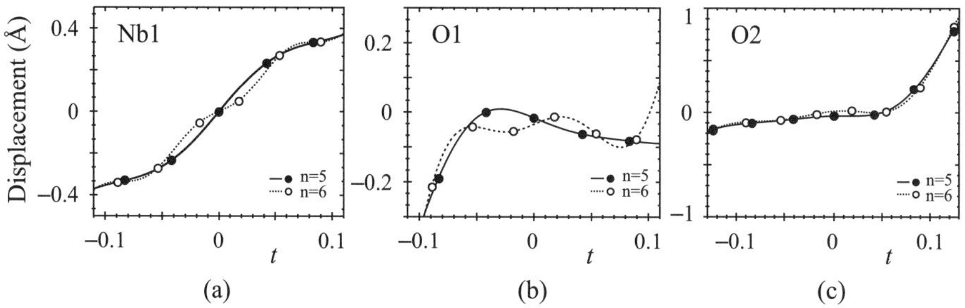

FIG. 11.13. Displacement modulation functions for selected atoms in the superspace structure models of $Ca_n(Nb,Ti)_nO_{3n+2}$ ($n = 5, 6$). (a) Nb1 atom; (b) O1 atom of the ABO layers; and (c) O2 atom of the O layers. t-Plots are given of the components $u_2(\bar{x}_{s4})$ for $n = 5$ (filled circles and solid lines) and for $n = 6$ (open circles and dashed lines). Reprinted with permission from Guevarra (2006), copyright (2006) by the IUCr.

incommensurate or commensurate. A (3+2)-dimensional superspace approach would be necessary (Elcoro *et al.*, 2003). Restricting the analysis to commensurate modulations with $\sigma_1 = \frac{1}{2}$, allows a (3+1)-dimensional superspace model to be used, with a basic structure given by $\mathbf{a}_1 = 2\mathbf{a}'_1$, $\mathbf{a}_2$ and $\mathbf{a}_3^0$. The appropriate subgroup of $F''mmm(0\,0\,\sigma_3)00\bar{1}$ is monoclinic with the symbol $X2_1/d(\sigma_1\,0\,\sigma_3)\bar{1}0$ of a non-standard setting with centring translations $(\frac{1}{2}, 0, 0, \frac{1}{2})$ $(\frac{1}{4}, \frac{1}{2}, 0, \frac{3}{4})$ and $(\frac{3}{4}, \frac{1}{2}, 0, \frac{1}{4})$, and with 'accidentally' $\sigma_1 = 0$. Superspace refinements have been presented for the $n = 5, 6$ members of the homologous series $Ca_n(Nb,Ti)_nO_{3n+2}$ (Guevarra, 2006). The unified superspace model has been found to apply to the crystal structures of both compounds. Displacement modulation functions follow similar trends, but they differ in details (Fig. 11.13). This is especially true for the displacement modulation functions of the oxygen atoms in the ABO layers, that cannot be represented by a single function for both compounds $n = 5, 6$ [Fig. 11.13(b)].

APPENDIX A

COMPUTER PROGRAMS

A.1 Data collection and data processing (Chapter 6)

Measurements of the intensities of Bragg reflections of incommensurate crystals requires measurements at setting angles corresponding to scattering vectors that cannot be indexed by three integers. The computer program DIF4 by K. Eichhorn (University of Karlsruhe) computes setting angles of reflections from indexings with up to six integers. Other, commercial diffractometer software sometimes allows the collection of data at a series of scattering vectors $(h_1\, h_2\, h_3)$, where real-valued indices h_i can be given. The latter can easily be calculated from the four-integer indexing by *ad hoc* software [eqn (2.5)]. Data processing is possible with HELENA (Spek, 1997) and REDUCE (Eichhorn, 1995).

The measurement with area detectors involves identical procedures for periodic and aperiodic crystals. Software for indexing and integration of measured images need to account for the particular positions of satellite reflections. Several commercially available software packages incorporate options for integration of data from aperiodic crystals, including EVAL by Bruker, CRYSALIS by Oxford Diffraction and X-AREA by STOE.

Refinements of the components of modulation wave vectors on area-detector data can be performed with NADA (Schönleber *et al.*, 2001). Automatic indexing and refinement of lattice parameters and modulation wave vectors is possible with BAYINDEX (Pilz *et al.*, 2002).

A.2 Structure refinements (Chapter 7)

The first general-purpose computer program for the refinement of modulated crystal structures against single-crystal X-ray diffraction data was REMOS by A. Yamamoto. REMOS82 was initially released in 1982 (Yamamoto, 1982*c*). The present version, REMOS95.1_3, is available from A. Yamamoto, Advanced Materials Laboratory, Tsukuba, 305-0044, Japan, or by download from the web site `http://quasi.nims.go.jp/yamamoto/`. It allows the refinement of modulated and composite crystals against either single-crystal or powder diffraction data.

The second, widely used computer program for structure refinement is JANA, originally written by Petricek *et al.* (1985). The present version, JANA2000, can be obtained from V. Petricek, Department of Crystallography of the Institute of Physics of the Academy of Sciences of the Czech republic, Cukrovarnicka 10, 162 53 Praha, Czech Republic, or by download from the web site `http://www-xray.fzu.cz/jana/Jana2000/jana.html`. A new major release, JANA2006, is currently under development. JANA2000 allows the refinement of modulated and composite crystals against single-crystal or powder diffraction data. Important feature of JANA2000 is the plethora of possibilities for special functions and restrictions between parameters, including rigid-body modulations, non-crystallographic site symmetries, block-wave and saw-tooth-shaped modulation functions, anharmonic temperature parameters, and direct refinement of the parameters of the TLS formalism.

Other computer programs for structure refinements are of less general scope than the two programs introduced above, or support for them has been discontinued. They include

1. SIMREF Version 2.8 for Rietveld refinement against powder diffraction data, written by U. Amann, H. Ritter, J. Ihringer, J.K. Maichle and W. Prandl, Institut für Angewandte Physik der Universität Tübingen Auf der Morgenstelle 10, D-72070 Tübingen, Germany. SIMREF is available from `http://www.uni-tuebingen.de/uni/pki/simref/simref.html`.
2. XND for Rietveld refinement of incommensurately modulated structures against X-ray powder diffraction data. Written by J.–F. Bérar, Laboratoire de Cristallographie – CNRS, BP 166, 38042 Grenoble CEDEX 09, France; Email: berar@polycnrs-gre.fr.
3. MSR for the refinement of incommensurately modulated structures was developed around 1985 by W. A. Paciorek. More recent versions include efficient algorithms for computing the modulated atomic scattering factor (Paciorek and Uszynski, 1987; Paciorek and Chapuis, 1992; Paciorek and Chapuis, 1994).
4. A computer program for refinement of structures of crystals with one-dimensional, displacive modulations by Hogervorst (1986).
5. MINREF by Elsenhans (1990), especially developed for the refinement of incommensurate magnetic structures against neutron powder diffraction data.

A.3 Fourier maps and the maximum entropy method (Chapter 8)

Fourier maps and difference Fourier maps can be computed with the refinement programs REMOS95.1 and JANA2000, then producing maps based on observed structure factor amplitudes and calculated structure factors from the refinement (Section A.2).

Electron density maps according to the maximum entropy method (MEM) can be computed with BAYMEM, written by van Smaalen *et al.* (2003). BAYMEM can be obtained from S. van Smaalen, Laboratory of Crystallography, University of Bayreuth, D-95440 Bayreuth, Germany, or from the web site `http://www.crystal.uni-bayreuth.de/BayMEM.html`. BAYMEM allows MEM maps to be calculated for periodic and aperiodic crystals of arbitrary dimensions. It includes modules for the generation of PRIOR densities, for the extraction of modulation functions from electron density maps, and for the analysis of density maps according to Bader's Atoms in Molecules theory.

A.4 Structure solution (Chapter 10)

A computer program, DIMS, for the application of direct methods to diffraction data of incommensurately modulated structures and composite crystals has been written by Fu and Fan (1994). DIMS can be obtained from H.–F. Fan at the web site `http://cryst.iphy.ac.cn/VEC/Tutorials/DIMS/main.html`. The most recent version has been described in the newsletter of the Commission on Crystallographic Computing of the IUCr (Fan, 2005).

Charge-flipping in superspace can be performed with the computer programs SUPERFLIP (Palatinus, 2004) and BAYMEM (Section A.3). SUPERFLIP can be obtained from L. Palatinus or from the web site `http://superspace.epfl.ch/superflip/`.

A.5 Structural analysis and plotting

t-Plots of many quantities can be made with the refinement programs JANA2000 and REMOS95 (Section A.2), and by MISTEK, available from S. van Smaalen.

Basic structures are usually plotted by one of many plotting programs for atomic structures of periodic crystals. The most recent version of DRAWXTL allows basic structures to be plotted of all subsystems of a composite crystal in a single plot (Finger *et al.*, 2007).

APPENDIX B

GLOSSARY OF SYMBOLS

$\mathbf{a}_i$ — $(i = 1, 2, 3)$ Basis vectors of a lattice in physical space.

$\mathbf{A}_i$ — $(i = 1, 2, 3)$ Basis vectors of the superlattice in physical space corresponding to the setting $\mathbf{q}_r = 0$ (Section 3.9.1).

$\mathbf{a}_{\nu i}$ — $(i = 1, 2, 3)$ Basis vectors of the lattice of subsystem ν of a composite crystal [eqn (1.11)].

$\mathbf{a}_{sk}$ — $(k = 1, \cdots, 3{+}d)$ Basis vectors of the lattice Σ in superspace [eqn (2.9)].

$\mathbf{a}_{\nu sk}$ — $(k = 1, \cdots, 3{+}d)$ Basis vectors of the lattice Σ_ν in superspace [eqn (4.20)].

$\mathbf{a}^*_k$ — $(k = 1, \cdots, 3{+}d)$ Reciprocal vectors in physical space that provide an integer indexing of the Bragg reflections [eqn (2.4)]. $\{\mathbf{a}^*_1, \mathbf{a}^*_2, \mathbf{a}^*_3\}$ form a reciprocal lattice in physical space; $\mathbf{q}^j = \mathbf{a}^*_{3+j}$.

$\mathbf{A}^*_i$ — $(i = 1, 2, 3)$ Reciprocal basis vectors of the superlattice in physical space corresponding to the setting $\mathbf{q}_r = 0$ (Section 3.9.1).

$\mathbf{a}^*_{\nu k}$ — $(k = 1, \cdots, 3{+}d)$ Reciprocal vectors in physical space that provide an integer indexing of the Bragg reflections [eqn (4.12)]. $\{\mathbf{a}^*_{\nu 1}, \mathbf{a}^*_{\nu 2}, \mathbf{a}^*_{\nu 3}\}$ form the reciprocal lattice of the basic structure of subsystem ν of a composite crystal [eqn (4.6)].

$\mathbf{a}^*_{sk}$ — $(k = 1, \cdots, 3{+}d)$ Basis vectors of the reciprocal lattice Σ^* in superspace [eqn (2.7)].

$\mathbf{a}^*_{\nu sk}$ — $(k = 1, \cdots, 3{+}d)$ Basis vectors of the reciprocal superspace lattice Σ^*_ν of subsystem ν of a composite crystal [eqn (4.19)].

$\mathbf{A}^n(\mu)$ — n^{th}-order harmonic sine coefficients for displacive modulation of atom μ [eqn (1.10)].

$A^n_i(\mu)$ — $(i = 1, 2, 3)$ n^{th}-order harmonic sine coefficients along $\mathbf{a}_i$.

A^{sw}_i — $(i = 1, 2, 3)$ Amplitude of a saw-tooth-shaped displacive modulation function for components relative to the basis vectors $\mathbf{a}_i$.

b — $= 0.37$ Å. Universal constant in the bond-valence method [eqn (11.1)].

$\mathbf{B}^n(\mu)$ — n^{th}-order harmonic cosine coefficients for displacive modulation of atom μ [eqn (1.10)].

$B^n_i(\mu)$ — $(i = 1, 2, 3)$ n^{th}-order harmonic cosine coefficients along $\mathbf{a}_i$.

B — Weighted average temperature parameter [eqn (6.23)].

B^μ — Isotropic temperature parameter of atom μ [eqn (6.9)].

B_ϕ	Phason temperature parameter [eqn (6.19)].
$B^\mu(\bar{x}_4)$	Modulation function for the isotropic temperature parameter B^μ [eqn (6.20)].
$B^{cn}(\mu)$	n^{th}-order harmonic cosine coefficient of modulation function $B^\mu(\bar{x}_4)$.
$B^{sn}(\mu)$	n^{th}-order harmonic sine coefficient of modulation function $B^\mu(\bar{x}_4)$.
C_{F^2}	F-constraint in the MEM [eqn (8.10)].
$C_{F^2}^{\mathrm{PDC}}$	Prior-derived F-constraint (PDC) in the MEM [eqn (8.27)].
C_G	G-constraint in the MEM [eqn (8.26)].
d	Number of independent modulation waves; the dimension of superspace is 3+d.
E, E_s	Identity operator.
$E(\mathbf{H}_s)$	Normalized structure factor of Bragg reflection $\mathbf{H}_s$ [eqn (6.53)].
$f_j(S)$	Atomic scattering factor of a free spherical atom [eqn (1.20)].
$f_j^0(S)$	Atomic scattering factor of a free spherical atom excluding the contributions of anomalous scattering [eqn (6.3)].
$F(\mathbf{H}_s)$	Structure factor of a modulated or composite structure of Bragg reflection $\mathbf{H}_s$ [eqns (2.34a) and (4.32)].
$F_\nu(\mathbf{H}_{\nu s})$	Contribution of subsystem ν to the structure factor of a composite crystal of Bragg reflection $\mathbf{H}_{\nu s} = \mathbf{H}_s(W^\nu)^{-1}$ [eqn (4.29)].
$F(\mathbf{S})$	Structure factor of a periodic structure [eqn (1.26)].
$F(\mathbf{S}; m)$	Structure factor of a modulated or composite structure [eqn (1.44a)].
$F_{\mathrm{cal}}(\mathbf{H}_s)$	Calculated structure factor of Bragg reflection $\mathbf{H}_s$ [eqn (6.2)].
$F_{\mathrm{cal}}^{\mathrm{elec}}(\mathbf{H}_s)$	Calculated structure factor of Bragg reflection $\mathbf{H}_s$ excluding contributions from anomalous scattering [eqn (8.21)].
$F_{\mathrm{MEM}}(\mathbf{H}_s)$	Structure factor of Bragg reflection $\mathbf{H}_s$, obtained by discrete Fourier transform of an electron density $\{\rho_k\}$ in the MEM [eqn (8.10)].
$F_{\mathrm{obs}}(\mathbf{H}_s)$	Observed structure factor of Bragg reflection $\mathbf{H}_s$ [eqn (6.1)].
$F_{\mathrm{obs}}^{\mathrm{elec}}(\mathbf{H}_s)$	Observed structure factor of Bragg reflection $\mathbf{H}_s$ as used in the MEM [eqn (8.23)].
$g_\mu(\mathbf{H_s})$	Atomic modulation scattering factor of Bragg reflection $\mathbf{H}_s$ [eqn (2.34b)].
$g_\mu(\mathbf{S}; \mathbf{m})$	Atomic modulation scattering factor [eqn (1.44b)].
$\mathbf{G}$	Reciprocal lattice vector in three-dimensional space [eqn (1.14)].
$G_{1,2}$	Product of normalized structure factors of a triplet structure invariant [eqn (10.15)].
G_ν	Subsystem space group of the basic structure of subsystem ν of a composite crystal.
G_s^ν	Subsystem superspace group of subsystem ν of a composite crystal.
$G(H; m)$	Average diffracted intensity of satellite reflections of order m at length of scattering vectors H [eqn (6.37)].

$\mathbf{H}$	Scattering vector in physical space of a Bragg reflection, with integer indices h_k relative to $\mathbf{a}_k$ $(k = 1, \cdots, 3{+}d)$ [eqn (1.37) or eqn (1.47)], or with real-valued indices with respect to $\{\mathbf{a}_1^*, \mathbf{a}_2^*, \mathbf{a}_3^*\}$ [eqn (2.5)].
$\mathbf{H}_\nu$	Scattering vector in physical space of a Bragg reflection, with integer indices $(h_{\nu k} \cdots h_{\nu\,3+d})_\nu$ relative to $\mathbf{a}_{\nu k}$ [eqn (4.14)] .
$\mathbf{H}_s$	Scattering vector in superspace of a Bragg reflection with integer components h_k relative to $\mathbf{a}_{sk}$ $(k = 1, \cdots, 3{+}d)$ [eqn (2.6)].
$\mathbf{H}_{\nu s}$	Scattering vector in superspace of a Bragg reflection with integer components $h_{\nu k}$ relative to $\mathbf{a}_{\nu sk}$ $(k = 1, \cdots, 3{+}d)$ [eqn (4.14)].
h_k	$(k = 1, \cdots, 3{+}d)$ Integer reflection indices with respect to $\mathbf{a}_k^*$ or $\mathbf{a}_{sk}^*$ [eqn (1.36)].
H_k	$(k = 1, \cdots, 3{+}d)$ Integer reflection indices with respect to $\mathbf{A}_i^*$ and $\mathbf{q}_i$ (Section 3.9.1).
$h_{\nu k}$	$(k = 1, \cdots, 3{+}d)$ Integer reflection indices with respect to $\mathbf{a}_{\nu k}^*$ or $\mathbf{a}_{\nu sk}^*$ [eqn (4.14)].
$I(\mathbf{S})$	Intensity of scattered X-rays [eqn (1.21)].
$I_{\text{obs}}(\mathbf{H}_s)$	Observed integrated intensity of Bragg reflection $\mathbf{H}_s$ [eqn (6.1)].
J_ν	Jacobian defining the relative volumes of the basic-structure unit cells of subsystems ν [eqn (4.31)].
k_E	Correction factor for secondary extinction [eqn (7.1)].
k_s	Scale factor between observed and calculated structure factors [eqn (7.1)].
K	Scale factor in the Wilson plot [eqn (6.24)].
$\mathbf{L}$	Lattice vector in physical space [eqn (1.2)].
$\mathbf{L}_\nu$	Lattice vector of subsystem ν in physical space [eqn (4.3)].
$\mathbf{L}_s$	Lattice vector in superspace [eqn (3.20)].
l_i	$(i = 1, 2, 3)$ Integer components of $\mathbf{L}$ relative to the basis vectors $\mathbf{a}_i$.
$l_{\nu i}$	$(i = 1, 2, 3)$ Integer components of $\mathbf{L}_\nu$ relative to the subsystem basis vectors $\mathbf{a}_{\nu i}$.
l_{sk}	$(k = 1, \cdots, 3{+}d)$ Integer components of $\mathbf{L}_s$ relative to the basis vectors $\mathbf{a}_{sk}$.
$\mathbf{m}^*$	Reciprocal lattice vector as part of $(R_s)^{-1}$ [eqn (3.3)].
m	Satellite order of a Bragg reflection [Section 1.5 and eqn (4.18)].
m_k^ρ	Multiplicity of pixel k in the unit cell [eqn (8.18)].
M	Set of reciprocal vectors $\mathbf{a}_k^*$ $(k = 1, \cdots, 3{+}d)$ [eqn (2.4)].
M_ν	Set of reciprocal vectors $\mathbf{a}_{\nu k}^*$ $(k = 1, \cdots, 3{+}d)$ [eqn (4.12)].
$\mathbf{n}^*$	Reciprocal lattice vector as part of R_s [eqn (3.14)].
N	Number of atoms in the basic-structure unit cell.
N_{cell}	Number of unit cells in a crystal [eqn (1.25)].

N_{el}	Number of electrons in the unit cell or in the superspace unit cell [eqn (8.9)].
N_F	Number of reflections.
N_{pix}	$= N_1 \times \cdots \times N_{3+d}$ Number of pixels in a grid over the unit cell in physical space ($d = 0$) or in superspace [eqns (8.5) and (8.17)].
N_{super}	Superperiod of a commensurate modulation [eqn (5.1)].
N_{vol}	Number of atoms in a crystal [eqn (1.23)].
$p^{bw\,\mu}(\bar{x}_{s4})$	Block-wave function with centre $x^{bw}(\mu)$ and width $\Delta(\mu)$ [eqn (7.27)].
$p^{\mu}(\bar{x}_{s4})$	Occupational modulation function of atom μ [eqn (7.23)].
${}_1P_{\lvert E\rvert}(\lvert E\rvert)$	Acentric probability distribution function of normalized structure factor amplitudes [eqn (6.51)].
${}_{\bar{1}}P_{\lvert E\rvert}(\lvert E\rvert)$	Centrosymmetric probability distribution function of normalized structure factor amplitudes [eqn (6.52)].
${}_1P_{\lvert F\rvert}(\lvert F\rvert)$	Acentric probability distribution function of structure factor amplitudes [eqn (6.48)].
$P_+(\frac{1}{2}G_{1,2})$	Probability of a triplet phase of a centrosymmetric crystal to be positive [eqn (10.16)].
$P_\Phi(\Phi)$	Cochran probability distribution of values of the triplet phase of an acentric crystal [eqn (10.14)].
$P(\mathbf{z})$	Patterson function in physical space [eqn (10.3) or eqn (10.4)].
$P_{\text{ave}}(\mathbf{z})$	Average periodic Patterson function in physical space [eqn (10.5)].
$P_s(\mathbf{z}_s)$	Generalized Patterson function in superspace [eqn (10.6) or eqn (10.7)].
$P^0(\mu)$	Average occupation probability of atom μ at site μ [eqn (7.22)].
P^{n_p}	Average modulation amplitude of the ${n_p}^{th}$ harmonic of an occupational modulation function [eqn (6.40)].
$P^{c\,n}(\mu)$	n^{th}-order harmonic cosine coefficient of the occupational modulation function of atom μ at site μ [eqn (7.23)].
$P^{s\,n}(\mu)$	n^{th}-order harmonic sine coefficient of the occupational modulation function of atom μ at site μ [eqn (7.23)].
$\mathbf{q}$	Modulation wave vector [eqn (1.4)].
$\mathbf{q}_i$	Unrestricted (irrational) part of the modulation wave vector.
$\mathbf{q}_r$	Rational part of the modulation wave vector [eqn (3.4)].
$\mathbf{q}^j$	$= \mathbf{a}^*_{3+j}$ $(j = 1, \cdots, d)$ Modulation wave vectors of d independent modulation waves.
$\mathbf{q}^j_\nu$	$= \mathbf{a}^*_{\nu\,3+j}$ $(j = 1, \cdots, d)$ Modulation wave vectors of subsystem ν of d independent modulation waves [eqn (1.12)].
R	Point symmetry operator in physical space, as well as its 3×3 matrix representation [eqn (3.2)].

R^{-1}	Inverse of R.
R_0	Bond-valence parameter [eqn (11.1)].
R_{exp}	Expected R index [eqn (7.5)].
R_{int}	Internal R index between intensities of equivalent Bragg reflections [eqn (9.1)].
R_{Bragg}	Bragg R index in Rietveld refinement [eqn (7.37)].
R_F	R index between observed and calculated structure factor amplitudes [eqn (7.3)].
R_s	$= (R, \epsilon)$ Point symmetry operator in superspace, as well as its $(3{+}d) \times (3{+}d)$ matrix representation [eqn (3.11)].
$\{R\|\mathbf{v}\}$	Symmetry operator in physical space with rotational part R and translational part $\mathbf{v}$ [eqn (3.5)].
$\{R_s\|\mathbf{v}_s\}$	$= \{R, \epsilon\|\mathbf{v}_s\}$ Symmetry operator in superspace with rotational part R_s and translational part $\mathbf{v}_s$ [eqn (3.18)].
$\hat{\mathbf{R}}(\kappa)$	Polar vector in physical space defining the rotation of rigid body κ towards it final orientation [eqn (7.17b)].
$\mathbf{S}$	Scattering vector in physical space.
$\mathcal{S}_n$	Sum over one basic-structure unit cell of the nth power of atomic scattering factors, $[f_\mu(H)]^n$ [eqn (6.22)].
$\mathbf{S}_s$	Scattering vector in superspace.
S	Entropy of the discrete electron density or discrete generalized electron density $\{\rho_k\}$ relative to the prior $\{\tau_k\}$ [eqn (8.6)].
S_i	$(i = 1, 2, 3)$ Components of the scattering vector relative to $\mathbf{a}_i^*$.
S_{sk}	$(k = 1, \cdots, 3{+}d)$ Components of the superspace scattering vector relative to $\mathbf{a}_{sk}^*$.
t	Parameter representing (i) a physical-space section of superspace, and (ii) the phase of the modulation wave [eqn (2.23)].
t^ν	t-Parameter of subsystem ν; is completely defined by t [eqn (4.22)].
T^μ_{phason}	Phason Debye–Waller factor of atom μ [eqns (6.16)–(6.19)].
T^μ_{phonon}	Debye–Waller factor of atom μ [eqns (6.10) and (6.11)].
$\mathbf{T}(\kappa)$	Vector in physical space defining the translation of rigid body κ towards it final location [eqn (7.17a)].
$\mathbf{u}_i^{sw}(\bar{x}_4)$	$(i = 1, 2, 3)$ Saw-tooth function with amplitude A_i^{sw} and centre x_i^{sw} [eqn (7.6)].
$\mathbf{u}_R^\kappa[\bar{x}_{s4}(rb\,\kappa)]$	Modulation function of rotation of rigid body κ [eqn (7.20)].
$\mathbf{u}_T^\kappa[\bar{x}_{s4}(rb\,\kappa)]$	Modulation function of translation of rigid body κ [eqn (7.20)].
$\mathbf{u}^\mu(\bar{x}_4)$	Displacement modulation function of atom μ [eqn (1.8)].
$\mathbf{u}_\nu^\mu(\bar{x}_4)$	Displacement modulation function of atom μ of subsystem ν [eqn (4.4)].
$u_i^\mu(\bar{x}_4)$	$(i = 1, 2, 3)$ Components of $\mathbf{u}^\mu(\bar{x}_4)$ with respect to the basis vectors $\mathbf{a}_i$.

$u^{\mu}_{\nu i}(\bar{x}_4)$	$(i = 1,2,3)$ Components of $\mathbf{u}^{\mu}_{\nu}(\bar{x}_4)$ relative to the subsystem basis vectors $\mathbf{a}_{\nu i}$ [eqn (4.4)].
U	Average modulation amplitude of a single-harmonic displacive modulation function [eqn (6.35)].
U^{n_u}	Average modulation amplitude of the ${n_u}^{th}$ harmonic of a displacive modulation function [eqn (6.38)].
U^{μ}	3×3 tensor of temperature parameters of atom μ [eqn (6.12)].
U^{μ}_{ij}	Components of U^{μ} with respect to the basis vectors $\mathbf{a}_i$.
$U^{\mu}_{ij}(\bar{x}_4)$	Modulation function for the temperature tensor U^{μ} .
$\mathbf{v}$	Translational part of a symmetry operator in physical space, with components relative to $\mathbf{a}_i$ $(i = 1,2,3)$.
$\mathbf{v}^r$	Translation in physical space with rational components [eqn (5.11)].
$\mathbf{v}_s$	Translational part of a symmetry operator in superspace, with components relative to $\mathbf{a}_{sk}$ $(k = 1, \cdots, 3{+}d)$ [eqn (3.17)].
$\mathbf{v}^r_s$	Translation in superspace with rational components (Section 5.3.2).
V	Volume of a unit cell in physical space.
V_i	Valence of atom i as obtained by the bond-valence method [eqn (11.2)].
$V_i(t)$	t-Plot of the valence of atom i of the basic structure, as computed by the bond-valence method for an incommensurate crystal [eqn (11.5)].
w_i	Weight of Bragg reflection $\mathbf{H}_i$ in χ^2 [eqn (7.2)].
wR_{F^2}	Weighted R index between observed and calculated structure factor amplitudes [eqn (7.4)].
W^{ν}	$(3{+}d) \times (3{+}d)$ Integer matrix defining the subsystem reciprocal basis vectors and subsystem modulation wave vectors in terms of the reciprocal basis M [eqn (4.13)], and defining a coordinate transformation in superspace between superspace (lattice Σ and reciprocal lattice Σ^*) and standard superspace of subsystem ν (Σ_{ν} and Σ^*_{ν}) [eqn (4.14)].
$\mathbf{x}$	Vector in physical space.
$\mathbf{x}(j)$	Position of atom j in physical space.
$\mathbf{x}_{\nu}$	Vector in physical space with components relative to the basis vectors of subsystem ν.
x_i	$(i = 1,2,3)$ Components of $\mathbf{x}$ relative to the basis vectors $\mathbf{a}_i$.
$x^{bw}(\mu)$	Centre on the $\bar{x}_{s4}$-axis of a block-wave-shaped occupational modulation function of atom μ [eqn (7.27)].
x^{sw}_i	$(i = 1,2,3)$ Centre on the $\bar{x}_{s4}$-axis of a saw-tooth-shaped displacive modulation function for components relative to the basis vectors $\mathbf{a}_i$.
$x_{\nu i}$	$(i = 1,2,3)$ Components of $\mathbf{x}_{\nu}$ relative to the basis vectors $\mathbf{a}_{\nu i}$.
$\mathbf{x}_s$	Vector in superspace with components relative to the basis Σ.
$\mathbf{x}_s(j)$	Vector in superspace describing the position of atom j [eqn (2.21)].

$\mathbf{x}_{\nu s}$ Vector in superspace with components relative to the basis Σ_ν of subsystem ν.

x_{sk} $(k = 1, \cdots, 3{+}d)$ Components of $\mathbf{x}_s$ relative to the basis vectors $\mathbf{a}_{si}$.

x_{rsk} $(k = 1, \cdots, 3{+}d)$ Components of $\mathbf{x}_{rs}$ relative to the basis Σ_r [eqn (2.27)].

$x_{\nu sk}$ $(k = 1, \cdots, 3{+}d)$ Components of $\mathbf{x}_{\nu s}$ relative to the basis vectors $\mathbf{a}_{\nu sk}$ [eqn (4.21)].

$\mathbf{x}^0(\mu)$ Vector with three components $x_i^0(\mu)$ describing the basic-structure position of atom μ relative to the unit cell [eqn (1.1)].

$\mathbf{x}^0(rb\,\kappa)$ $= \mathbf{T}(\kappa)$. Reference point or centre of rigid body κ [eqn (7.18)].

$\mathbf{x}^0_{rb}(\kappa;\mu)$ Vector with three components $x^0_{rb\,i}(\kappa;\mu)$ describing the basic-structure position of atom μ relative to the centre of rigid body κ [eqn (7.16)].

$\mathbf{x}^0_\nu(\mu)$ Vector with three components $x^0_{\nu i}(\mu)$ describing the basic-structure position of atom μ of subsystem ν relative to the unit cell [eqn (4.2)].

$x_i^0(\mu)$ $(i = 1, 2, 3)$ Components of $\mathbf{x}^0(\mu)$ relative to the basis vectors $\mathbf{a}_i$.

$x^0_{\nu i}(\mu)$ $(i = 1, 2, 3)$ Components of $\mathbf{x}^0_\nu(\mu)$ relative to the subsystem basis vectors $\mathbf{a}_{\nu i}$.

$\bar{\mathbf{x}}$ $= \mathbf{L} + \mathbf{x}^0(\mu)$. Basic-structure position in physical space [eqn (1.3)].

$\bar{x}_4$ $= t + \mathbf{q}\cdot\bar{\mathbf{x}}$. Argument of the modulation functions [eqn (1.5)].

$\bar{x}_{s4}(rb\,\kappa)$ $= t + \mathbf{q}\cdot[\mathbf{L} + \mathbf{x}^0(rb\,\kappa)]$ Fourth basic-structure coordinate corresponding to the reference point of rigid body κ [eqn (7.19)].

$\bar{x}_{sk}$ $= \bar{x}_k$ $(k = 1, \cdots, 3{+}d)$ Superspace coordinates of the basic structure relative to the basis vectors $\mathbf{a}_{sk}$ [eqn (2.20)].

$\bar{x}_{\nu 4}$ $= t_\nu + \mathbf{q}^\nu\cdot\bar{\mathbf{x}}_\nu$. Argument of the modulation functions of subsystem ν of a composite crystal [eqn (4.21)].

$\bar{x}_{\nu sk}$ $= \bar{x}_{\nu k}$ $(k = 1, \cdots, 3{+}d)$ Superspace coordinates of the basic structure relative to the basis Σ_ν of subsystem ν of a composite crystal [eqn (4.21)].

$\bar{x}'_{\nu sk}$ $(k = 1, \cdots, 3{+}d)$ Superspace coordinates of the basic structure of subsystem ν relative to a common origin [eqn (4.23)].

$Y(H;m)$ Modulation contribution to the average diffracted intensity of satellite reflections of order m at length of scattering vector H, for an arbitrary modulation wave [eqn (6.42)].

$\mathbf{z}$ Vector in physical space as argument of the Patterson function (Section 10.3).

$\mathbf{z}_s$ Vector in superspace as argument of the Patterson function (Section 10.3).

Z_μ Atomic number of atom μ.

$Z(H; m, U)$	Modulation contribution to the average diffracted intensity of satellite reflections of order m at length of scattering vector H, for a single-harmonic displacive modulation with average modulation amplitude U [eqn (6.34)].
$Z(\lambda_1)$	Partition function as a function of the Lagrange multiplier corresponding to the F- or G-constraint in the MEM [eqn (8.12)].
ϵ	Symmetry operator in d-dimensional additional space, as well as its $d \times d$ matrix representation [eqn (3.16)].
λ	Wavelength of radiation.
λ_l	Lagrange multiplier of constraint C_l in the MEM [eqn (8.7)].
Λ	Lattice in physical space spanned by basis vectors $\{\mathbf{a}_1, \mathbf{a}_2, \mathbf{a}_3\}$ [eqn (2.10)].
Λ_ν	$(\nu = 1, 2 \cdots)$. Basic-structure lattice of subsystem ν of a composite crystal [eqn (1.11)].
Λ^*	Reciprocal lattice of Λ [eqn (2.2)].
Λ^*_ν	Reciprocal lattice of Λ_ν of subsystem ν [eqn (4.6)].
$\rho(\mathbf{x})$	Electron density distribution in physical space. $\rho(\mathbf{x})\,\mathrm{d}\mathbf{x}$ is the number of electrons in an infinitesimal volume $\mathrm{d}\mathbf{x}$ centred on the position $\mathbf{x}$ in physical space [eqn (1.17)].
$\rho^{\mathrm{FLIP}}(\mathbf{x})$	Electron density representing the solution of charge-flipping (Section 10.5).
$\rho^{\mathrm{MEM}}(\mathbf{x})$	Electron density representing the solution to the MEM [eqn (8.12)].
ρ_k	Electron density at point $\mathbf{x}_k$ or point $\mathbf{x}_{s,k}$ of a grid over the unit cell with pixels $k = 1, \cdots, N_{\mathrm{pix}}$ [eqns (8.5) and (8.17)].
$\rho_s(\mathbf{x}_s)$	Generalized electron density in superspace [eqn (2.12)].
$\rho_{sj}(\mathbf{x}_s)$	Generalized electron density of atom j in superspace [eqn (2.22)].
$\rho_\infty(\mathbf{x})$	Electron density in physical space of an infinite crystal [eqn (1.28)].
$\hat{\rho}_\infty(\mathbf{x})$	Fourier transform of $\rho_\infty(\mathbf{x})$.
σ	$= (\sigma_1, \sigma_2, \sigma_3)$.
$\sigma(\mathbf{H}_i)$	Standard uncertainty of the observed structure factor amplitude $\lvert F_{\mathrm{obs}}(\mathbf{H}_i)\rvert$, of Bragg reflection $\mathbf{H}_i$, (Section 7.1).
σ_i	$(i = 1, 2, 3)$ Components of the modulation wave vector $\mathbf{q}$ with respect to reciprocal basis vectors $\mathbf{a}^*_i$ [eqn (1.4)].
σ_{ji}	$(i = 1, 2, 3)$, $(j = 1, \cdots, d)$ Components of $\mathbf{q}^j$ with respect to $\mathbf{a}^*_i$.
σ^ν	$= (\sigma^\nu_1, \sigma^\nu_2, \sigma^\nu_3)$ [eqn (4.16)].
σ^ν_i	$(i = 1, 2, 3)$ Components of the modulation wave vector $\mathbf{q}_\nu$ relative to $\mathbf{a}^*_{\nu i}$ [eqn (1.12)].
σ_V	Crystal form function [eqn (1.27)].
$\hat{\sigma}_V(\mathbf{S})$	Fourier transform of the crystal form function.

Σ	Direct lattice in superspace [eqn (2.10)].
Σ_ν	Direct lattice in superspace of subsystem ν [eqn (4.20)].
Σ_r	Rectangular direct lattice in superspace [eqn (2.26)].
Σ^*	Reciprocal lattice in superspace [eqn (2.7)].
Σ^*_ν	Reciprocal lattice in superspace of subsystem ν [eqn (4.19)].
τ_k	Value of the prior electron density at point $\mathbf{x}_k$ or point $\mathbf{x}_{s,k}$ of a grid over the unit cell with pixels $k = 1, \cdots, N_{\text{pix}}$ [eqns (8.6)].
μ	Enumerates independent atoms in the basic structure.
ν	Enumerates subsystems of a composite crystal.
ν_{ij}	Contribution of a pair of atoms i and j to the atomic valence of both i and j, as computed by the bond-valence method [eqn (11.1)].
$\nu_{ij}(t)$	t-Plot of the contribution to the atomic valence of a single bond between atoms i and j, as computed by the bond-valence method [eqn (11.3)].
$\delta(x-a)$	Dirac delta function [eqn (1.31)].
$\Delta F(\mathbf{H}_s)$	Difference between observed and calculated structure factor amplitudes of Bragg reflection $\mathbf{H}_s$ [eqn (8.4)].
$\Delta F_{\text{M}}(\mathbf{H}_j)$	$= F_{\text{obs}}(\mathbf{H}_j) - F_{\text{MEM}}(\mathbf{H}_j)$ [eqn (8.13a)].
$\Delta(\mu)$	Width of the interval on the $\bar{x}_{s4}$-axis with value one of a block-wave function describing occupational modulation of atom μ [eqn (7.27)].
$\Delta\rho_s(\mathbf{x}_s)$	Difference Fourier map in superspace [eqn (8.4)].
2θ	Scattering angle [eqn (1.15)].
$\varphi(\mathbf{H}_s)$	Phase of the structure factor of reflection $\mathbf{H}_s$ [eqn (6.21)].
χ^2	Sum of squares of differences between observed and calculated structure factors [eqn (7.1)].

REFERENCES

Abrahams, J. P. (1997). Bias reduction in phase refinement by modified interference functions: introducing the γ correction. *Acta Crystallogr. D*, **53**, 371–376.

Aramburu, I., Friese, K., Perez-Mato, J. M., Morgenroth, W., Aroyo, M., Breczewski, T., and Madariaga, G. (2006). Modulated structure of Rb_2ZnCl_4 in the soliton regime close to the lock-in phase transition. *Phys. Rev. B*, **73**, 014112.

Aramburu, I., Madariaga, G., Grebille, D., Perez-Mato, J. M., and Breczewski, T. (1996). High-order diffraction satellites in Rb_2ZnCl_4: experimental evidence against overall phason thermal factors. *Europhys. Lett.*, **36**, 515–520.

Aramburu, I., Madariaga, G., Grebille, D., Perez-Mato, J. M., and Breczewski, T. (1997). High-order diffraction satellites and temperature variation of the modulation in the incommensurate phase of Rb_2ZnCl_4. *J. Phys. I France*, **7**, 371–383.

Axe, J. D. (1980). Debije–Waller factors for incommensurate structures. *Phys. Rev. B*, **21**, 4181–4190.

Babkevich, A. Y. and Cowley, R. A. (1999). The absence of a discommensuration lattice in the incommensurate phase of Rb_2ZnCl_4. *J. Phys.: Condens. Matter*, **11**, 1639–1655.

Bader, R. F. W. (1994). *Atoms in Molecules.* Oxford University Press, Oxford.

Baudour, J. L. and Sanquer, M. (1983). Structural phase transition in polyphenyls. VIII. The modulated structure of phase III of biphenyl (T = 20 K) from neutron diffraction data. *Acta Crystallogr. B*, **39**, 75–84.

Blinc, R., Apih, T., Dolinsek, J., Prelovsek, P., Slak, J., Ailion, D. C., and Ganesan, K. (1994). NMR in substitutionally disordered incommensurate $(Rb_{1-x}K_x)_2ZnCl_4$. *Phys. Rev. B*, **50**, 2827–2832.

Boeyens, J. C. A. and Comba, P. (2001). Molecular mechanics: theoretical basis, rules, scope and limits. *Coord. Chem. Rev.*, **212**, 3–10.

Böhm, H. (1977). *Eine erweiterte Theorie der Satellitenreflexe und die Bestimmung der modulierten Struktur des Natriumnitrits.* Habilitation Thesis, University of Münster, Münster, Germany.

Brese, N. E. and O'Keeffe, M. (1991). Bond-valence parameters for solids. *Acta Crystallogr. B*, **47**, 192–197.

Bricogne, G. (1988). A Bayesian statistical theory of the phase problem. I. A multichannel maximum-entropy formalism for constructing generalized joint probability distributions of structure factors. *Acta Crystallogr. A*, **44**, 517–545.

Brouwer, R. and Jellinek, F. (1977). Multiple order in sulfides and selenides. *J. Physique*, **38 Colloque C7**, 36–41.

Brown, H., Bülow, R., Neubüser, J., Wondratschek, H., and Zassenhaus, H.

(1978). *Crystallographic Groups of Four-Dimensional Space.* John Wiley, New York.

Brown, I. D. (2002). *The Chemical Bond in Inorganic Chemistry, The Bond Valence Method.* Oxford University Press, Oxford.

Buck, B. and Macaulay, V. A. (ed.) (1991). *Maximum Entropy in Action.* Clarendon Press, Oxford.

Burkov, S. E. (1991). Structure model of the Al–Cu–Co decagonal quasicrystal. *Phys. Rev. Lett.*, **67**, 614–617.

Caracas, R. (2002). A database of incommensurate phases. *J. Appl. Crystallogr.*, **35**, 120–121.

Cornier-Quiquandon, M., Quivy, A., Lefebvre, S., Elkaim, E., Heger, G., Katz, A., and Gratias, D. (1991). Neutron diffraction study of icosahedral Al–Cu–Fe single quasicrystals. *Phys. Rev. B*, **44**, 2071–2084.

Coxeter, H. S. M. (1973). *Regular Polytopes, third edition.* Dover publications, New York.

Cummins, H. Z. (1990). Experimental studies of structurally incommensurate crystal phases. *Phys. Rep.*, **185**, 211–409.

Currat, R., Bernard, L., and Delamoye, P. (1986). Incommensurate phase in β-$ThBr_4$. In *Incommensurate Phases and Dielectrics 2. Materials* (ed. R. Blinc and A. P. Levanyuk), pp. 161–204. North–Holland, Amsterdam.

Daniels, P., Tamazyan, R., Kuntscher, C. A., Dressel, M., Lichtenberg, F., and van Smaalen, S. (2002). The incommensurate modulation of the structure of $Sr_2Nb_2O_7$. *Acta Crystallogr. B*, **58**, 970–976.

David, W. I. F., Shankland, K., McCusker, L. B., and Baerlocher, C. (ed.) (2002). *Structure Determination from Powder Diffraction Data.* Oxford University Press, Oxford.

de Gelder, R., Israel, R., Lam, E., Beurskens, P., van Smaalen, S., Fu, Z., and Fan, H. (1996). Direct methods for incommensurately modulated structures. On the applicability of normalized structure factors. *Acta Crystallogr. A*, **52**, 947–954.

de Vries, R. Y., Briels, W. J., and Feil, D. (1994). Novel treatment of the experimental data in the application of the maximum-entropy method to the determination of the electron-density distribution from X-ray experiments. *Acta Crystallogr. A*, **50**, 383–391.

de Vries, R. Y., Briels, W. J., and Feil, D. (1996). Critical analysis of non-nuclear electron-density maxima and the maximum entropy method. *Phys. Rev. Lett.*, **77**, 1719–1722.

de Wolff, P. M. (1974). The pseudo-symmetry of modulated crystal structures. *Acta Crystallogr. A*, **30**, 777–785.

de Wolff, P. M., Janssen, T., and Janner, A. (1981). The superspace groups for incommensurate crystal structures with a one-dimensional modulation. *Acta Crystallogr. A*, **37**, 625–636.

Degtyareva, O., Gregoryanz, E., Mao, H. K., and Hemley, R. J. (2005). Crystal structure of sulfur and selenium at pressures up to 160 GPa. *High Pressure*

Res., **25**, 17–33.

Duisenberg, A. J. M. (1992). Indexing in single-crystal diffractometry with an obstinate list of reflections. *J. Appl. Crystallogr.*, **25**, 92–96.

Dusek, M., Chapuis, G., Meyer, M., and Petricek, V. (2003). Sodium carbonate revisited. *Acta Crystallogr. B*, **59**, 337–352.

Dusek, M., Petricek, V., Wunschel, M., Dinnebier, R. E., and van Smaalen, S. (2001). Refinement of modulated structures against X-ray powder diffraction data with JANA2000. *J. Appl. Crystallogr.*, **34**, 398–404.

Eichhorn, K. D. (1995). REDUCE *Software*. HASYLAB, DESY, Hamburg.

Elcoro, L., Perez-Mato, J. M., Darriet, J., and El Abed, A. (2003). Superspace description of trigonal and orthorhombic $A_{1+x}A'_xB_{1-x}O_3$ compounds as modulated layered structures; application to the refinement of trigonal $Sr_6Rh_5O_{15}$. *Acta Crystallogr. B*, **59**, 217–233.

Elcoro, L., Perez-Mato, J. M., and Withers, R. L. (2001). Intergrowth polytypoids as modulated structures: A superspace description of the $Sr_n(Nb,Ti)_nO_{2n+2}$ compound series. *Acta Crystallogr. B*, **57**, 471–484.

Elsenhans, O. (1990). Minref – a new computer program for neutron refinement of incommensurate multiphase nuclear and magnetic structures. *J. Appl. Crystallogr.*, **23**, 73–76.

Elser, V. (2003). Solution of the crystallographic phase problem by iterated projections. *Acta Crystallogr. A*, **59**, 201–209.

Engel, P. (1986). *Geometric Crystallography.* D. Reidel, Dordrecht.

Estermann, M. A. and Steurer, W. (1998). Diffuse scattering data acquisition techniques. *Phase Transitions*, **67**, 165–195.

Fabry, J. and Perez-Mato, J. M. (1994). Some stereochemical criteria concerning the structural bility of A_2BX_4 compounds of type β-K_2SO_4. *Phase Transitions*, **49**, 193–229.

Fan, H.-f. (2005). DIMS (Direct-methods program for solving incommensurate modulated structures) on the VEC platform. *IUCr Compcomm Newsletter*, **5**, 16–23.

Fan, H. F., Wang, Z. H., Li, J. Q., Fu, Z. Q., Mo, Y. D., Li, Y., Sha, B. D., Cheng, T. Z., Li, F. H., and Zhao, Z. X. (1998). Multi-dimensional electron crystallography of Bi-based superconductors. In *Electron Crystallography* (ed. D. L. Dorset, S. Hovmöller, and X. Zou), pp. 285–294. Kluwer Academic Publishers, Dordrecht.

Favre-Nicolin, V. (1999). *Développement de la diffraction anomale dispersive, application à l'étude de structures modulées inorganiques et de macromolécules biologiques.* Ph. D. thesis, University Joseph Fourier – Grenoble I, France.

Finger, L. W., Kroeker, M., and Toby, B. H. (2007). DRAWxtl, an open-source computer program to produce crystal structure drawings. *J. Appl. Crystallogr.*, **40**, 188–192.

Folcia, C. L., Ortega, J., Etxebarria, J., and Breczewski, T. (1993). Optical properties and symmetry restrictions in the incommensurate phase of $[N(CH_3)_4]_2ZnCl_4$. *Phys. Rev. B*, **48**, 695–700.

Fu, Z. Q. and Fan, H. F. (1994). DIMS – a direct-method program for incommensurate modulated structures. *J. Appl. Crystallogr.*, **27**, 124–127.

Fu, Z. Q., Li, Y., Cheng, T. Z., Zhang, Y. H., Gu, B. L., and Fan, H. F. (1995). Bi-2212. *Science in China A*, **38**, 210–216.

Fujishita, H., Sato, M., and Hoshino, S. (1984). Incommensurate superlattice reflections in quasi one dimensional conductors, $(MSe_4)_2I$ (M = Ta and Nb). *Solid State Commun.*, **49**, 313–316.

Gao, Y. and Coppens, P. (1989). Structure of modulated molecular crystals. VI. Lattice-energy analysis of the modulated phase of thiourea. *Acta Crystallogr. B*, **45**, 298–303.

Gardner, M. (1976). Mathematical games: extraodinary nonperiodic tiling that enriches the theory of tiles. *Sci. Amer.*, **236**, 110–119.

Giacovazzo, C. (1998). *Direct Phasing in Crystallography.* Oxford University Press, Oxford.

Giacovazzo, C., Monaco, H., Artioli, G., Viterbo, D., Ferraris, G., Gilli, G., Zanotti, G., and Catti, M. (2002). *Fundamentals of Crystallography, second edition.* Oxford University Press, Oxford.

Gilmore, C., Donga, W., and Bricogne, G. (1999). A multisolution method of phase determination by combined maximization of entropy and likelihood. VI. The use of error-correcting codes as a source of phase permutation and their application to the phase problem in powder, electron and macromolecular crystallography. *Acta Crystallogr. A*, **55**, 70–83.

Gilmore, C. J. (1996). Maximum entropy and Bayesian statistics in crystallography: a review of practical applications. *Acta Crystallogr. A*, **52**, 561–589.

Goldstein, H. (1980). *Classical Mechanics, second edition.* Addison-Wesley, Reading, Massachusetts.

Gourdon, O., Petricek, V., Dusek, M., Bezdicka, P., Durovic, S., Gyepesova, D., and Evain, M. (1999). Determination of the modulated structure of $Sr_{41/11}CoO_3$ through a $(3+1)$-dimensional space description and using non-harmonic ADPs. *Acta Crystallogr. B*, **55**, 841–848.

Gourdon, O., Petricek, V., and Evain, M. (2000). A new structure type in the hexagonal perovskite family; structure determination of the modulated misfit compound $Sr_{9/8}TiS_3$. *Acta Crystallogr. B*, **56**, 409–418.

Grebille, D., Lambert, S., Bouree, F., and Petricek, V. (2004). Contribution of powder diffraction for structure refinements of aperiodic misfit cobalt oxides. *J. Appl. Crystallogr.*, **37**, 823–831.

Grebille, D., Weigel, D., Veysseyre, R., and Phan, T. (1990). Crystallography, geometry and physics in higher dimensions. VII. The different types of symbols of the 371 mono-incommensurate superspace groups. *Acta Crystallogr. A*, **46**, 234–240.

Grünbaum, B. and Shephard, G. C. (1987). *Tilings and Patterns.* W. H. Freeman, New York.

Guevarra, J. S. (2006). *Crystal structures of perovskite-related* $Ca_n(Nb,Ti)_nO_{3n+2}$ *(n = 5 and 6).* Ph. D. thesis, University of Bayreuth,

Bayreuth, Germany.

Guinier, A. (1994). *X-ray Diffraction in Crystals, Imperfect Crystals and Amorphous Bodies.* Dover publications, New York.

Gull, S. F. and Skilling, J. (1999). *Quantified Maximum Entropy,* MEMSYS5 *Users' Manual.* Maximum Entropy Data Consultants Ltd., Suffolk, U.K.

Gummelt, P. (1996). Penrose tilings as coverings of congruent decagons. *Geom. Dedic.*, **62**, 1–17.

Hagiya, K., Ohmasa, M., and Iishi, K. (1993). The modulated structure of synthetic Co-åkermanite, $Ca_2CoSi_2O_7$. *Acta Crystallogr. B*, **49**, 172–179.

Hahn, T. (ed.) (2002). *International Tables for Crystallography Vol. A, fifth edition.* Kluwer Academic Publishers, Dordrecht.

Hao, Q., Liu, Y. W., and Fan, H. F. (1987). Direct methods in superspace. I. preliminary theory and test on the determination of incommensurate modulated structures. *Acta Crystallogr. A*, **43**, 820–824.

Heilmann, I. U., Axe, J. D., Hastings, J. M., Shirane, G., Heeger, A. J., and Macdiarmid, A. G. (1979). Neutron investigation of the dynamical properties of the mercury-chain compound $Hg_{3-\delta}AsF_6$. *Phys. Rev. B*, **20**, 751–762.

Hogervorst, A. C. R. (1986). *Comparative study of the modulated structures in Rb_2ZnBr_4 and in related compounds.* Ph. D. thesis, Technical University Delft, Delft.

Iizumi, M., Axe, J. D., Shirane, G., and Shimaoka, K. (1977). Structural phase transformation in K_2SeO_4. *Phys. Rev. B*, **15**, 4392–4411.

Janner, A. and Janssen, T. (1979). Superspace groups. *Physica A*, **99A**, 47–76.

Janner, A. and Janssen, T. (1980*a*). Symmetry of incommensurate crystal phases. I. Commensurate basic structure. *Acta Crystallogr. A*, **36**, 399–408.

Janner, A. and Janssen, T. (1980*b*). Symmetry of incommensurate crystal phases. II. Incommensurate basic structure. *Acta Crystallogr. A*, **36**, 408–415.

Janner, A., Janssen, T., and de Wolff, P. M. (1983). Bravais classes for incommensurate crystal phases. *Acta Crystallogr. A*, **39**, 658–666.

Janot, C. (1994). *Quasicrystals a Primer, second edition.* Clarendon Press, Oxford.

Janssen, T. (1986). Microscopic theories of incommensruate crystal phases. In *Incommensurate Phases in Dielectrics. 1. Fundamentals* (ed. R. Blinc and A. P. Levanyuk), pp. 67–142. North–Holland, Amsterdam.

Janssen, T., Janner, A., Looijenga-Vos, A., and de Wolff, P. M. (1995). Incommensurate and commensurate modulated structures. In *International Tables for Crystallography Vol. C* (ed. A. J. C. Wilson), pp. 797–835. Kluwer Academic Publishers, Dordrecht.

Jauch, W. (1994). The maximum-entropy method in charge-density studies. II. General aspects of reliability. *Acta Crystallogr. A*, **50**, 650–652.

Jaynes, E. T. (2003). *Probability Theory: The Logic of Science.* Cambridge University Press, Cambridge.

Jobst, A. (2003). *Bestimmung der modulierten Strukturen von $(LaS)_{1.13}TaS_2$ und $(LaS)_{1.14}NbS_2$ mittels hochauflösender Röntgenbeugung, und Charakter-*

isierung der Bindung zwischen den Teilsystemen. Ph. D. thesis, University of Bayreuth, Bayreuth.

Jobst, A. and van Smaalen, S. (2002). Intersubsystem chemical bonds in the misfit layer compounds $(LaS)_{1.13}TaS_2$ and $(LaS)_{1.14}NbS_2$. *Acta Crystallogr. B*, **58**, 179–190.

Katrych, S., Weber, T., Kobas, M., Massger, L., Palatinus, L., Chapuis, G., and Steurer, W. (2007). New stable decagonal quasicrystal in the system Al–Ir–Os. *J. Alloys Compounds*, **428**, 164–172.

Kobayashi, J. (1990). Incommensurate phase transitions and optical activity. *Phys. Rev. B*, **42**, 8332–8338.

Koehler, W. C. (1972). Magnetic structures of rare earth metals and alloys. In *Magnetic Properties of Rare Earth Metals*, pp. 81–128. Plenum Press, London.

Koritsanszky, T. S. and Coppens, P. (2001). Chemical applications of X-ray charge-density analysis. *Chem. Rev.*, **101**, 1583–1627.

Kumazawa, S., Takata, M., and Sakata, M. (1995). On the single-pixel approximation in maximum-entropy analysis. *Acta Crystallogr. A*, **51**, 47–53.

Lam, E., Beurskens, P., and van Smaalen, S. (1992). Scaling of x-ray diffraction intensities for crystals with a one-dimensional, incommensurate, displacive modulation. *Solid State Commun.*, **82**, 345–349.

Lam, E., Beurskens, P., and van Smaalen, S. (1993). Intensity statistics and normalized structure factors for crystals with an incommensurate, one-dimensional modulation. *Acta Crystallogr. A*, **49**, 709–721.

Lam, E., Beurskens, P., and van Smaalen, S. (1994). A more general expression for the average X-ray diffraction intensity of crystals with an incommensurate one-dimensional modulation. *Acta Crystallogr. A*, **50**, 690–703.

Lam, E. J. W., Beurskens, P. T., Smits, J. M. M., van Smaalen, S., de Boer, J. L., and Fan, H. F. (1995). Determination of the incommensurately modulated structure of (Perylene)Co(mnt)$_2$-$(CH_2Cl_2)_{0.5}$ by direct methods. *Acta Crystallogr. B*, **51**, 779–789.

Lander, G. H., Fischer, E. S., and Bader, S. D. (1994). The solid-state properties of uranium. A historical perspective and review. *Adv. Phys.*, **43**, 1–111.

Lee, C. H., Matsuhata, H., Yamaguchi, H., Sekine, C., Kihou, K., Suzuki, T., Noro, T., and Shirotani, I. (2004). Charge-density-wave ordering in the metal-insulator transition compound $PrRu_4P_{12}$. *Phys. Rev. B*, **70**, 153105.

Lichtenberg, F., Herrnberger, A., Wiedenmann, K., and Mannhart, J. (2001). Synthesis of perovskite-related layered $A_nB_nO_{3n+2} = ABO_x$ type niobates and titanates and study of their structural, electric and magnetic properties. *Prog. Solid State Chem.*, **29**, 1–70.

Lüdecke, J., Jobst, A., and van Smaalen, S. (2000). The incommensurate structure of the CDW state of the phosphate tungsten bronze $P_4W_8O_{32}$ ($m = 4$). *Europhys. Lett.*, **49**, 357–361.

Lüdecke, J., Jobst, A., van Smaalen, S., Morre, E., Geibel, C., and Krane, H. G. (1999). Acentric low-temperature superstructure of NaV_2O_5. *Phys. Rev. Lett.*, **82**, 3633–3636.

Mackay, A. L. (1982). Cystallography and the Penrose pattern. *Physica A*, **114**, 609–613.

Magnéli, A. (1953). Structures of the ReO_3-type with recurrent dislocations of atoms: 'homologous series' of molybdenum and tungsten oxides. *Acta Crystallogr.*, **6**, 495–500.

Maradudin, A. A., Montroll, E. W., and Weiss, G. H. (1963). *Theory of Lattice Dynamics in the Harmonic Approximation.* Academic Press, New York.

Maradudin, A. A. and Vosko, S. H. (1968). Symmetry properties of the normal vibrations of a crystal. *Rev. Mod. Phys.*, **40**, 1–37.

Marmeggi, J. C., Lander, G. H., van Smaalen, S., Bruckel, T., and Zeyen, C. M. E. (1990). Neutron-diffraction study of the charge-density wave in α-uranium. *Phys. Rev. B*, **42**, 9365–9376.

McMahon, M. and Nelmes, R. (2004). High-pressure structures of the metallic elements. *Z. Kristallogr.*, **219**, 742–748.

McMahon, M. I., Degtyareva, O., and Nelmes, R. J. (2000). Ba-IV-Type incommensurate crystal structure in group-V metals. *Phys. Rev. Lett.*, **85**, 4896–4899.

McMahon, M. I., Degtyareva, O., Nelmes, R. J., van Smaalen, S., and Palatinus, L. (2007). Incommensurate modulations of Bi-III and Sb-II. *Phys. Rev. B*, **in press**.

Mermin, N. D. (1992). Copernican crystallography. *Phys. Rev. Lett.*, **68**, 1172–1175.

Mermin, N. D. and Lifshitz, R. (1992). Bravais classes for the simplest incommensurate crystal phases. *Acta Crystallogr. A*, **48**, 515–532.

Meyer, M., Paciorek, W. A., Schenk, K. J., Chapuis, G., and Depmeier, W. (1994). Incommensurately modulated structure of γ-PAMC: new experimental evidence for amplitude and phase fluctuations. *Acta Crystallogr. B*, **50**, 333–343.

Miller, R., Gallo, S. M., Khalak, H. G., and Weeks, C. M. (1994). SnB: crystal structure determination via Shake-and-Bake. *J. Appl. Cryst.*, **27**, 613–621.

Mo, Y. D., Fu, Z. Q., Fan, H.-F., van Smaalen, S., Lam, E. J. W., and Beurskens, P. T. (1996). Direct methods for incommensurate intergrowth compounds. III. Solving the average structure in multidimensional space. *Acta Crystallogr. A*, **52**, 640–644.

Moncton, D. E., Axe, J. D., and DiSalvo, F. J. (1977). Neutron scattering study of the charge–density wave transitions in $2H$–$TaSe_2$ and $2H$–$NbSe_2$. *Phys. Rev. B*, **16**, 801–819.

Ollivier, J., Etrillard, J., Toudic, B., Ecolivet, C., Bourges, P., and Levanyuk, A. P. (1998). Direct observation of a phason gap in an incommensurate molecular compound. *Phys. Rev. Lett.*, **81**, 3667–3670.

Onoda, M., Saeki, M., Yamamoto, A., and Kato, K. (1993). Structure refinement of the incommensurate composite crystal $Sr_{1.145}TiS_3$ through the Rietveld analysis process. *Acta Crystallogr. B*, **49**, 929–936.

Opgenorth, J., Plesken, W., and Schulz, T. (1998). Crystallographic algorithms and tables. *Acta Crystallogr. A*, **54**, 517–531.

Oszlányi, G. and Sütő, A. (2004). Ab initio structure solution by charge flipping. *Acta Crystallogr. A*, **60**, 134–141.

Oszlányi, G. and Sütő, A. (2005). Ab initio structure solution by charge flipping. II. Use of weak reflections. *Acta Crystallogr. A*, **61**, 147–152.

Ottolenghi, A. and Pouget, J. P. (1996). Evidence of high critical temperature charge density wave transitions in the $(PO_2)_4(WO_3)_{2m}$ family of low dimensional conductors for $m \geq 8$. *J. Phys. I France*, **6**, 1059–1083.

Overhauser, A. W. (1971). Observability of charge-density waves by neutron diffraction. *Phys. Rev. B*, **3**, 3173–3182.

Paciorek, W. and Chapuis, G. (1992). A new algorithm for incommensurate structure refinement. *J. Appl. Crystallogr.*, **25**, 317–322.

Paciorek, W. and Chapuis, G. (1994). Generalized bessel functions in incommensurate structure analysis. *Acta Crystallogr. A*, **50**, 194–203.

Paciorek, W. and Uszynski, I. (1987). An efficient algorithm for incommensurate structure refinement. *J. Appl. Crystallogr.*, **20**, 57–59.

Palatinus, L. (2004). Ab initio determination of incommensurately modulated structures by charge flipping in superspace. *Acta Crystallogr. A*, **60**, 604–610.

Palatinus, L., Amami, M., and van Smaalen, S. (2004). Structure of incommensurate ammonium tetrafluoroberyllate studied by structure refinements and the maximum entropy. *Acta Crystallogr. B*, **60**, 127–137.

Palatinus, L., Dusek, M., Glaum, R., and El Bali, B. (2006). The incommensurately and commensurately modulated crystal structures of chromium(II) diphosphate. *Acta Crystallogr. B*, **62**, 556–566.

Palatinus, L., Lee, H., and van Smaalen, S. (2005*a*). Modulation waves in $Sr_{1.132}TiS_3$. Unpublished.

Palatinus, L., Schönleber, A., and van Smaalen, S. (2005*b*). Two-fold superstructure of titanium(III)-oxybromide at T = 17.5 K. *Acta Crystallogr. C*, **61**, i48–i49.

Palatinus, L. and van Smaalen, S. (2002). The generalized F-constraint in the Maximum Entropy Method – a study on simulated data. *Acta Crystallogr. A*, **58**, 559–567.

Palatinus, L. and van Smaalen, S. (2004*a*). The ferroelectric phase transition and modulated valence electrons in the incommensurate phase of ammonium tetrafluoroberyllate. *Ferroelectrics*, **305**, 49–52.

Palatinus, L. and van Smaalen, S. (2004*b*). Incommensurate modulations made visible by the Maximum Entropy Method in superspace. *Z. Kristallogr.*, **219**, 719–729.

Penrose, R. (1974). The role of aesthetics in pure and applied mathematical research. *Bull. Inst. Math. Appl.*, **10**, 266–271.

Perez-Mato, J. M. (1991). Superspace description of commensurately modulated structures. In *Methods of Structural Analysis of Modulated Structures and Quasicrystals* (ed. J. M. Perez-Mato, F. J. Zuniga, and G. Madariaga), pp. 117–128. World Scientific, Singapore.

Perez-Mato, J. M., Madariaga, G., and Elcoro, L. (1991). Influence of phason dynamics on atomic Debye–Waller factors of incommensurate modulated structures and quasicrystals. *Solid State Commun.*, **78**, 33–37.

Perez-Mato, J. M., Zakhour-Nakhl, M., Weill, F., and Darriet, J. (1999). Structure of composites $A_{1+x}(A'_x B_{1-x})O_3$ related to the $2H$ hexagonal perovskite: relation between composition and modulation. *J. Mater. Chem.*, **9**, 2795–2808.

Peschar, R., Israel, R., and Beurskens, P. T. (2001). The joint probability distributions of structure factor doublets in displacive incommensurately modulated structures and their applicability to direct methods. *Acta Crystallogr. A*, **57**, 461–472.

Peterkova, J., Dusek, M., Petricek, V., and Loub, J. (1998). Structures of fluoroarsenates $KAsF_{6-n}(OH)_n$, $n = 0, 1, 2$: application of the heavy-atom method for modulated structures. *Acta Crystallogr. B*, **54**, 809–818.

Petricek, V., Cisarova, I., de Boer, J. L., Zhou, W., Meetsma, A., Wiegers, G. A., and van Smaalen, S. (1993). The modulated structure of the commensurate misfit-layer compound $(BiSe)_{1.09}TaSe_2$. *Acta Crystallogr. B*, **49**, 258–266.

Petricek, V., Coppens, P., and Becker, P. (1985). Structure analysis of displacively modulated molecular crystals. *Acta Crystallogr. A*, **41**, 478–483.

Petricek, V., Dusek, M., and Palatinus, L. (2000). *The Crystallographic Computing System* JANA2000. Institute of Physics, Praha, Czech Republic.

Petricek, V., Gao, Y., Lee, P., and Coppens, P. (1990). X-ray analysis of the incommensurate modulation in the 2 : 2 : 1 : 2 Bi–Sr–Ca–Cu–O superconductor including the oxygen atoms. *Phys. Rev. B*, **42**, 387–392.

Petricek, V., van der Lee, A., and Evain, M. (1995). On the use of crenel functions of occupationally modulated structures. *Acta Crystallogr. A*, **51**, 529–535.

Phan, T., Veysseyre, R., Weigel, D., and Grebille, D. (1989). Crystallography, geometry and physics in higher dimensions. VI. Geometrical 'WPV' symbols for the 371 crystallographic mono-incommensurate space groups in four-dimensional space. *Acta Crystallogr. A*, **45**, 547–557.

Pilz, K., Estermann, M., and van Smaalen, S. (2002). Automatic indexing of area-detector data of periodic and aperiodic crystals. *J. Appl. Crystallogr.*, **35**, 253–260.

Plesken, W. and Schulz, T. (2000). Counting crystallographic groups in low dimensions. *Experimental Mathematics*, **9**, 407–411.

Ravy, S., Requardt, H., Le Bolloc'h, D., Foury-Leylekian, P., Pouget, J. P., Currat, R., Monceau, P., and Krisch, M. (2004). Inelastic X-ray scattering study of charge-density-wave dynamics in the $Rb_{0.3}MoO_3$ blue bronze. *Phys. Rev. B*, **69**, 115113.

Sakata, M., Mori, T., Kumazawa, S., and Takata, M. (1990). Electron-density distribution from X-ray powder data by use of profile fits and the Maximum-Entropy Method. *J. Appl. Crystallogr.*, **23**, 526–534.

Schmicker, D. and van Smaalen, S. (1996). Dynamical behavior of aperiodic intergrowth crystals. *Int. J. Mod. Phys. B*, **10**, 2049–2080.

Schmicker, D., van Smaalen, S., de Boer, J. L., Haas, C., and Harris, K. D. M. (1995). Observation of the sliding mode in incommensurate intergrowth compounds: Brillouin scattering from the inclusion compound of urea and heptadecane. *Phys. Rev. Lett.*, **74**, 734–737.

Schomaker, V. and Trueblood, K. N. (1968). On the rigid-body motion of molecules in crystals. *Acta Crystallogr. B*, **24**, 63–76.

Schönleber, A., Meyer, M., and Chapuis, G. (2001). NADA — a computer program for the simultaneous refinement of orientation matrix and modulation vector(s). *J. Appl. Crystallogr.*, **34**, 777–779.

Schönleber, A., Zuniga, F. J., Perez-Mato, J. M., Darriet, J., and zur Loye, H.-C. (2006). Description of $Ba_{1+x}Ni_xRh_{1-x}O_3$ with $x = 0.1170$ (5) in superspace: modulated composite versus modulated-layer structure. *Acta Crystallogr. B*, **62**, 197–204.

Schutte, W. J. and de Boer, J. L. (1988). Valence fluctuations in the incommensurately modulated structure of calaverite $AuTe_2$. *Acta Crystallogr. B*, **44**, 486–494.

Schutte, W. J., Disselborg, F., and de Boer, J. L. (1993). Determination of the two-dimensional incommensurately modulated structure of Mo_2S_3. *Acta Crystallogr. B*, **49**, 787–794.

Schwarz, U., Takemura, K., Hanfland, M., and Syassen, K. (1998). Crystal structure of Cesium-V. *Phys. Rev. Lett.*, **81**, 2711–2714.

Senechal, M. (1995). *Quasicrystals and Geometry.* Cambridge University Press, Cambridge.

Shechtman, D., Blech, I., Gratias, D., and Cahn, J. W. (1984). Metallic phase with long-range orientational order and no translational symmetry. *Phys. Rev. Lett.*, **53**, 1951–1953.

Shmueli, U. and Weiss, G. H. (1995). *Introduction to Crystallographic Statistics.* Oxford University Press, Oxford.

Skilling, J. and Bryan, R. K. (1984). Maximum entropy image reconstruction: general algorithm. *Mon. Not. R. Astr. Soc.*, **211**, 111–124.

Spek, A. L. (1997). HELENA *Software.* Utrecht University, The Netherlands.

Steinhardt, P. J., Jeong, H. C., Saitoh, K., Tanaka, M., Abe, E., and Tsai, A. P. (1998). Experimental verification of the quasi-unit-cell model of quasicrystal structure. *Nature*, **396**, 55–57.

Steller, I., Bolotovsky, R., and Rossmann, M. G. (1997). An algorithm for automatic indexing of oscillation images using fourier analysis. *J. Appl. Crystallogr.*, **30**, 1036–1040.

Steurer, W. (1987). (3+1)-Dimensional Patterson and Fourier methods for the determination of one-dimensionally modulated structures. *Acta Crystallogr. A*, **43**, 36–42.

Steurer, W. (1991). The n-dim maximum-entropy method. In *Methods of Structural Analysis of Modulated Structures and Quasicrystals* (ed. J. M. Perez-Mato, F. J. Zuniga, and G. Madariaga), pp. 344–349. World Scientific, Singapore.

Steurer, W. (2004). Twenty years of structure research on quasicrystals. Part I. Pentagonal, octagonal, decagonal and dodecagonal quasicrystals. *Z. Kristallogr.*, **219**, 391–446.

Stitzer, K. E., Darriet, J., and zur Loye, H. C. (2001). Advances in the synthesis and structural description of $2H$-hexagonal perovskite-related oxides. *Curr. Opin. Sol. State Mater. Sci.*, **5**, 535–544.

Stitzer, K. E., Henley, W. H., Claridge, J. B., zur Loye, H. C., and Layland, R. C. (2002). Sr_3NiRhO_6 and Sr_3CuRhO_6—two new one-dimensional oxides. Magnetic behavior as a function of structure: commensurate vs incommensurate. *J. Sol. State Chem.*, **164**, 220–229.

Takata, M., Nishibori, E., Kato, K., Sakata, M., and Moritomo, Y. (1999). Direct observation of orbital order in manganites by MEM charge-density study. *J. Phys. Soc. Jpn.*, **68**, 2190–2193.

Takemura, K., Sato, K., Fujihisa, H., and Onoda, M. (2004). Structural phase transitions in iodine under high pressure. *Z. Kristallogr.*, **219**, 749–754.

Taye, A., Michel, D., and Petersson, J. (2004). Phason and amplitudon dynamics in the incommensurate phase of bis(4-chlorophenyl)sulphone. *Phys. Rev. B*, **69**, 224206.

Terwilliger, T. C. (2000). Maximum-likelihood density modification. *Acta Crystallogr. D*, **56**, 965–972.

Tsinde, B. P. F. and Doert, T. (2005). The ternary rare-earth polychalcogenides $LaSeTe_2$, $CeSeTe_2$, $PrSeTe_2$, $NdSeTe_2$, and $SmSeTe_2$: syntheses, crystal structures, electronic properties, and charge-density-wave-transitions. *Solid State Sci.*, **7**, 573–587.

Tuinstra, F. and Fraase Storm, G. (1972). *Internal report No. MSB KR 47 a.* Technical University Delft, Delft.

Tun, Z. and Brown, I. D. (1986). The low-temperature structures of $Hg_{3-\delta}SbF_6$ and $Hg_{3-\delta}TaF_6$. *Acta Crystallogr. B*, **42**, 209–213.

van Aalst, W., den Hollander, J., Peterse, W. J. A. M., and de Wolff, P. M. (1976). The modulated structure of γ-Na_2CO_3 in a harmonic approximation. *Acta Crystallogr. B*, **32**, 47–58.

van Smaalen, S. (1987). Superspace-group description of short-period commensurately modulated crystals. *Acta Crystallogr. A*, **43**, 202–207.

van Smaalen, S. (1988). Consequences of the monoclinic superspace group description of triclinic NbS_3. *Phys. Rev. B*, **38**, 9594–9600.

van Smaalen, S. (1991). Symmetry of composite crystals. *Phys. Rev. B*, **43**, 11330–11341.

van Smaalen, S. (1992*a*). Renormalization of bond valences: Application to incommensurate intergrowth crystals. *Acta Crystallogr. A*, **48**, 408–410.

van Smaalen, S. (1992*b*). Superspace description of incommensurate intergrowth compounds and the application to inorganic misfit layer compounds. *Materials Science Forum*, **100–101**, 173–222.

van Smaalen, S. (1995). Incommensurate crystal structures. *Crystallogr. Rev.*, **4**, 79–202.

van Smaalen, S. (1999). Atomic valences in aperiodic crystals studied by the bond valence method. In *Computational Studies of New Materials* (ed. D. A. Jelski and T. F. George), pp. 273–294. World Scientific, Singapore.

van Smaalen, S. (2005). The Peierls transition in low-dimensional electronic crystals. *Acta Crystallogr. A*, **61**, 51–61.

van Smaalen, S., de Boer, J. L., Meetsma, A., Graafsma, H., Sheu, H.-S., Darovskikh, A., Coppens, P., and Levy, F. (1992). Determination of the structural distortions correspondingto the q^1- and q^2-type modulations in niobium triselenide $NbSe_3$. *Phys. Rev. B*, **45**, 3103–3106.

van Smaalen, S. and George, T. F. (1987). Determination of the incommensurately modulated structure of α-uranium below 37 K. *Phys. Rev. B*, **35**, 7939–7951.

van Smaalen, S. and Harris, K. D. M. (1996). Superspace group descriptions of the symmetries of incommensurate urea inclusion compounds. *Proc. Roy. Soc. London A*, **452**, 677–700.

van Smaalen, S., Lam, E. J., and Lüdecke, J. (2001). Structure of the charge-density wave in $(TaSe_4)_2I$. *J. Phys.: Condens. Matter*, **13**, 9923–9936.

van Smaalen, S. and Palatinus, L. (2004). Applications of the Maximum Entropy Method in superspace. *Ferroelectrics*, **305**, 57–62.

van Smaalen, S., Palatinus, L., and Schneider, M. (2003). The Maximum Entropy Method in superspace. *Acta Crystallogr. A*, **59**, 459–469.

Wang, B. C. (1985). Resolution of phase ambiguity in macromolecular crystallography. *Methods Enzymol.*, **115**, 90–112.

Weigel, D., Phan, T., and Veysseyre, R. (1987). Crystallography, geometry and physics in higher dimensions. III. Geometrical symbols for the 227 crystallographic point groups in four-dimensional space. *Acta Crystallogr. A*, **43**, 294–304.

Welberry, T. R. (2004). *Diffuse X-ray Scattering and Models of Disorder.* Oxford University Press, Oxford.

Wiegers, G. A. (1996). Misfit layer compounds: structures and physical properties. *Prog. Solid State Chem.*, **24**, 1–139.

Wilson, A. J. C. (ed.) (1995). *International Tables for Crystallography Vol. C.* Kluwer Academic Publishers, Dordrecht.

Yamada, N., Ono, Y., and Ikeda, T. (1984). A structural study of the incommensurate-to-ferroelectric phase transition in K_2SeO_4. *J. Phys. Soc. Jpn.*, **53**, 2565–2574.

Yamamoto, A. (1982*a*). Modulated structure of CuAu II (one-dimensional modulation). *Acta Crystallogr. B*, **38**, 1446–1451.

Yamamoto, A. (1982*b*). Modulated structure of Wustite ($Fe_{1-x}O$) (three-dimensional modulation). *Acta Crystallogr. B*, **38**, 1451–1456.

Yamamoto, A. (1982*c*). Structure factor of modulated crystal structures. *Acta Crystallogr. A*, **38**, 87–92.

Yamamoto, A. (1993). Determination of composite crystal structures and superspace groups. *Acta Crystallogr. A*, **46**, 831–846.

Yamamoto, A. (1996). Crystallography of quasiperiodic crystals. *Acta Crystallogr. A*, **52**, 509–560.

Yamamoto, A. (2001). Toward automatic analysis of modulated and composite crystals. *Ferroelectrics*, **250**, 139–142.

Yamamoto, A. (2005). *Superspace Groups for One-, Two- and Three-Dimensionally Modulated Structures.* Advanced Materials Laboratory at `http://quasi.nims.go.jp/yamamoto/spgr.new.html`, Tsukuba, Japan.

Yamamoto, A. and Nakazawa, H. (1982). Modulated structure of the NC-type ($N = 5.5$) pyrrhotite, $Fe_{1-x}S$. *Acta Crystallogr. A*, **38**, 79–86.

Yamamoto, A., Nakazawa, H., Kitamura, M., and Morimoto, N. (1984). The modulated structure of intermediate plagioclase feldspar $Ca_xNa_{1-x}Al_{1+x}Si_{3-x}O_8$. *Acta Crystallogr. B*, **40**, 228–237.

Yamamoto, A., Onoda, M., Takayama-Muromachi, E., Izumi, F., Ishigaki, T., and Asano, H. (1990). Rietveld analysis of the modulated structure in the superconducting oxide $Bi_2(Sr,Ca)_3Cy_2O_{8+x}$. *Phys. Rev. B*, **42**, 4228–4239.

Yan, Y. and Pennycook, S. J. (2001). Chemical ordering in $Al_{72}Ni_{20}Co_8$ decagonal quasicrystals. *Phys. Rev. Lett.*, **86**, 1542–1545.

Yeo, L. and Harris, K. D. M. (1997). Definitive structural characterization of the conventional low-temperature host structure in urea inclusion compounds. *Acta Crystallogr. B*, **53**, 822–830.

Young, R. A. (ed.) (1995). *The Rietveld Method.* Oxford University Press, Oxford.

Zeyher, R. and Finger, W. (1982). Phason dynamics of incommensurate crystals. *Phys. Rev. Lett.*, **49**, 1833–1837.

INDEX